Nuclear Fuel
Reprocessing and Waste Management

Modern Nuclear Energy Analysis Methods

Series Editors: Tunc Aldemir *(The Ohio State University, USA)*
Steven A. Arndt *(University of Tennessee, USA)*

This series will focus on the analysis methods used in all areas of modern nuclear energy modeling and evaluation. It will provide an up to date review of all of the tools and processes needed to design and model, systems, structures and components used today in nuclear power plant. The series will include nuclear both design basis method and risk analysis methods that are being used by designers of both current generation and next generation nuclear power plants.

A second part of the series will be on non-reactor radiation safety and would include radiation safety for the medial industry, industrial radiation safety and mining and waste management radiation safety.

Published

Vol. 2   *Nuclear Fuel Reprocessing and Waste Management*
edited by Jinsuo Zhang

Vol. 1   *Advanced Concepts in Nuclear Energy Risk Assessment and Management*
edited by Tunc Aldemir

Forthcoming

Design Basis Accident Analysis Methods for Light-Water Nuclear Power Plants
edited by Robert Martin & Cesare Frepoli

Risk Analysis for Nuclear Plants: Methods, Tools, and Applications
by Vinod Mubayi & Robert W. Youngblood

Nuclear Fuel
Reprocessing and Waste Management

Editor

Jinsuo Zhang

Virginia Tech, USA

World Scientific

NEW JERSEY · LONDON · SINGAPORE · BEIJING · SHANGHAI · HONG KONG · TAIPEI · CHENNAI · TOKYO

Published by

World Scientific Publishing Co. Pte. Ltd.

5 Toh Tuck Link, Singapore 596224

USA office: 27 Warren Street, Suite 401-402, Hackensack, NJ 07601

UK office: 57 Shelton Street, Covent Garden, London WC2H 9HE

Library of Congress Cataloging-in-Publication Data

Names: Zhang, Jinsuo, editor.

Title: Nuclear fuel reprocessing and waste management / editor, Jinsuo Zhang
(Virginia Tech, USA).

Description: [Hackensack] New Jersey : World Scientific, 2018. |
Series: Modern nuclear energy analysis methods ; volume 2 |
Includes bibliographical references and index.

Identifiers: LCCN 2018014816 | ISBN 9789813271364 (hc : alk. paper)

Subjects: LCSH: Reactor fuel reprocessing. | Radioactive waste disposal. |
Radioactive wastes--Storage.

Classification: LCC TK9360 .N8525 2018 | DDC 621.48/38--dc23

LC record available at https://lccn.loc.gov/2018014816

British Library Cataloguing-in-Publication Data

A catalogue record for this book is available from the British Library.

For any available supplementary material, please visit
http://www.worldscientific.com/worldscibooks/10.1142/11017#t=suppl

Desk Editors: Dr. Sree Meenakshi Sajani/Amanda Yun

Typeset by Stallion Press
Email: enquiries@stallionpress.com

Preface

Nuclear power is increasingly recognized and acknowledged as clean, safe, reliable, and economic. Multiple new generating facilities as well as advanced nuclear reactors are being developed around the world. However, efforts on the development of innovative nuclear energy are encountering significant resistance. There is widespread public skepticism of the benefits and concerns about potential problems. These concerns include the methods used for disposal of discharged nuclear fuels (also known as spent nuclear fuel or used nuclear fuel) and other radioactive wastes generated by the reactor. How to manage the spent nuclear fuel is a major impediment to increasing nuclear power usage and advanced nuclear reaction development.

There are two types of nuclear fuel cycles: open fuel cycle also known as once-through fuel cycle and closed fuel cycle. In the open fuel cycle, the spent nuclear fuel is treated as waste and is disposed in a deep geological repository directly without additional reprocessing. The once-through fuel cycle approach is now the preferred strategy in U.S. for various reasons including cost considerations. While there is considerable validity to this approach, there are also countervailing considerations. The long-term high radioactive level and high decay heat may lead to both engineering and scientific issues such as structural material degradation that will influence the performance and long-term stability of the repository. In the closed fuel cycle, the useful fissionable materials as well as the radioactive isotopes with long-term half-life are separated from the spent fuel using a reprocessing technology before permanent disposal. Therefore, the closed fuel cycle has significant benefits over open fuel cycle in terms of the waste volume, long-term radioactive level, and decay heat.

There are books available that cover the entire nuclear fuel cycle from uranium mining to radioactive waste disposal. However, considering that the nuclear fuel cycle is a broad area including ore mining, milling, enrichment, fuel fabrication, nuclear reactor, spent fuel storage, reprocessing technology, and radioactive waste disposal, deep discussions on each topic in one book that covers the entire range of process should not be expected. There are also books available that focus on particular topics; however, some the topics need to be updated because of the most recent development in science, engineering, and technology. The present book focuses on the back-end processes of a closed nuclear fuel cycle, including spent fuel interim storage, reprocessing technology, waste form design, and radioactive waste disposal. For each topic, there are general instruction as well as the most recent research progresses. Therefore, the book addresses readers with professional research interests or with a little background information, such as college students.

In principle, there are two types of nuclear fuel reprocessing technology: aqueous reprocessing such as PUREX (Plutonium-Uranium-EXtraction) and dry processing such as preprocessing based on electrochemical separation. The book covers both of the technologies. Chapter 1 is an introduction chapter on PUREX and its variant processes such as UREX (Uranium EXtraction) and NUEX (New Uranium EXtraction). The separation components, fundamentals, and chemical reactions are addressed in this chapter. Chapter 2 presents an introduction on pyroprocessing that addresses history, background, and fundamentals of the process. The chapter covers the entire process including oxide fuel treatment, electrochemical separation, electrode processing, and salt waste processing.

Aqueous-based process is a mature technology, and reprocessing plants based on this technology are commercially available. There are books that particularly focus on aqueous-based process. While pyroprocessing is considered as an under-developed technology and commercial plants based on the technology are not available, there are also no books particularly on this topic. Therefore, Chapter 3 and Chapter 4 address two research topics on pyroprocessing: electrochemical studies on uranium and safeguarding for pyroprocessing.

In the electrorefiner of the pyroprocesing system, molten LiCl-KCl (40.5 at%) salt was selected as the electrolyte. Thermochemical properties of actinides and fission products in the salt must be known for designing a pyroprocessing system. Chapter 3 focuses on available data of uranium properties in the salt. These properties include activity coefficient, potential, diffusion coefficient, as well as the exchange current. Electrochemical methods on how to measure these properties are also discussed in Chapter 3.

Safeguarding approaches for PUREX have been well developed. However, this is not the case for pyroprocessing. Compared with PUREX, pyroprocessing does not have an accountability tank, which make it almost impossible to accurately measure the amount of nuclear materials that enter into separation system, such as electrorefiner, which is heart of the pyproprocessing system. Therefore, Chapter 4 focuses on the safeguarding approaches for pyroporcessing. The chapter first discusses the available electrochemical methods to measure the materials in operation for safeguarding purpose. Then the chapter discusses a modeling approach for safeguarding a pyroprocessing facility. The model has the capabilities for predicting material distribution in the system; therefore, the model can be applied to track the nuclear materials in operation. The chapter also conducts a critical review on available fundamental data for model input.

Through reprocessing the spent nuclear fuel, the quantities (e.g. mass, volume, radiotoxicity) are expected to reduce. Then, the radioactive waste can be ultimately disposed. Chapter 5 focuses on a few technical, safe, and environmentally sound options for ultimate disposal of high-level waste resulting from reprocessing activities. The chapter discusses alternatives to mined repositories including interim storage and geological alternatives. Then the chapter focuses on geological disposal. The principles, performance assessment, as well as issues and challenges are discussed.

About half the spent nuclear fuel around the world is stored for eventual treatment. Dry storage is one type of interim storage. Chapter 6 focuses on dry storage. In dry storage, one of the major concerns is the material degradation by environment which may lead to radioactive material release during storage or transportation of the storage canisters. Chapter 6 discusses the potential structural material degradation mechanisms including general corrosion, pitting corrosion, and chloride-induced stress corrosion cracking. Experimental methods on how to measure the corrosion and results are also discussed.

The stability for ultimate disposal depends on the disposed wasteform. Chapter 7 discusses the crystalline wasteform phases for ^{137}Cs and ^{90}Sr. The chapter focuses on the structure and stability of the wasteform. The experimental techniques, microporous ion-exchanges for Cs and Sr separation, and the wasteform phases for Cs and Sr immobilization are discussed in the chapter.

Finally, I would like to thank all the contributors: Dr. Jack D. Law, Prof. Michael F. Simpson, Dr. Dalsung Yoon, Prof. Supathorn Phongika-roon, Prof. Wentao Zhou, Dr. Jean-Francois Lucchini, Dr. Yi Xie, and Dr. Hongwu Xu. The authors of Chapter 3 and Chapter 4 would like acknowledge the funding support for the original research on the topic by U.S. Department of Energy (DE-NE0000710). The authors of Chapter 6 would like to acknowledge the funding support for the original research on the topic by U.S NRC (NRC-HQ-11-G-39-0036).

About the Authors

Jack D. Law is a Chemical Engineer with 33 years of experience at Idaho National Laboratory (INL), USA in the areas of nuclear fuel reprocessing and the nuclear waste treatment. His work has primarily focused on flowsheet design and testing for separation of actinides, lanthanides, and fission products from radioactive solutions, solvent extraction equipment testing, and facility design support. Jack is currently Manager of the Aqueous Separation and Radiochemistry Department at INL. He has contributed to over 50 peer reviewed publications, and has been awarded 10 US patents and 3 Russian patents. He was awarded the AICHE Nuclear Engineering Division Robert E. Wilson Award in 2016 and the INL Laboratory Director's Award for Lifetime Achievement in 2015.

Jean-Francois (Jef) Lucchini has been a scientist at Los Alamos National Laboratory Carlsbad Operations, USA for over 15 years. He is a technical expert on difficult transuranic waste to the Department of Energy Carlsbad Field Office and generator sites in support of the Waste Isolation Pilot Plant. He holds a PhD in radiochemistry from the University of Paris XI Orsay, France. He is author or co-author of more than 25 peer-reviewed publications on different topics of radioactive waste research and management, and an active member of the American Nuclear Society.

Supathorn Phongikaroon earned his PhD and BS degrees in chemical engineering and nuclear engineering from University of Maryland, College Park, USA in 2001 and 1997, respectively. Prior to joining the Virginia Commonwealth University, USA in January 2014, he held academic and research positions at University of Idaho in Idaho Falls, USA; Idaho National Laboratory, USA; and the United States Naval Research Laboratory, Washington, D.C. He has established the chemical and electrochemical separation of used nuclear fuel through pyroprocessing technology and extended his expertise toward reactor physics and material detection, and accountability for safeguarding applications. His work has been published in over 30 papers in peer-reviewed journals and presented at over 90 international and national conferences and workshops.

Michael F. Simpson is Professor of Metallurgical Engineering at the University of Utah, USA. After earning a BS in chemical engineering from California Institute of Technology, USA in 1991 and a PhD in chemical engineering from Princeton University, USA in 1996, he joined Argonne National Laboratory — West (later Idaho National Laboratory), USA as a research engineer. His career at Idaho National Laboratory notably included developing extensive research collaborations with the Republic of Korea in the area of pyroprocessing. His accomplishments in the field of pyroprocessing include the development of high accuracy, *in situ* molten salt analysis methodology, measurement of kinetics of redox reactions with soluble metals in molten salts, development of durable zeolite-based waste forms for salts, and high efficiency actinide extraction from molten salts.

Yi Xie has been on the postdoc of the Mechanical Engineering department at Virginia Tech, USA since 2017. She received her PhD in Nuclear Engineering (materials) from Ohio State University, USA in 2016, and BS in Nuclear Engineering from University of Science and Technology of China in 2012. She received the Nuclear Engineering Achievement Award from Ohio State University in 2016. Her research interests include nuclear materials degradation, advanced nuclear fuel development, electrochemical analysis, and multi-dimensional materials characterization. She has published about 10 journal papers.

Hongwu Xu is a Senior Scientist in the Earth and Environmental Sciences Division of Los Alamos National Laboratory, USA. He received his PhD and MA in geosciences from Princeton University, USA, and his MS and BS in crystallography, mineralogy, petrology, and geochemistry from Nanjing University, China. His research interests focus on the determination of structure-stability relationships of both natural minerals and synthetic materials at high-pressure variable-temperature conditions using a combined approach of synchrotron X-ray/neutron scattering and calorimetric techniques. He is a Fellow of the Mineralogical Society of America, serves as an Associate Editor for *American Mineralogist*, and has published more than 130 papers in peer-reviewed journals.

Dalsung Yoon joined Korea Atomic Energy Research Institute (KAERI), South Korea in August 2017. He held a postdoctoral position at the Virginia Commonwealth University (VCU), USA after receiving his PhD in mechanical and nuclear engineering from VCU in December 2017, focusing on uranium electrochemical studies in molten LiCl–KCl eutectic salts. He earned his MS in quantum energy chemical engineering from the University of Science and Technology in South Korea on February

2010. He has been focusing on determining electrochemical/kinetic parameters of nuclear materials via different electrochemical techniques,and has authored/co-authored over 10 published papers in peer-reviewed journals and 6 presentations at international and national conferences.

Wentao Zhou is currently Assistant Professor of the Nuclear Engineering Program at Shanghai Jiao Tong University, China. He received his PhD from The Ohio State University in 2017 and BS degree from the University of Science and Technology of China in 2013, both in nuclear engineering. His research mainly focuses on the pyroprocessing of spent nuclear fuel, electrochemical separation, and facility safeguards. He specializes in investigating the material properties in molten salt systems and developing models to monitor the electrochemical separation process; predicting the separation performance; and detecting possible nuclear material diversions.

About the Editor

Jinsuo Zhang is currently Professor of the Nuclear Engineering and Science Program at Virginia Tech, USA. Before that, Professor Zhang was an Associate Professor and the Director of the Center of Nuclear Materials and Fuel Cycle Research in the Nuclear Engineering Program at The Ohio State University (2012–2017). Through 2004–2012, Professor Zhang was a staff scientist at the Los Alamos National Laboratory (LANL), USA, and he was a postdoc research associate of the same laboratory from 2001 to 2004. Professor Zhang received his PhD from Zhejiang University, China in 2001, and his BS from the same university in 1997. Professor Zhang focuses on studies of advanced used nuclear fuel reprocessing, safeguards and non-proliferation, nuclear materials, material compatibility and materials corrosion in advanced and current nuclear reactors.

Contents

Chapter 2.   Fundamentals of Spent Nuclear Fuel Pyroprocessing

27

Michael F. Simpson

Chapter 5. Waste Disposal   145

Jean-Francois Lucchini

Chapter 6. Spent Fuel Interim Dry Storage System and Chloride-Induced Stress Corrosion Cracking   177

Yi Xie and Jinsuo Zhang

Chapter 7. Crystalline Wasteform Phases for ^{137}Cs and ^{90}Sr: Structure and Stability 237

Hongwu Xu

Chapter 1

Aqueous Reprocessing of Used Nuclear Fuel

Jack D. Law

Idaho National Laboratory
Idaho Falls, ID 83415-3870, USA
jack.law@inl.gov

1. Introduction

Aqueous technologies for used nuclear fuel reprocessing have historically been utilized to recover uranium (U) and plutonium (Pu) from irradiated nuclear reactor fuel for recycle. The purpose for reprocessing used fuel has been to recover unused U and Pu in the used fuel elements. This results in gaining more energy from the original U and contributes to the national energy security of the country reprocessing their fuel. Also, reprocessing results in a reduction in volume of high-level waste for disposal, and the radiotoxicity is lower and reduces more rapidly than with used nuclear fuel.

With aqueous technologies, the used nuclear fuel is typically mechanically chopped into small pieces and leached into an acidic solution. The resulting dissolver product is chemically processed to separate actinides for recycle to a reactor. The remaining metals and fission products are treated for disposal as a high-level waste. The primary aqueous separation method utilized to accomplish the required separation is solvent extraction, although precipitation has been used initially in the US defense industry for the recovery of Pu for weapons.

The plutonium uranium reduction extraction (PUREX) process is the most common solvent extraction technology utilized in the USA and internationally for the separation of U and Pu from used nuclear fuel. Variations in the PUREX process are being developed and implemented to prevent the separation of pure Pu. Additionally, advanced aqueous separation technologies are being developed in the USA and internationally for the separation and recycle/transmutation of minor actinides.

2. Solvent Extraction

Solvent extraction is the primary technology utilized in aqueous used nuclear fuel reprocessing. First, used nuclear fuel from a reactor, after some amount of decay storage, is leached with an acidic solution. The resulting aqueous solution is separated from the remaining fuel cladding and then chemically processed via solvent extraction to separate the components of interest, typically U and/or Pu.

Solvent extraction within used nuclear fuel reprocessing utilizes an organic phase containing an extractant, in contact with the aqueous dissolver product via mixing, to extract the components of interest into the organic phase. Typically, this process is carried out by intimately mixing the two immiscible phases, allowing for the selective transfer of solute(s) from one phase to the other, then allowing the two phases to separate. The component of interest is subsequently removed from the organic phase to an aqueous phase via back-extraction. In order for effective processing, the two phases must be immiscible, have enough of a density difference to allow rapid disengagement, be of an appropriate viscosity to be transported through process equipment, limited solubility of the organic phase in the aqueous phase to maintain the extractant concentration over long-term use, and the organic phase must have sufficient hydrolytic and radiolytic stability to allow for long-term reuse to minimize organic waste volumes.

One of the primary advantages of solvent extraction processes is the ability to operate in a continuous, countercurrent manner with multiple contacting/separating stages to achieve the desired removal efficiency of the components being extracted. This allows for continuous operation instead of batch operation. Countercurrent operation is shown graphically for an extraction section of a U/Pu separation flowsheet in Figure 1.

In this flow diagram, the aqueous dissolver product feed stream containing the components to be extracted enters at one end of the process (A_{N+1}), and the fresh solvent (organic) stream enters the other end (O_0). The aqueous and organic steams flow countercurrently from stage to stage where they are intimately mixed and separated, and the final products are the solvent loaded with the desired components (e.g., U and Pu), O_N, leaving stage N and the aqueous raffinate, A_1, depleted in U and Pu leaving stage 1.

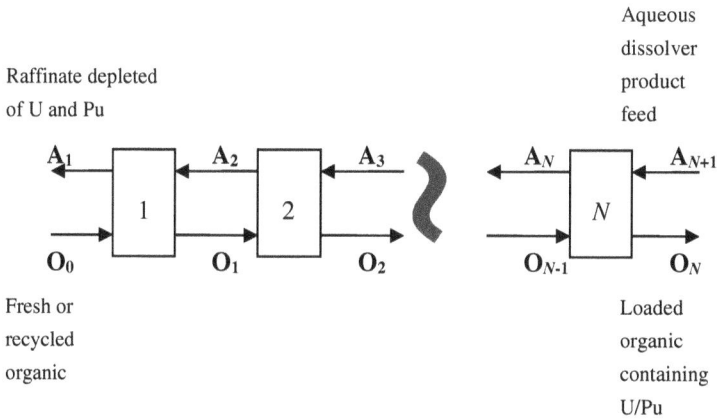

Figure 1. Countercurrent — Multistage extraction process flow diagram.

Following an extraction section, scrub, strip, and solvent wash sections are typically employed. The purpose of the scrub section is to back extract (scrub) to the extraction section any metals/fission products (e.g., Zr) that may have co-extracted along with the desired components. The strip section is utilized to back-extract the desired components (e.g., U and Pu) into an aqueous phase where they will ultimately be re-enriched and converted into an oxide form for recycle to a reactor, typically as MOX fuel. The solvent wash section is utilized to remove any hydrolytic or radiolytic degradation products from the solvent and prepare the solvent for recycle to the extraction section, thus minimizing organic waste volumes.

3. Solvent Extraction Equipment

Solvent extraction equipment has been utilized in the nuclear industry since the implementation of the PUREX process for the separation of Pu and U from fission products for weapons production in the 1950s [1]. This technology was later implemented for the reprocessing of used nuclear fuel from commercial power reactors. PUREX-based solvent extraction processing has been and continues to be utilized worldwide including France, Japan, Russia, the UK, and the USA. Additionally, considerable research in the development of advanced separation processes is ongoing, requiring the use of solvent extraction equipment. In these efforts, solvent extraction equipment is utilized to efficiently mix and separate the aqueous and organic phases to facilitate separation of the species of interest. The solvent extraction equipment is designed for countercurrent flow to facilitate high separation efficiencies and continuous operation.

To support operation in a radioactive environment, solvent extraction equipment must be remotely operable and maintainable, resistant to high radiation fields, capable of continuous countercurrent operation, and be critically safe for certain applications. In addition, desirable attributes include the ability to accommodate solids, a small process footprint, and operational flexibility (continuous long-term operation or frequent start–stop operation). The three main types of solvent extraction equipment used in industrial-scale reprocessing facilities and in fuel cycle research laboratories include (1) columns, (2) mixer-settlers, and (3) centrifugal contactors. A brief description of each type of equipment follows. More detailed description of these type of equipments can be found in literature [2, 3].

3.1. *Columns*

Packed columns and pulse columns with sieve plates or trays are the typical type of solvent extraction column utilized in the nuclear industry. Packed columns contain packing material (e.g., Raschig Rings) to create mixing of the two phases as they flow counter currently. The phases disengage at either the top or the bottom of the column depending upon whether the column is operated in an aqueous or organic continuous mode. Pulsed columns with trays or plates were developed to increase the mixing intensity of the phases and, thus, decrease the height requirement of the column. The mechanical energy applied to the column via pulsing with air facilitates the formation of small droplets for mass transfer. A photograph of an operating pulsed column used to support laboratory testing is shown in Figure 2.

3.2. *Mixer-settlers*

A mixer-settler contains a small mixing chamber and a larger settling chamber. The two immiscible phases enter the mixing chamber containing an impeller. This dispersion that forms flows into the settling chamber where the two phases separate by gravity. A system of weirs allow for the light phase to flow over the higher weir and the heavy phase to flow under the lower weir. Multiple mixer-settlers are configured for countercurrent flow to achieve high separation efficiency. A photo of several mixer-settlers is shown in Figure 3.

Figure 2. Operating lab-scale pulsed column at Idaho National Laboratory.

Figure 3. Industrial and lab-scale mixer-settlers.

3.3. *Centrifugal contactors*

Centrifugal contactors consist of a hollow rotor that rotates within a cylindrical housing utilizing a motor mounted above the housing. The aqueous and organic feed solutions enter a stage near the top of the housing and are mixed as they flow downward. For annular centrifugal contactors, the

Figure 4. Operating 5 cm centrifugal contactor with plastic housing.

shear forces between the spinning rotor and the stationary housing mix the two phases. Other contactor designs utilize alternative methods of mixing such as mixing pins attached to the bottom of the spinning rotor. When the mixture enters the spinning rotor through an opening at its base, the spinning rotor acts as a centrifuge with the heavy phase going to the rotor wall. A series of weirs allow the separated heavy and light phases to flow out of the contactors as separate streams. Multiple centrifugal contactors are configured for countercurrent flow to achieve high separation efficiency. An operating centrifugal contactor is shown in Figure 4.

4. Reprocessing History

Aqueous reprocessing technologies were first utilized at the Hanford Site during the Manhattan Project for the recovery of Pu-239 by using a bismuthate phosphate precipitation process [1]. Recovery transitioned to solvent extraction processes that allowed for continuous instead of batch operation, as well as the concurrent recovery of U and Pu. Also, the purpose of the reprocessing of used fuel transitioned from defense purposes to commercial use for the recycle of U and Pu.

The first solvent extraction process implemented was the reduction oxidation (REDOX) process [1]. The process was developed and tested at Argonne National Laboratory and Oak Ridge National Laboratory, and the first REDOX plant operated at Hanford. The REDOX process utilizes Methyl isobutyl ketone (hexone) to extract uranyl nitrate and Pu nitrate. U(VI) and Pu(IV) are co-extracted then the Pu is reduced to Pu(III) by using ferrous sulfamate and selectively stripped from the U. One disadvantage of this process is the required addition on aluminum nitrate to increase the nitrate concentration. Also, hexone is a flammable and volatile solvent.

The BUTEX process was developed by British scientists at Chalk River Laboratory [4]. This process utilized dibutyl carbitol as the extractant and used nitric acid as a salting agent instead of aluminum nitrate, thus reducing the problem of larger waste volumes due to the addition of $Al(NO_3)_3$. This process was implemented at the Windscale plant at the Sellafield Site in the UK for the reprocessing of used fuel.

Aqueous reprocessing transitioned to the PUREX process in the 1950s [1]. The PUREX process was developed by Knolls Atomic Power Laboratory and tested at ORNL. A modified PUREX production-scale plant was used in Idaho at the Idaho Chemical Processing Plant beginning in 1953. This process utilized hexone as an extractant in the first cycle and PUREX in the second and third cycles. The PUREX process was utilized on a production scale at Savannah River in the F-Canyon in 1954 [5]. It has been used at the Savannah River Site since 1955 in the H-Canyon facility [5]. It replaced the REDOX process at Hanford in 1956. From here, the use of the PUREX process expanded within the USA and internationally.

5. PUREX Process

The PUREX process has historically been the primary aqueous separation process utilized worldwide for the reprocessing of used nuclear fuel. Table 1 lists historical, current, and in-construction reprocessing facilities utilizing the PUREX process or a variant.

The front-end of the PUREX-based reprocessing consists of chopping the used fuel into small pieces; leaching the used fuel from the cladding using a nitric acid solution; separation of the fuel cladding pieces, spacers, and other fittings; chemical adjustment; and filtration of this dissolver product. This dissolver product is then fed to a first cycle of solvent extraction where the U and Pu are separated from the dissolver product and subsequently separated from each other. Several additional cycles of solvent extraction are used to further purify the U and Pu. The resulting U and/or Pu product

Table 1. Historical PUREX reprocessing facilities.

Facility	Country	Operation	Capacity (MTHM/yr)	Comments
Savannah River F, H-Canyon [5]	USA	1955–present	5–10 MT/yr	Recovered Pu, U, and Np from weapon production reactors
Hanford [6]	USA	1956–1972, 1983–1988	8.3 MT/day design up to 20 actual	Recovered Pu, U, and Np from Hanford production reactors
UP1 Magnox-Marcoule [7]	France	1958–1997	960 MT/yr	Fuel from Pu production reactors — later, power reactors
B20 — Sellafield [8, 9]	UK	1964–present	1500 MT/yr	Magnox fuel [6]
Trombay [8, 9]	India	1964–1973, 1983–present	60 MT/yr	Research reactor fuel [10]
West Valley [8, 10]	USA	1966–1972	300 MT/yr	First plant in USA to process commercial fuel
Morris, IL [8, 10]	USA	NA	300 MT/yr	Construction halted prior to ops. F volatility polishing
Barnwell, SC [8, 10]	USA	NA	1500 MT/yr	Shutdown during startup testing US reprocessing policy change
UP2 Magnox-LaHague [7]	France	1966–1976	800 MT/yr	Magnox fuel then converted into oxide fuel
Tokai [8]	Japan	1975–2014	210 MT/yr design, 100 actual	Boiling water reactor (BWR) and pressurized water reactor (PWR) fuel, adv. thermal reactor fuel

UP2 Oxide-LaHague [5]	France	1976–present	800 MT/yr design, 400 actual	Commercial light water reactor (LWR) fuel
RT1 Mayak [8]	Russia	1976–present	400 MT/yr	VVER–440 fuel, preparing to expand to VVER-1000 fuel
UP3 Oxide-LaHague [7]	France	1989–present	800 MT/yr	Commercial LWR fuel for foreign countries
Thorp [8]	UK	1994–present	1200 MT/yr	Foreign and domestic fuel. Planned shutdown in 2018
KARP-Kalpakkam [9]	India	1996–present	100 MT/yr	Reprocesses oxide fuels from PHWRs
Jiuquan [11]	China	2010–present	50 MT/yr	Pilot scale facility
PREFRE-2 Tarapur [9]	India	2011–present	100 MT/yr	Oxide fuels from PHWRs
Rokkasho [6]	Japan	2018	800 MT/yr	Planned startup in 2018. PUREX process variant
New [12]	China	2025–2030	800 MT/yr	AREVA contract to design, 2030 target

is converted into an oxide form. In commercial fuel reprocessing, the Pu oxide product, with or without U, is recycled to a reactor as mixed oxide (MOX) fuel in light water reactors (LWRs). About 50 reactors in Europe and Japan are licensed to use MOX fuel with around 30 actually doing so.

The PUREX process utilizes 20–40 vol% tributyl phosphate (TBP) in a hydrocarbon diluent (typically kerosene or n-dodecane) to extract U and Pu from the dissolver product. The PUREX process is effective in extracting actinides in the +4 and +6 oxidation state. U, as U(VI) is extracted as follows:

$$UO_2^{2+} + 2NO_{3-} + 2TBP \leftrightarrow UO_2(NO_3)_2 \cdot 2TBP \tag{1}$$

The chemical equilibria for the actinides in the +4 oxidation state is

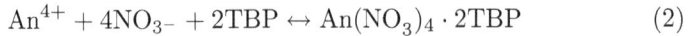

$$An^{4+} + 4NO_{3-} + 2TBP \leftrightarrow An(NO_3)_4 \cdot 2TBP \tag{2}$$

In the PUREX process, Pu is present as Pu(IV) and is co-extracted with the U. Other actinides and lanthanides present in the +3 or lower oxidation state are not extracted with the PUREX process, thus allowing for an efficient separation. Neptunium is another actinide that can be maintained as Np(VI), if desired, and co-extracted with the U and Pu.

6. PUREX Process Variants

Numerous modifications to the standard PUREX process have been or are currently being developed. The primary purpose of these variations is to prevent the production of a pure Pu stream due to concerns with proliferation. With the standard PUREX process, a pure U stream and a pure Pu stream are produced. Theft or diversion of the reactor-grade Pu for weapons use has been a concern under these conditions. While no separations process can be made proliferation-proof, modifications to the PUREX process can effectively reduce the proliferation risk. Table 2 lists several PUREX process variants that have been or are currently being developed.

6.1. COEX^{TM} process

The co-extraction (COEX^{TM}) process, developed in France by AREVA, utilizes variants of the PUREX process chemistry to allow some of the extracted U to follow the Pu, resulting in three streams; a U/Pu product, a U product, and a raffinate waste solution containing fission products and minor actinides [13]. This is accomplished using a reducing agent, such as U(IV) nitrate or hydroxylamine nitrate to reduce the Pu(IV) to

Table 2. PUREX process variants.

Technology	Country	Variation
COEX™	CEA/AREVA France	No separation of pure Pu
UREX	Department of Energy (DOE), USA	Single cycle U and technetium (Tc) separation, lower acidity
Co-decontamination	DOE, USA	No separation of pure Pu
NUEX	Energy Solutions	No separation of pure Pu
NNL advanced process	National Nuclear Laboratory, UK	Single cycle, no separation of pure Pu
NEXT process	JAEA, Japan	Combines crystallization with PUREX process, no separation of pure Pu
Simplified PUREX	Russia	Thermochemical decladding of the used fuel assemblies instead of chopping. Operated at lower acidity
PARC process	India	Tc and Np are extracted by TBP along with the U and Pu. Single cycle

Pu(III) and back-extract this Pu while only partially reducing and back-extracting the U(VI). Additionally, this multi-cycle process includes the use of a co-conversion process to produce a U/Pu oxide product. At no point during the process is pure Pu separated.

6.2. *UREX*

The uranium extraction (UREX) process is a variant of the PUREX process that was developed by the US Department of Energy (DOE) as a method to separate U, without co-extracting Pu [14, 15]. Co-extraction of Pu and Np is prevented by introduction of a complexant/reductant, such as acetohydroxamic acid (AHA) in the scrub feed. The complexation of Pu and Np are enhanced in the UREX process through the use of low-acidity feed and scrub solutions, as well as to enhance the extractability of technetium. Development of the UREX process has progressed to the point of demonstrations with actual used nuclear fuel using 2-cm centrifugal contactors, with good results [14].

6.3. *Co-decontamination process*

The co-decontamination process is a variant of the PUREX process that is being developed by the US DOE as a method of co-extracting U and Pu, with no separation of pure Pu [16]. There are several variants of this process with similar goals that utilize various reductants. Previously, a co-decontamination process has been demonstrated with actual used nuclear fuel at the laboratory scale. With these tests, it has been demonstrated that a U–Pu–Np product can be produced which contains approximately 10% Pu–Np and, in a second test, which contains approximately 30% Pu–Np [15]. Recent focus of development of the co-decontamination process is being pursued through a multi-year experimental study to evaluate the technological capability to control the preparation of U/Pu product [18]. To this end, the process being developed, after co-extraction of U and Pu via standard PUREX process chemistry, uses hydrazine-stabilized U(IV) as a Pu reductant. An excess of the reductant will produce a U/Pu nitrate solution at a flexible range of U/Pu ratios, dependent upon process goals. This ratio is determined by adjusting the flow rate in the back-extraction section, the purpose of which is to remove excess U in the solvent and produce the desired product ratio. The co-decontamination process is under active development by the US DOE, with flowsheet testing planned in the 2018 timeframe [18].

6.4. *NUEX process*

The new uranium extraction (NUEX) process is a PUREX process variant that was designed by EnergySolutions as a potential near-term reprocessing flowsheet for application in the USA for used LWR fuel [17, 19]. This process modifies the PUREX process flowsheet utilized in the Thorp reprocessing facility at the Sellafield Site in the UK to produce a U/Pu/Np product and prevent the production of a pure Pu stream. This is accomplished through modifications to the three cycle PUREX process by replace the $U(IV)/N_2H_4$ reductant from the THORP flowsheet with AHA. This would result in partitioning U from U/Pu by complexation rather than reduction, resulting in U and U/Pu/Np products. Technetium is also separated and recovered with this process.

6.5. *NNL advanced process*

The National Nuclear Laboratory in the UK is developing a simplified flowsheet variant of the PUREX process for the reprocessing of GenIV fuel [17]. The process consists of a single cycle flowsheet that utilizes a hydroxamic acid complexant, much like the NUEX process described previously. Research and development to date have indicated that the use of AHA (1) results in a high decontamination factor for Pu in the U product stream, (2) the U decontamination factor in the Pu product stream can be maintained low enough to result in a U/Pu product instead of pure Pu, (3) Np is separated with the U/Pu product stream, (4) technetium (Tc) mostly follows the U but more R&D is needed, (5) process kinetics allow the use of centrifugal contactors as the separation equipment, and (6) AHA degradation to acetic acid is an issue relative to recycle of nitric acid within a facility.

6.6. *NEXT process*

The new extraction system for transuranic (TRU) recovery (NEXT) is being developed by the Japan Atomic Energy Agency (JAEA) for the reprocessing of fast reactor fuel [20, 21]. The process combines PUREX process chemistry with crystallization. First, U is partially recovered (approximately 70%) by crystallization of uranyl nitrate hexahydrate from the used fast reactor fuel dissolver product accomplished by lowering the temperature. The resulting uranyl nitrate hexahydrate crystals are washed with nitric acid to further decontaminate them from TRU elements and fission products. The crystals can then be further treated in a crystal purification step [21]. U/Pu/Np is recovered from the mother liquor via PUREX process chemistry without the need for multiple cycles of purification. Further recovery of Am and Cm is obtained through the use of extraction chromatography. The PUREX solvent extraction portion of this process has been demonstrated with actual fast reactor (JOYO) irradiated MOX fuel dissolved in nitric acid by using laboratory-scale centrifugal contactors. With this test, it was shown that Np is recovered with the U and Pu [20].

6.7. *Simplified PUREX process*

The simplified PUREX process is being developed in Russia as a next-generation reprocessing with the goal of reducing the volume of low-level waste generated [22]. The simplified PUREX process, compared with classical PUREX, implements thermo-chemical (dry) operations in the head-end of the process. This is accomplished through thermochemical decladding of the used fuel assemblies instead of chopping. Low-temperature voloxidation of the fuel is performed to release volatile fission products prior to dissolution and to generate a more quickly dissolvable U_3O_8 form. The end result is a more concentrated dissolver product, relative to U concentration in a lower nitric acid concentration feed, resulting the generation of less liquid waste. One variant of the simplified PUREX process dissolves the SNF, after voloxidation, with a TBP $\cdot$ nHNO$_3$ adduct in a supercritical fluid for extraction [22].

6.8. *PARC process*

The partitioning conundrum key (PARC) process is a PUREX process variant being developed by JAEA [23]. With this process, Tc and Np are extracted by TBP along with the U and Pu in a co-extraction step. Extracted Np and Tc are separated from the U/Pu stream via selective reduction of Np(VI) to Np(V) by using normal-butyraldehyde in the presence of U(VI) and Pu(IV) and high acid scrubbing of technetium. The U and Pu can then be separated from each other resulting in four product streams — Tc, Np, Pu, and U products. Second and third cycles of PUREX purification are not required. Further processing of the raffinate to separate Am and Cm can then be performed using adsorption techniques with TODGA and alkyl-BTP solvents [23].

7. Advanced Reprocessing Technologies

Reprocessing technologies for the separation of minor actinides from lanthanides are expected to be an important part of future advanced reprocessing for used nuclear fuel. The recovery and transmutation, in a fast reactor or potentially LWR or Boiling water reactors (BWRs) [24], of long-lived minor actinides to short-lived fission products would reduce the long-term heat load and radiotoxicity of used fuel or high-level waste resulting from reprocessing. Transmutation is accomplished by irradiation in an intense neutron field to form short-lived fission products. The main focus for minor actinide recycle is the separation of Am and Cm from used fuel. There are two main approaches, homogenous and heterogeneous recycles. In the homogeneous recycle, the minor actinides are combined with the U and Pu in fast reactor nuclear fuel. In the heterogeneous recycle, the minor actinides are manufactured into targets, which are subsequently transmutated in the reactor. Minor actinide separation has become a key R&D area worldwide in the development of advanced reprocessing technologies. Many, but not all, of the areas of development are presented below.

7.1. *TRUEX/TALSPEAK process*

The Transuranic Extraction/Trivalent Actinide Lanthanide from Separation with Phosphorus Reagent Extraction from Aqueous Komplexes (TRUEX/TALSPEAK) process is a two cycle solvent extraction process in which the trivalent actinides and lanthanides are first co-extracted from the high-acidity raffinate of a PUREX-based U/Pu separation process raffinate by using the transuranic extraction (TRUEX) process. The actinide/lanthanide product stream from the TRUEX process is then treated using the TALSPEAK process to separate the actinides from the lanthanides.

The TRUEX process was developed by Argonne National Laboratory [25] and consists of an octyl(phenyl)-N,N-diisobutylcarbamoylmethylphosphine oxide (CMPO) extractant in an aliphatic diluent. The CMPO extracts the trivalent actinides and lanthanides from the acidic raffinate of the upfront U/Pu separation process. TBP is added as a solvent modifier to prevent the third phase formation. The TRUEX solvent typically consists of 0.2 mol/L CMPO and 1.4 mol/L TBP in n-dodecane. In the TRUEX process, trivalent actinides and lanthanides are co-extracted from the PUREX raffinate. Once extracted, the actinides and lanthanides are effectively stripped using low-acidity nitric acid. Alternatively, an aqueous solution consisting of diethylenetriaminepentaacetic acid (DTPA) in a lactate buffer at pH 5 is used to facilitate separation in the subsequent TALSPEAK process. This actinide/lanthanide strip product can then be fed directly into the TALSPEAK process with only minor adjustments.

The TALSPEAK process was developed at Oak Ridge National Laboratory (ORNL) [26]. This process functions based on the higher affinity of polyaminocarboxylate ligands for trivalent actinides compared with the trivalent lanthanides. DTPA is added to the aqueous phase, resulting in selective complexation of the actinides, holding them in the aqueous phase while the lanthanides are extracted into the organic phase by using bis-(2-ethylhexyl)phosphoric acid (HDEHP). The aqueous phase is buffered with lactic acid to control the pH and also to improve the extraction kinetics [27]. The TRUEX/TALSPEAK process has been demonstrated at the laboratory-scale by using actual used nuclear fuel. In these tests, the TRUEX and TALSPEAK testing followed an upfront UREX process [28]. Results from these tests indicated >99.99% recovery of Pu, Np, and Cm and 99.97% recovery of Am [28].

7.2. *TRUEX/advanced TALSPEAK*

The TALSPEAK process is very sensitive to aqueous solution pH. In order to reduce the dependence of the process performance on the pH, and thereby obtain more predictable extraction behavior, and, additionally, to obtain more rapid extraction kinetics, the advanced TALSPEAK process is currently being developed in the USA by the DOE. For the advanced TALSPEAK process, some adjustments are made to the TRUEX process. Primarily, the DTPA/lactate stripping solution is replaced with a (HEDTA)/citric acid stripping solution which is compatible with the advanced TALPSEAK process. The advanced TALSPEAK flowsheet is a modified version to the TALSPEAK flowsheet that utilizes a 2-ethylhexylphosphonic acid mono-2-ethylhexyl ester acid (HEH[EHP]) extractant instead of HDEHP and utilizes an aqueous feed composition of (HEDTA)/citric acid instead of DTPA/lactate [29].

Recently, as a collaboration between the US DOE and Forschungszentrum (FZ) Jülich in Germany, an advanced TALSPEAK flowsheet test was performed at Jülich facilities by using a radio-traced feed simulant and 24 stages of 1-cm annular centrifugal contactor manufactured by the Institute of Nuclear Energy Technology (INET) in Beijing, China. Results of these recent tests are yet to be published but separation goals were met.

7.3. *ALSEP process*

The most recent focus in the USA for development of a minor actinide separation process center on simplification to a single process as opposed the two process TRUEX/TALSPEAK system. The primary simplified process being developed is the actinide–lanthanide separation (ALSEP) concept which consists of N,N,N',N'-tetraoctyldiglycolamide (TODGA) or N,N,N',N'-tetra(2-ethylhexyl)diglycolamide (T2EHDGA) as extractants combined with (mono-2-ethylhexyl ester [2-ethylhexylphosphonic acid] (HEH[EHP]) for the co-extraction of trivalent actinides and lanthanides [30]. Scrub sections with nitric acid and citrate are used to back-extract Mo. A citrate-buffered DTPA solution in the pH range of 2.5–4 is used to selectively strip the actinides from the solvent, and a solution of tetraethyldiglycolamide (TEDGA) in nitric acid is used to strip the lanthanides from the solvent.

Proof-of-principle testing of the ALSEP concept has been completed using radio-traced feed streams [30]. These tests have resulted in separation factors of the minor actinides from the lanthanides in the range of 20–40. ALSEP development continues with the development of stripping agents (i.e., modified polyaminocarboxylate chelates) that exhibit enhanced kinetics [31] with a near-term goal of laboratory-scale flowsheet testing in centrifugal contactor equipment using radio-traced simulant leading to a laboratory-scale demonstration with actual dissolved used nuclear fuel.

7.4. *Am(VI) extraction*

Another approach being developed in the USA by the DOE-NE Sigma Team for Advanced Actinide Recycle (STAAR) is through the exploitation of higher oxidation states of Am. The approach is to use a strong oxidant (standard potential of 1.7 V for the Am(III)/Am(VI) redox couple) to oxidize Am(III) to Am(VI), leaving the lanthanides as Ln(III) (with the exception of Ce) and extracting the Am(VI). The most mature process being developed in the USA uses sodium bismuthate [32]. The solid sodium bismuthate oxidizes Am(III) to Am(VI) and 1 M diamylamylphosphonate (DAAP) in *n*-dodecane extracts the Am(VI). The sodium bismuthate also oxidizes Ce(III) to Ce(IV) and the DAAP co-extracts the Ce(IV). A selective strip using dilute HNO_3 or H_2O_2 is accomplished based on the large difference in stability between Am(VI), which is unstable, and Ce(VI), which is stable. This process has been tested using a radio-traced simulant in 5-cm centrifugal contactors with a resulting removal efficiency of 62% compared with a batch contact removal efficiency of 65% obtained immediately prior to the radio-tracer test. These results demonstrated that Am(VI) can remain oxidized long enough to accomplish a separation by using solvent extraction in engineering-scale equipment. Recent research focuses on developing the selective stripping and testing of alternative extractants such as butyramides [33].

7.5. *Group hexavalent actinide precipitation*

Another promising approach being developed by the US DOE-NE STAAR program focuses on a group separation of all actinides from U to Am in the hexavalent form [34]. This process would forgo a required upfront PUREX-type process to separate and Pu prior to the minor actinide separation process, thus further simplifying the overall reprocessing flowsheet. The main challenge to such a process is the oxidation and stability of Am(VI). Sodium bismuthate has been shown to be an effective oxidant for Am [32]. As such, the concept behind this group hexavalent actinide separation process is to oxidize U, Pu, Np, and Am to the hexavalent state using sodium bismuthate. The solution is then cooled to 2C, resulting in co-crystallization of the hexavalent actinides. Results to date indicate near proportional removal in the range of 61–71% with a single crystallization [34]. Multiple recrystallizations have the potential to significantly increase the recovery of the hexavalent actinides. Study of the fission products Zr, Cs, Ce, and Nb indicates that Nb is separated significantly (12%). The fission products Nb, and to a lesser extent Zr, Cs, and Ce showed some separation under the highly oxidizing conditions created with the sodium bismuthate [34].

7.6. *DIAMEX/SANEX process*

The diamide extraction (DIAMEX)/selective actinide (SANEX) process, developed in France in collaboration with European researchers, is a two-step process which would follow a PUREX type U/Pu separation process. First, the lanthanides and trivalent actinides are separated from the remaining fission products (DIAMEX), and then the trivalent actinides are separated from the lanthanides in the SANEX process [35].

The DIAMEX process uses a malonamide, such as 1 M N,N'-dimethyl-N,N'-dioctyl-hexylethoxy-malonamide (DMDOHEMA) in a hydrogenated tetrapropylene, to co-extract the lanthanides and trivalent actinides [36]. After scrubbing with a HNO_3/oxalic acid/HEDTA scrub and a second 1 M HNO_3 scrub, the actinides and lanthanides are back-extracted into a 0.3 HNO_3 stream. This strip product is feed for the SANEX process.

The SANEX process uses an extractant consisting of 15 mM 6,6'-bis(5,5,8,8-tetramethyl-5,6,7,8-tetrahydro-benzo[1,2,4]triazin-3-yl)-[2,2']bi-pyridine (CyMe4-BTBP) and 0.25 M DMDOHEMA in an octanol diluent [35]. The acidity of the aqueous feed from the DIAMEX process is increased to 2 M HNO_3. A dilute HNO_3 scrub is used to back-extract the lanthanides and 0.5 M glycolic acid at pH 4 is used to strip the actinides from the solvent.

The DIAMEX and SANEX processes are relatively well developed. The DIAMEX process has been tested with simulated and actual used nuclear fuel solutions in a variety of equipment — mixer-settlers, pulse columns, and centrifugal contactors [17, 36]. The SANEX process has been demonstrated in laboratory-scale centrifugal contactors with actual used fuel solutions [35].

7.7. *DIAMEX-SANEX/HDEHP process*

A single-step process has been developed by the French CEA by combining bis-(2-ethylhexyl)phosphoric acid (HDEHP) with the DMDOHEMA from the DIAMEX/SANEX process to co-extract the lanthanides and trivalent actinides and selectively strip first the actinides and then the lanthanides [17, 37]. The HEDTA holds the lanthanides in the organic phase during the actinide strip. The actinide strip solution uses HEDTA and citric acid to back-extract the actinides followed by dilute nitric acid to back-extract the lanthanides. Zr, Mo, and Fe co-extract with this process, requiring a citric acid strip prior to the actinide strip, to remove the Mo, and an oxalic acid/nitric acid strip to remove the Zr and Fe [17]. Simulant testing and hot testing with actual used fuel solution (ATALANTE facility) have been successfully performed using laboratory-scale equipment [17].

7.8. *I-SANEX*

The innovative selective actinide separation (I-SANEX) process uses TODGA and octanol in a TPH diluent to extract the minor actinides and lanthanides from a PUREX raffinate [38]. A series of two scrub solutions back-extract Mo, Zr, Sr, and HNO_3 from the solvent. Trans-1,2-diaminocyclohexane-N,N,N',N'-tetraacetic acid (CDTA) is added to the feed and scrub as a complexing agent for Zr and Pd. Selective stripping of the actinides from the lanthanides is accomplished using SO_3–Ph–BTP in HNO_3 to back-extract the actinides and citric acid solution buffered to pH 3 to back-extract the lanthanides. This process has been demonstrated at the laboratory scale by using centrifugal contactors and radio-traced PUREX raffinate simulant [38]. Results of the testing indicate >99.9% recovery of the Am(III), Cm(III), and Ln(III) with <0.1% contamination of the actinide product with lanthanides. The scrub section was effective in back-extraction of Mo, Zr, and Sr. Ruthenium was found to extract (16%) and mostly remained in the solvent, requiring further research.

7.9. *1-Cycle SANEX*

The one-cycle selective actinide separation (1-Cycle SANEX) process uses $CyMe_4BTBP$ and TODGA in a TPH/1-octanol diluent to extract the minor actinides from a PUREX raffinate [38]. Scrub solutions consisting of oxalic acid and HNO_3 are used to back-extract Zr and residual lanthanides and -cysteine in HNO_3 is used to back-extract Pd(II). The minor actinides are then stripped using a solution of glycolate at pH 4. This process has been demonstrated at the laboratory scale by using centrifugal contactors and radio-traced PUREX raffinate simulant [38]. Results of the testing indicate >99.8% recovery of Am(III), >99.4% recovery for Cm(III), and satisfactory decontamination of the actinide product from fission products and lanthanides. Slow extraction kinetics and limited loading capacity of the organic phase are issues requiring further research [38].

7.10. *CEA-GANEX and Euro-GANEX processes*

The group actinide extraction (GANEX) process was developed by the French CEA. This process supports homogenous recycle of actinides by first separating a pure U product in the first cycle solvent extraction process that uses DEHiBA (*N,N*-di-(ethyl-2-hexyl)isobutyramide) as a U(VI) extractant, followed by the second cycle in which the TRU is separated together for recycle. DEHiBA is used instead of TBP due to increased selectivity for U(VI) over Pu(IV) and its high loading capacity for U. This portion of the CEA-GANEX process has been demonstrated at the Atalante facility, using actual used nuclear fuel dissolved in nitric acid, with good results [39].

The second cycle of the CEA-GANEX process uses N,N'-dimethyl-N,N'-dioctylhexylethoxymalonamide (DMDOHEMA) and HDEHP extractants to co-extract the actinides and lanthanides, and a few other extractable fission products (Mo, Ru, Tc) [40]. The Mo, Ru, and Tc are stripped from the solvent prior in a series of two strip sections. The actinides and lanthanide are then selectively stripped using a mixture of HEDTA and citric acid at pH 3 for the actinides and a TEDGA/oxalic acid/nitric acid solution to strip the lanthanides. This second cycle of the CEA-GANEX process was demonstrated at the Atalante facility, using actual used nuclear fuel dissolved in nitric acid, with good actinide recovery but higher than expected contamination of the actinide product with lanthanides [17, 40].

As part of the European Union Actinide Recycling by Separation and Transmutation (ACSEPT) and Safety of Actinide Separation processes (SACSESS) programs, the EURO-GANEX process was developed [41]. This process consists of a first cycle as described above, followed by a second cycle that utilizes TODGA and N,N'-dimethyl-N,N'-dioctyl-2-(2-hexyloxyethyl) malonomide (DMDOHEMA) extractants. CDTA is used in the extraction/scrub aqueous phase to suppress Zr and Pd extraction. The TRU and lanthanides are selectively stripped using SO$_3$–Ph–BTP, AHA, and HNO$_3$ to back-extract the actinides and dilute nitric acid to back-extract the lanthanides [17, 41]. Testing of the process with fast reactor carbide fuel that was oxidized and dissolved in nitric acid, and processed in a U-extraction cycle, was performed at the European Joint Research Centre Institute of TransUranium elements (ITU). Results were positive — good recovery of Am, Np, and Pu (>99%) with very little lanthanide contamination (<0.1%) [41].

7.11. *Japan — DGA extraction process*

Diglycolamide (DGA) extractants are being developed in Japan for the separation of minor actinides from used nuclear fuel [42]. One of the more promising DGA's is N,N,N',N'-tetraoctyldiglycolamide (TODGA).

TODGA effectively extracts Am, Cm, and rare earth elements at acidities greater than 1 M HNO_3. Back-extraction can be accomplished at low acidity. The third phase formation was noted with the extraction of Nd. To alleviate this issue, N,N,N',N'-tetradodecyldiglycolamide (TDdDGA) was developed [42]. Alternative methods to address the third phase formation include the addition of N,N-dihexyloctanamide (DHOA) to the TODGA/dodecane solvent, or through the use of TBP or octanol as a phase modifier [43, 44]. Also, Russian researchers at the Khlopin Radium Institute have developed several alternative polar-fluorinated diluents for DGA extractants that increase the extraction capability of TODGA for Am and Eu [17].

Countercurrent flowsheet testing with HLW simulant in mixer-settler equipment has been performed in Japan by using TDdDGA in n-dodecane [45]. With the flowsheet tested, minor actinides and lanthanides are extracted by the TDdDGA and back-extracted, together, using dilute HNO_3. Zirconium and Pd extraction is suppressed through the use of HEDTA and H_2O_2 in the feed and scrub. Separation of the minor actinides from the lanthanides is then accomplished using extraction chromatography. The countercurrent flowsheet test of the TRU recovery step achieved >99.99% recovery of Am and 62% recovery of Np [45].

7.12. *Diamides of dipicolinic acid*

Diamides of dipicolinic acid (DPA) are being investigated in Russia and other countries for the separation of actinides and lanthanides from used fuel [46, 47]. Mixtures of cobalt dicarbollide and DPA in an FS-13 diluent have been investigated in Russia. The study included N,N,N',N'-tetrabutyl-dipicolinic acid (TBDPA) and the *ortho*, *meta*, and *para* isomers of N,N-diethyl-N',N'-ditolyl diamides. Results indicate that the *ortho*-position is the most favorable and the ethyl-tolyl isomer Et(o)TDPA has good selectivity between the heavy and light lanthanides [46]. In the Czech Republic, N,N'-diethyl-N,N'-di-meta-tolyldipicolinamide (Et(m)TDPA) was found to have the high Am extractability while maintaining good Am(III)/Eu(III) selectivity [47].

Alternatively, at the Khlopin Radium Institute, dipyridyl-dicarboxylic acid in polar fluorinated diluents has been studied [48]. Diamides of 2,2-dipyridyl-6,6-dicarboxylic acid can be used for the separation of minor trivalent actinides from lanthanides with good separation factors (>10) obtained.

7.13. *EXAm*

The extraction of americium (EXAm) process is being developed in France by the CEA for the recovery of only Am, as the main contributor to long-term heat generation and radiotoxicity, from PUREX raffinate [49, 50]. The solvent consists of DMDOHEMA and HDEHP in TPH. TEDGA is used as a complexing agent to maintain Cm and heavy lanthanides in the aqueous phase, thus improving the Am/Cm selectivity. Molybdenum, Pd, and Ru are stripped from the solvent by using citric acid and NaOH prior to stripping of the Am. A selective strip is then employed which first back-extracts Am using DTPA at a low acidity and then back-extracts the lighter lanthanides by using oxalic acid and TEDGA.

Hot testing of the EXAm process has been performed using actual PUREX raffinate [49]. Greater than 99% of the Am was extracted with a decontamination factor of >500 relative to Cm. Approximately 0.7% of the extracted Am was lost to the Mo strip effluent. Good decontamination from the light lanthanides was obtained for the Am product.

Recent efforts have focused on operation of the EXAm process by using a concentrated PUREX raffinate [50]. These efforts have included countercurrent flowsheet testing with simulants and actual PUREX raffinate to produce AmO_2 pellets for irradiation testing. These tests resulted in an acceptable Am product; however, there was a 10% loss of Am to the raffinate.

8. Summary

Historically, there is a great deal of experience with aqueous separation technologies for the separation of U and Pu from used nuclear fuel. Much of this experience is based on the use of the PUREX process on an industrial-scale for decades. As countries move forward with development of advanced separation processes, considerable effort has focused on modified PUREX processes in which there is not a pure Pu product. Additionally, a great deal of progress has been made in development of aqueous separation processes for the separation and recycle of minor actinides from used nuclear fuel. This is the primary area of ongoing research and development in the nuclear separations community with several processes having been demonstrated with actual used nuclear fuel in laboratory-scale equipment.

References

1. M. S. Gerber, The plutonium production story at the Hanford site: Processes and facilities history, report WHC-MR-0532, Westinghouse Hanford Company (1998).

2. M. F. Simpson and J. D. Law, Nuclear fuel reprocessing, In *Encyclopedia of Sustainability Science and Technology*, R. A. Meyers ed., Springer, New York, pp. 7142–7156 (2012).

3. F. Drain, R. Vinoche, and J. Duhamet, 40 years of experience with liquid-liquid extraction equipment in the nuclear industry, In *Proc. Waste Management 2003*, Tucson, AZ, USA (2003).

4. C. H. Castano, Nuclear fuel reprocessing, In *Nuclear Energy Encyclopedia: Science, Technology, and Applications*, T. B. Kingrey ed., Wiley, Hoboken, New Jersey, pp. 121–126 (2011).

5. T. F. Severynse, Nuclear material processing at the Savannah River Site, report WSRC-MS-98-00515, Westinghouse Savannah River Company (1998).

6. M. S. Gerber, A brief history of the PUREX and UO_3 facilities, report WHC-MR-0437, Westinghouse Hanford Company (1993).

7. M. Schneider and Y. Marignac, Spent nuclear fuel reprocessing in France, A research report of the International Panel on Fissile Materials, Research Report No. 4, International Panel on Fissile Materials (2008).

8. A. G. Croff, R. G. Wymer, L. L. Tavlarides, J. H. Flack, and H. G. Larson, Background, status, and issues related to the regulation of advanced spent nuclear fuel recycle facilities: ACNW&M white paper (Nureg-1909), Advisory Committee on Nuclear Waste and Materials, U.S. Nuclear Regulatory Commission (2008).

9. R. Natarajan and B. Raj, Technology development of fast reactor fuel in India, *Curr. Sci.* **108**(1), 30–38 (2015).

10. J. H. Saling and A. W. Fentiman eds., *Radioactive Waste Management*, 2nd edition, CRC Press, New York (2001).

11. H. Zhang, Chinese reprocessing and nuclear security issues, In *Proc. Institute of Nuclear Materials Management 55th Annual Meeting*, Atlanta, GA, USA (2014).

12. F. Fiori and Z. Zhou, Sustainability of the Chinese nuclear expansion: The role of ADS to close the nuclear fuel cycle, *Prog. Nucl. Energy* **83**, 123–134 (2015).

13. P. Baron, Process for reprocessing a spent nuclear fuel and of preparing a mixed uranium-plutonium oxide, US Patent US 2007/0290178 Al (2007).

14. C. Pereira, G. F. Vandegrift, M. C. Regalbutto, A. Bakel, D. Bowers, A. V. Gelis, A. S. Hebden, L. E. Maggos, D. Stepinski, Y. Tsai, and J. J. Laidler, Lab-scale demonstration of the UREX+1a process using spent fuel, In *Proc. Waste Management 2007*, Tucson, AZ, USA (2007).

15. T. S. Rudisill and M. C. Thompson, Demonstration of the use of formohydroxamic acid in the UREX process, *Separ. Sci. Technol.* **50**(18), 2823–2831 (2015).

16. E. D. Collins, D. E. Benker, W. D. Bond, D. O. Campbell, and B. B. Spencer, Development of the UREX+ co-decontamination solvent extraction process, In *Proc. Global 2003*, New Orleans, LA, USA (2003).

17. P. Baron *et al.*, Organization for Economic Co-operation and Development Nuclear Energy Agency, State-of-the-art report on progress on separation

chemistry, minor actinide separation and perspectives for future R&D, OECD/NEA, Paris, France, in press, https://www.oecd-nea.org/science/pubs/2018/7267-soar.pdf (2018).

18. J. Bresee, P. Paviet, and T. Todd, Nuclear separations process control and accountability, In *Proc. 14th Information Exchange Meeting on Partitioning and Transmutation (IEMPT'14)*, San Diego, CA, USA (2016).

19. S. T. Arm and E. J. Butcher, Treatment and disposal of process wastes arising from the recycle of spent nuclear fuel in the USA, In *Proc. Waste Management 2009*, Phoenix, AZ, USA (2009).

20. M. Nakahara, Y. Sano, Y. Koma, M. Kamiya, A. Shibata, T. Koizumi, and T. Koyama, Separation of actinide elements by solvent extraction using centrifugal contactors in the NEXT process, *J. Nucl. Sci. Technol.* **44**(3), 373–381 (2007).

21. M. Nakahara, Separation of uranyl nitrate hexahydrate crystal from dissolver solution of irradiated fast neutron reactor fuel, In *Advances in Crystallization Processes*, Y. Mastai ed., InTech (2012).

22. V. V. Bondin, P. M. Gavrilov, Y. A. Revenko, B. Y. Zilberman, B. N. Romanovskij, Y. S. Fedorov, A. Y. Shadrin, E. G. Kudryavcev, and A. V. Haperskaja, Simplified PUREX process — Perspective SNF reprocessing technology for the plant of the next generation, In *Proc. GLOBAL 2007*, Boise, ID, USA, pp. 1484–1489 (2007).

23. G. Uchiyama, H. Mineo, T. Asakura, S. Hotoku, M. Llzuka, S. Fujisaki, H. Isogai, Y. Itoh, M. Sato, and N. Hosoya, Long-lived nuclide separation for advancing back-end fuel cycle process, *J. Nucl. Sci. Technol.* **39**(Suppl. 3), 925–928 (2002).

24. IAEA, Status of minor actinide fuel development, IAEA Nuclear Energy Series No. NF-T-4.6 (2010).

25. E. P. Horwitz, D. G. Kalina, H. Diamond, G. F. Vandegrift, and W. W. Schulz, The TRUEX process — A process for the extraction of the transuranic elements from nitric acid wastes utilizing modified PUREX solvent, *Solv. Extr. Ion Exch.* **3**(1–2), 75–109 (1985).

26. B. F. Weaver and F. A. Kappelmann, Preferential extraction of lanthanides over trivalent actinides by monoacidic organophosphates from carboxylic acids and from mixtures of carboxylic acids and aminopolyacetic acids, *J. Inorg. Nucl. Chem.* **30**(1), 263–272 (1968).

27. P. R. Danesi and C. Cianetti, Kinetics and mechanism of the interfacial mass transfer of Eu(III) in the system: Bis(2-ethylhexyl)phosphoric acid, n-dodecane-NaCl, lactic acid, polyaminocarboxylic acid, water, *Sep. Sci. Technol.* **17**(7), 969–984 (1982).

28. C. Pereira, G. F. Vandegrift, M. C. Regalbuto, A. Bakel, D. Bowers, A. V. Gelis, A. S. Hebden, L. E. Maggos, D. Stepinski, Y. Tsai, and J. J. Laidler, Lab-scale demonstration of the UREX+1a process using spent fuel, In *Proc. Waste Management 2007*, Tucson, AZ, USA (2007).

29. G. J. Lumetta, A. J. Casella, B. M. Rapko, T. G. Levitskaia, N. K. Pence, J. C. Carter, C. M. Niver, and M. R. Smoot, An advanced TALSPEAK concept using 2-ethylhexylphosphonic acid mono-2-ethylhexyl ester as the extractant, *Solv. Extr. Ion Exch.* **33**(3), 211–223 (2015).

30. G. J. Lumetta, A. V. Gelis, J. C. Carter, C. M. Niver, and M. R. Smoot, The actinide-lanthanide separation concept, *Solv. Extr. Ion Exch.* **32**(4), 333–347 (2014).

31. T. Grimes, C. Heathman, S. Jansone-Popova, V. Bryantsev, S. Srinivasan, M. Nakase, and P. Zalupski, Thermodynamic, spectroscopic, and computational studies of f-element complexation by N-hydroxyethyl-diethylenetriamine-N,N′,N″,N‴-tetraacetic Acid, *Inorg. Chem.* **56**(3), 1722–1733 (2017).

32. B. J. Mincher, R. D. Tillotson, T. G. Garn, V. Rutledge, J. D. Law, and N. C. Schmitt, The solvent extraction of Am(VI) using centrifugal contactors, *J. Radioanal. Nucl. Chem.* **307**(3), 1833–1836 (2016).

33. B. J. Mincher, T. S. Grimes, R. D. Tillotson, and J. D. Law, Am(VI) extraction final report: FY16, report FCRD-MRWFD-2016-000327, Idaho National Laboratory (2016).

34. J. D. Burns and B. A. Moyer, Group hexavalent actinide separations: A new approach to used nuclear fuel recycling, *Inorg. Chem.* **55**(17), 8913–8919 (2016).

35. D. Magnusson, B. Christiansen, M. R. S. Foreman, A. Geist, J.-P. Glatz, R. Malmbeck, G. Modolo, D. Serrano-Purroy, and C. Sorel, Demonstration of a SANEX process in centrifugal contactors using the CyMe4-BTBP molecule on a genuine fuel solution, *Solv. Extr. Ion Exch.* **27**(2), 97–106 (2009).

36. G. Modolo, H. Vijgen, D. Serrano-Purroy, B. Christiansen, R. Malmbeck, C. Sorel, and P. Baron, DIAMEX counter-current extraction process for recovery of trivalent actinides from simulated high active concentrate, *Separ. Sci. Technol.* **42**(3), 439–452 (2007).

37. C. Hill, Overview of recent advances in An(III)/Ln(III) separation, In *Ion Exchange and Solvent Extraction: A Series of Advances, Volume 19*, B. A. Moyer edn., CRC Press (2009).

38. G. Modolo, A. Wilden, P. Kaufholz, D. Bosbach, and A. Geist, Development and demonstration of innovative partitioning processes (i-SANEX and 1-cycle SANEX) for actinide partitioning, *Prog. Nucl. Energy* **72**, 107–114 (2014).

39. M. Miguirditchian, C. Sorel, I. Bisel, B. Cames, P. Baron, D. Espinoux, J. N. Calor, C. Viallesoubranne, C. Lorrain, and M. Masson, HA demonstration in the Atalante facility of the GANEX 1st cycle for the selective extraction of Uranium from HLW, In *Proc. GLOBAL 2009*, Paris, France (2009).

40. M. Miguirditchian, H. Roussel, L. Chareyre, P. Baron, D. Espinoux, C. N. Calor, C. Viallesoubranne, B. Lorrain, and M. Masson, HA demonstration in the Atalante facility of the Ganex 2nd cycle for the grouped TRU extraction, In *Proc. GLOBAL 2009*, Paris, France (2009).

41. R. Taylor, M. Carrott, H. Galan, A. Geist, X. Hères, C. Maher, C. Mason, R. Malmbeck, M. Miguirditchian, G. Modolo, C. Rhodes, M. Sarsfield, and A. Wilden, The EURO-GANEX process: Current status of flowsheet development and process safety studies, *Procedia Chem.* **21**, 524–529 (2016).

42. Y. Sasaki, Y. Kitatsuji, Y. Tsubata, Y. Sugo, and Y. Morita, Separation of Am, Cm and lanthanides by solvent extraction with hydrophilic and lipophilic organic ligands, *Solv. Extr. Res. Dev. Jpn.* **18**, 93–101 (2011).

43. G. Modolo, H. Asp, H. Vijgen, R. Malmbeck, D. Magnusson, and C. Sorel, Demonstration of a TODGA/TBP process for recovery of trivalent actinides and lanthanides from a PUREX raffinate, In *Proc. GLOBAL 2007*, Boise, ID, USA, pp. 1111–1116 (2007).

44. A. Geist and G. Modolo, TODGA process development: an improved solvent formulation, In *Proc. GLOBAL2009*, Paris, France (2009).

45. Y. Morita and T. Kimura, Development of separation process for transuranium elements and some fission products using new extractants and adsorbents, actinide and fission product partitioning and transmutation, In *Proc. 11th OECD/NEA Information Exchange Meeting (IEMPT)*, San Francisco, USA, pp. 235–244 (2012).

46. M. Y. Alyapyshev, V. A. Babain, R. S Herbst, J. D Law, and A. Paulenova, Extraction of lanthanoids with diamides of dipcolinic acid from nitric acid solutions II. Synergistic effect of ethyl-tolyl derivates and dicarbollide cobalt, *Solv. Extr. Ion Exch.* **31**(2), 184–197 (2013).

47. M. Bubenıkova, J. Rais, P. Selucky, and J. Kvıcala, Am(III) separation from acidic solutions by diamides of dipicolinic acid, *Radoichim. Acta* **101**, 753–759 (2013).

48. M. Y. Alyapyshev, V. A. Babain, L. I. Tkachenko, A. Paulenova, A. A. Popova, and N. E. Borisova, New diamides of 2,2′-dipyridyl-6,6′-dicarboxylic acid for actinide-lanthanide separation, *Solv. Extr. Ion Exch.* **32**(2), 138–152 (2014).

49. C. Rostaing, C. Poinssot, D. Warin, P. Baron, and B. Lorrain, Development and validation of the EXAm separation process for single Am recycling, *Procedia Chem.* **7**, 367–373 (2012).

50. V. Vanel, M-J. Bollesteros, C. Marie, M. Montuir, V. Pacary, F. Antegnard, S. Costenoble, and V. Boyer-Deslys, Consolidation of the EXAm process: Towards the reprocessing of a concentrated PUREX raffinate, *Procedia Chem.* **21**, 190–197 (2016).

Chapter 2

Fundamentals of Spent Nuclear Fuel Pyroprocessing

Michael F. Simpson

Department of Metallurgical Engineering, University of Utah
135 South 1460 East, Salt Lake City, UT 84112, USA
michael.simpson@utah.edu

1. Introduction

1.1. *Background*

Pyroprocessing is a high-temperature chemical process that has been demonstrated for treating spent nuclear fuel in the USA at Argonne National Laboratory (ANL, currently called Idaho National Laboratory (INL)). It is designed to recover actinides from spent fuel for the purpose of fabricating fresh fuel [1]. Its original development for spent fuel processing was under the Integral Fast Reactor (IFR) program led by ANL [2, 3]. The IFR program used the Experimental Breeder Reactor-II (EBR-II) and the Fuel Cycle Facility at Argonne National Laboratory-West (ANL-West) to develop and demonstrate its technology and functionality. EBR-II used a sodium-bonded metal (U–Zr) fuel. Driver fuel elements contained high-enriched uranium alloyed with zirconium, and blanket fuel elements contained depleted uranium metal. In 1994, the IFR program was canceled and soon replaced with a program to treat and dispose of irradiated fuel from EBR-II. ANL executed a 3-year demonstration project from 1996 to 1999 to prove the viability of pyroprocessing for treating the irradiated EBR-II fuel in addition to irradiated fuel from Enrico Fermi (Fermi-1) and Fast Flux Test Facility (FFTF) reactors [4]. Since the successful demonstration, ANL (now INL) has been processing EBR-II and FFTF fuel in the Fuel Conditioning Facility (FCF) at the site currently known as the Material and Fuels Complex (MFC). In the years since the demonstration was completed,

ANL/INL carried out research intended to evaluate and update the process for possible application to processing commercial light water reactor fuel. The USA has had a policy against reprocessing of spent fuel since 1977, but the US Department of Energy has considered various schemes and technologies for closing the nuclear fuel cycle and reducing the amount of high level waste that needs to be disposed of in a deep geologic repository. The US Department of Energy (DOE) planned to build three demonstration facilities under the Global Nuclear Energy Partnership (GNEP) program. This would have included two prototype fuel processing facilities and a demonstration fast reactor. The technology development portion of that program was canceled in 2008. Still the DOE has stated its interest in studying fuel processing technology such as pyroprocessing since 2008 in order to collect data and information to make a future decision about closing the nuclear fuel cycle. They currently administer this research under the DOE Fuel Cycle Technologies (FCT) program.

Meanwhile, interest in pyroprocessing in South Korea has increased to the point where it is widely considered to be the default plan for processing spent commercial fuel from the Korean pressurized water reactors starting within the next 10–20 years [5]. The Korea Atomic Energy Research Institute (KAERI) has a well-funded R&D program dedicated to scaling up pyroprocessing and designing facilities to use the technology to process the Korean light water reactor (LWR) fuel. In 2011, the Republic of Korea (ROK) and USA agreed to a 10-year study of the feasibility of pyroprocessing for commercial fuel treatment. Currently, a significant portion of the US government funded research in pyroprocessing is in support of the US–ROK Joint Fuel Cycle Study (JFCS).

Japan has also engaged in a significant research effort in pyroprocessing, led by the Central Research Institute for the Electric Power Industry (CRIEPI) [6]. Japan's involvement in pyroprocessing research dates back to the 1990s when they collaborated with ANL in the IFR program. Pyroprocessing has long been considered a backup to the plutonium uranium reduction extraction (PUREX)-based aqueous processing that has been scaled up and implemented for commercial fuel processing at the Rokkasho Reprocessing Plant.

Russia also has a significant program for developing pyroprocessing but has used an alternative approach that is ideally suited for mixed oxide (MOX) fuel. It utilizes different materials and process steps and will not be described in this chapter. An excellent overview of the Research Institute of Atomic Reactors (RIAR) developed process has been written by Vavilov *et al.* [7].

Other countries that have had pyroprocessing research programs include India, Russia, Great Britain, and France.

It is essential to keep in mind while reading this chapter that pyroprocessing has yet to be scaled up for commercial implementation. The largest known implementation with spent fuel is in INL's FCF, which contains two electrorefiners (ERs) that combined could hypothetically process up to 5 MT of spent fuel per year. This processing rate has never been demonstrated in FCF. The most likely first commercial facility will be built in the ROK, but that is largely hinging on revisions to the nuclear cooperation agreement between the USA and ROK. Currently, the ROK is forbidden from using any reprocessing technologies on US-origin fuel. The vast majority of ROK spent fuel is US origin, thus this agreement is very important.

1.2. *Flowsheet*

There is no single, definitive flowsheet for pyroprocessing. Depending on the input fuel type and the waste management objectives, a variety of options can be considered. The flowsheet shown in Figure 1 is intended to represent the most generic system with essential unit operations and minimal complexity. It includes the option to process oxide or metal fuel. It also includes the option to recover uranium (U) or U/transuranic (TRU) metal from the ER. A number of different salt waste processing schemes

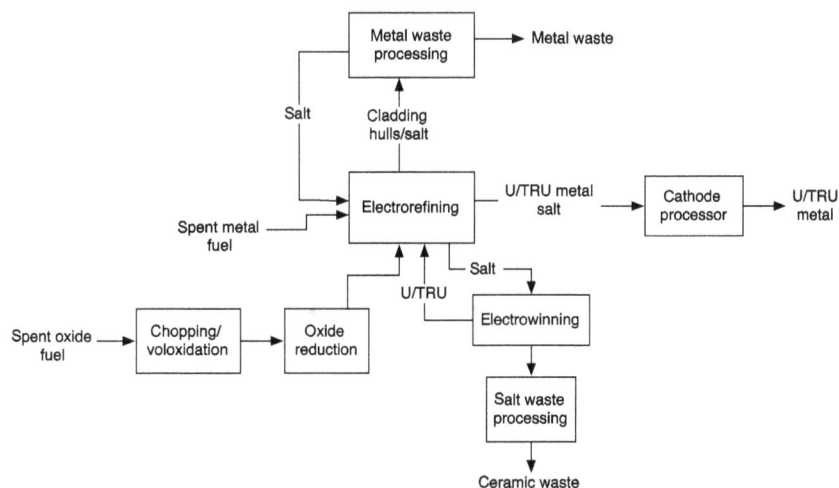

Figure 1. Generic flowsheet for spent nuclear fuel pyroprocessing.

have been developed and tested, but this groups them all into a single generic unit operation.

In each of the following sections, the individual unit operations shown in Figure 1 will be explained in terms of their fundamental chemical basis, functionality, equipment requirements, and process conditions. An attempt has been made to cite key references, while there are countless additional references that can be readily found in the open literature. Chopping/voloxidation is not discussed, as it is not unique to pyroprocessing. The function of that unit operation is to declad the fuel to enable further processing.

2. Oxide Fuel Treatment

2.1. *Lithium reduction*

Actinide separation occurs in the ER, which requires that the fuel be in the metallic state prior to processing. The IFR project used metallic fuel (U–Zr alloy) and, thus, could be used directly in the ER. For oxide fuel, a pre-conditioning is necessary to reduce the oxides to metals. Originally, pyroprocessing was not envisioned for processing commercial light water reactor fuel, which is composed primarily of uranium oxide. Early development of an oxide reduction process was based on the reaction of lithium metal with uranium oxide [8–10].

$$4Li + UO_2 \rightarrow U + 2Li_2O$$

Other actinides such as plutonium are also reduced at least partially. This reaction can be carried out via a solid–liquid reaction. The liquid solvent in this case is molten LiCl at a nominal temperature of 650°C. This system works well, because both metallic lithium and lithium oxide are partially soluble in molten LiCl. Lithium metal solubility has been measured to be 0.49–0.73 mol.% at approximately 650°C [11]. Solubility of Li_2O in LiCl has been reported to be 10–12 mol.% [12]. Lithium metal melts at 180°C and can be dissolved in the molten LiCl via simple liquid–liquid contact. Spent fuel is loaded into a basket with permeable steel walls to allow rapid mass transport of reactants and products. With each batch of fuel reduced, the Li_2O concentration increases. Reaction of the soluble lithium with the oxides in the spent fuel is limited thermodynamically. Refer to Table 1 that lists the free energy of formation of key oxides in this process at 650°C. In this table, the free energy values have been normalized per mole of oxygen for direct comparison. All of the oxides listed above

Table 1. Standard free energy of formation of oxides [13].

Oxide	ΔG_f (kcal/mol O) at 650°C
Y_2O_3	−130.28
CaO	−128.67
ThO_2	−125.92
Ce_2O_3	−123.27
Nd_2O_3	−123.19
La_2O_3	−121.94
MgO	−120.07
SrO	−119.23
Am_2O_3	−115.94
Li_2O	−113.70
Pu_2O_3	−112.51
BaO	−110.77
UO_2	−110.51
ZrO_2	−110.41
NpO_2	−109.04
MoO_2	−50.29

Li_2O are stable in the presence of Li metal. They will not reduce or will only partially reduce. This includes rare earths and minor actinides that are found in spent fuel. Uranium oxide spontaneously reduces because of its higher free energy of formation compared with lithium oxide. But several major components of spent fuel will remain as oxides when transferred from the oxide reduction process to the ER. The problem with having residual oxides in the fuel in the ER is that they will eventually react with UCl_3 in the ER-molten salt and form insoluble uranium oxides or oxychlorides that precipitate out of the ER salt. This problem was the primary motivation for developing an alternative oxide reduction method.

The density of metallic uranium ($19.1\,g/cm^3$) is much higher than that of uranium dioxide ($10.97\,g/cm^3$), which results in formation of pores in the solid fuel. This is a key phenomenon, as it opens up the pathway for the liquid reactant to fully penetrate the fuel and achieve complete reduction. A shrinking core model has been proposed for describing the process and deriving reaction rate expressions and has been shown to be consistent with the available kinetic data on lithium reduction of uranium oxide [10]. Specifically, the reduction rate is limited by diffusion of lithium from the surface of the pellet/particle through the layer of reduced fuel. This layer is akin to the ash layer described in the classical shrinking core model [14]. There is an unreacted core of uranium oxide that shrinks as the lithium diffuses to the surface and reacts to form uranium metal. In this model, it is assumed that the reaction rate at the surface is very fast compared with the mass transport rate across the ash layer. Understanding this kinetic mechanism is important for guiding process design and optimization. Particle size reduction is key, for example, to minimize the time scale for complete reduction of each fuel particle. As demonstrated via modeling/simulation in the paper by Simpson *et al.*, forced flow can play an important role in speeding up the reaction if the particle size is sufficiently low [10]. When the particle sizes are larger, forced flow through the fuel basket is less important as the reaction rate becomes dominated by diffusion through the particles/pellets.

2.2. *Direct electrolytic reduction*

The alternative oxide reduction process is called direct electrolytic reduction (DER) and is based on an electrochemical process variant of the lithium reduction process [15–22]. Many of the components of the system and the molten salt are the same between the lithium reduction and DER processes. A key difference is that the DER process maintains a low Li_2O concentration in the molten salt, which drives the reduction process for rare earths and minor actinides. Typically, the DER process can be run at 1 wt.% (1.4 mol.%) Li_2O in LiCl. Li_2O concentration does not increase during DER processing, because it is continuously decomposed at the anode in the DER cell. The DER reactions are as follows:

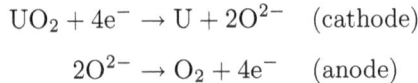

$$UO_2 + 4e^- \rightarrow U + 2O^{2-} \quad \text{(cathode)}$$
$$2O^{2-} \rightarrow O_2 + 4e^- \quad \text{(anode)}$$

The UO_2 reduction occurs at about $-1.8\,V$ relative to a Ni/NiO reference electrode [21]. If the cathode voltage goes much lower than this value, lithium oxide can also be reduced via the following reactions:

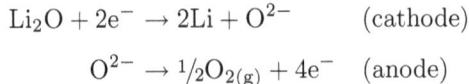

$$Li_2O + 2e^- \rightarrow 2Li + O^{2-} \quad \text{(cathode)}$$
$$O^{2-} \rightarrow {}^{1}\!/_{2}O_{2(g)} + 4e^- \quad \text{(anode)}$$

The only reaction that occurs at the anode, therefore, is oxidation of oxide ions to oxygen gas. Most metals are susceptible to reaction with oxygen and would corrode under these conditions, forming metal oxides. Platinum has been found to be stable and is the current anode material used in baseline designs for DER. A photograph of a platinum plate after an oxide reduction experiment is shown in Figure 2.

Figure 2. Platinum plate used for anode in oxide reduction experiment.

If lithium metal is formed in the reactions listed above, it will either react with the oxide fuel to form metals or will dissolve the platinum anode. Dual circuit designs have been tested for protecting the anode from interaction with lithium, as published by Herrmann *et al.* [18]. The dual circuit involves polarizing the cathode basket to a potential high enough to oxidize Li metal to Li^+. The reactions are as follows:

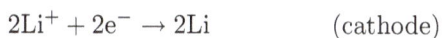

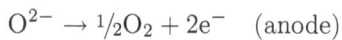

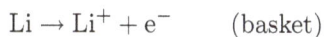

$$2Li^+ + 2e^- \rightarrow 2Li \qquad \text{(cathode)}$$

$$O^{2-} \rightarrow {}^1/_2O_2 + 2e^- \quad \text{(anode)}$$

$$Li \rightarrow Li^+ + e^- \qquad \text{(basket)}$$

M. F. Simpson

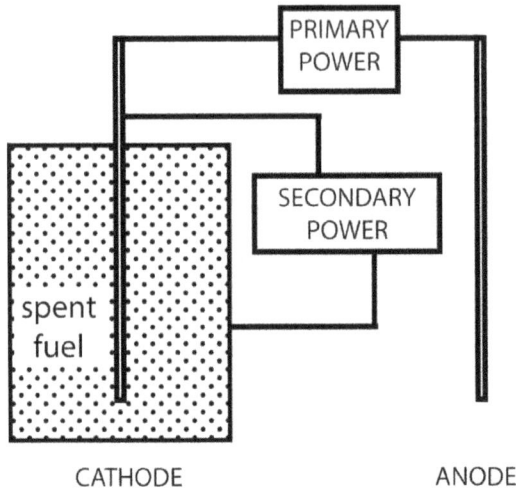

Figure 3. Electrochemical cell design for dual circuit DER consisting of a primary power supply and second power supply.

Uranium oxide has some electrical conductivity [22], but there is sufficient resistance between the cathode lead and the basket to allow for a significant potential gradient. Effectively charge transfer can be accomplished via Faradaic processes rather than electrical conduction.

A diagram of the dual circuit DER system is shown in Figure 3.

DER can be controlled via a number of different electrochemical process modes. A constant current can be applied between the cathode and anode, which is referred to as galvanostatic. Electric potential control is also an option, which can be subdivided into controlled cathode potential, controlled anode potential, or controlled cell potential. The last option refers to the difference between the cathode and anode potentials. The first two options require a reference electrode to provide a baseline (reference) potential. For DER in molten LiCl–Li$_2$O, a couple of options for reference electrodes have been reported in the literature. This includes Ni/NiO and Pb–Li [21, 23]. In both cases, the electrode is encased in a closed-ended magnesia tube. Low-density magnesia ceramics have sufficient porosity to support ionic conductivity and close the circuit. Cyclic voltammetry (CV) can identify the potentials relative to a given reference electrode that can drive different reactions. See the example in Figure 4. CV involved cycling the potential of an electrode at a constant rate over a specified range. In Figure 4, two different cycles are overlaid featuring

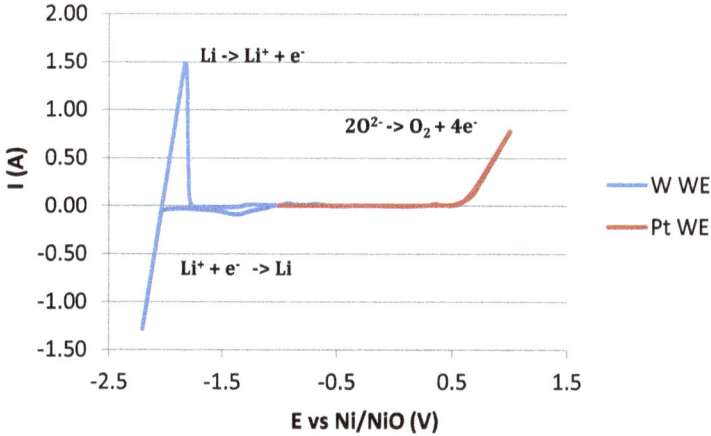

Figure 4. Cyclic voltammetry in molten LiCl–Li$_2$O at 650°C. High potential range used platinum working electrode, and low potential range used tungsten working electrode.

different electrodes — tungsten for the lower potentials and platinum for the higher potentials. Splitting the scans up in this way is necessary, because Pt will dissolve in the lithium that is generated at the lower potentials, and tungsten will oxidize at the higher potentials needed for O$_2$ generation. As can be seen in this figure, Li formation starts at about −1.9 V, and O$_2$ formation starts at about 0.5 V. Thus, one mode of operation that does not require a reference electrode is to set the potential difference between the cathode and anode at 2.4 V or higher. If the voltage difference is set too high, the anode potential may rise high enough to induce Pt oxidation followed by dissolution. Thus, potential control is key for the DER process.

Herrmann *et al.* reported CV data that demonstrated UO$_2$ reduces at about 0.1–0.2 V higher potential than Li$^+$ reduces [21]. An accurate and stable reference electrode can enable direct control of either the anode or cathode potential. Typically, anode potential control would be utilized to prevent the anode from dissolving. Attempts can be made to control the cathode potential to reduce UO$_2$ but not Li$^+$, but that is difficult to accomplish in practice, especially for large cathode baskets containing greater than a kilogram of spent fuel. The window between onset of UO$_2$ reduction and onset of Li$^+$ reduction is less than 200 mV, so this mode of control is difficult. The inevitable generation of Li metal at the cathode justifies the use of the dual circuit cell shown in Figure 3.

2.3. *Process outcome*

Reduction efficiencies for uranium and plutonium have been reported as high as 99.7% and 97.8%, respectively, in labscale experiments [21]. It was also reported by Herrmann *et al.* that Cs, Sr, and Ba partition into the molten salt phase. Presumably, this occurs via reaction of the metal oxides with LiCl, but this mechanism has yet to be proven. Another possibility is that there is solubility for Cs_2O, SrO, and BaO in molten LiCl. Either way, it is important to know that these fission products separate into the salt and are effectively separated from the rest of the spent fuel in the DER (or lithium reduction) process. This effectively lowers the radioactivity of the spent fuel and provides some flexibility on how to manage the radioactive waste. Cs and Sr are the key fission products that generate large amounts of heat during the first 200 years after removal from the reactor. This is because of the prominence of Cs-137 (half-life $= 30.2$ years, $0.154\,W/g$) and Sr-90 (half-life $= 28.8$ years, $0.536\,W/g$) as fission products. Much of the gamma radiation emitted from spent fuel comes from Cs-137, thus the oxide reduction process also substantially reduces the dose rate from the spent fuel. This can have proliferation consequences that should be considered when specifying facility safeguards requirements.

3. Electrorefining/Electrowinning

Key functionality for pyroprocessing is provided by the ER [24, 25]. It is where actinides are separated from fission products and other components of spent fuel via selective oxidation and reduction in molten salt. The electrolyte used for the ER is eutectic LiCl–KCl with a nominal composition of 58 mol.% LiCl and 42 mol.% KCl. The salt is anhydrous with very low water and oxygen content to minimize its tendency to cause corrosion. The ER uses a steel (2.5-Cr, 1-Mo) vessel [24], which is highly susceptible to corrosion at high temperatures in the presence of chloride salts and moisture. All of the pyroprocessing processes described in this chapter should be operated in an inert (argon or helium) atmosphere hot cell. The FCF at INL has an argon cell with an administrative limit of 100 ppm oxygen-based impurities ($O_2 + H_2O$). Small-scale experimental studies of electrorefining and other pyrochemical processes often utilize argon atmosphere glove boxes with <1 ppm for both O_2 and H_2O. In addition to LiCl and KCl, the electrolyte in the ER contains approximately 10 wt.% UCl_3, which serves to oxidize active metal fission products and enable fast electrotransport of U from the spent fuel basket (anode)

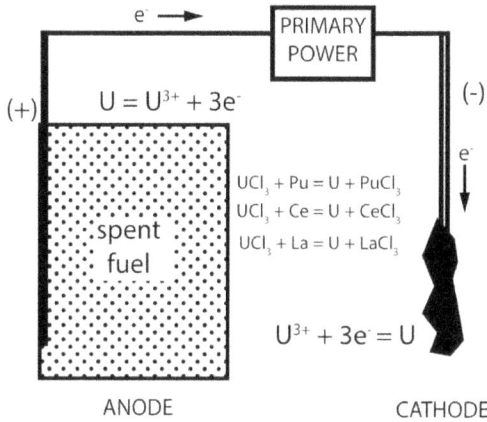

Figure 5. Electrorefiner configuration for processing metallic spent nuclear fuel.

to the product collector (cathode). A basic ER configuration is shown in Figure 5.

3.1. *Uranium electrorefining*

Two different chemical processes occur during uranium recovery via electrorefining, as illustrated in Figure 5. First, there is chemical oxidation of the active metal fission products and TRU metals via reaction with UCl_3. These active metals form chlorides, which have a relatively high solubility in eutectic LiCl–KCl. Examples of such reactions are listed below. Many more occur based on the relative values of the free energies of formation of the metal chlorides. Table 2 lists standard free energy of formation and standard reduction potential for all metal chlorides of interest at 500°C. Ions with standard reduction potentials less than that of U(III) are more stable as chlorides than uranium chloride and will, thus, tend to form as long as there is some UCl_3 available to react in the molten salt electrolyte:

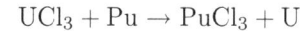

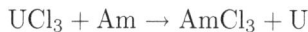

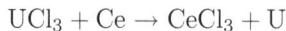

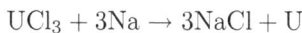

$$UCl_3 + Pu \rightarrow PuCl_3 + U$$

$$UCl_3 + Am \rightarrow AmCl_3 + U$$

$$UCl_3 + Ce \rightarrow CeCl_3 + U$$

$$UCl_3 + 3Na \rightarrow 3NaCl + U$$

Thus, they readily partition into the molten electrolyte phase and accumulate with each successive batch of spent fuel processed. Similar to oxide

Table 2. Standard free energy of formation and standard reduction potential for metals in spent fuel electrorefining systems [13].

Metal Ion	ΔG_f^0 (kJ/mole)	E^0 (V) vs. Cl$^-$/Cl$_2$	Metal Ion	ΔG_f^0 (kJ/mole)	E^0 (V) vs. Cl$^-$/Cl$_2$
Mo(III)	−172.2	−0.595	Mg(II)	−516.7	−2.678
Bi(III)	−229.9	−0.794	Pu(III)	−786.9	−2.719
Ni(II)	−188.3	−0.976	Am(III)	−795.3	−2.748
Co(II)	−206.3	−1.069	Pm(III)	−804.8	−2.780
Sn(II)	−226.3	−1.173	Y(III)	−817.1	−2.823
Fe(II)	−243.4	−1.261	Gd(III)	−818.5	−2.828
Cd(II)	−270.8	−1.403	Sm(III)	−838.2	−2.896
Cr(II)	−298.3	−1.546	Nd(III)	−852.7	−2.946
Zr(II)	−316.9	−1.642	Ce(III)	−861.5	−2.976
Mn(II)	−378.7	−1.962	Pr(III)	−866.1	−2.992
Ti(II)	−389.4	−2.018	La(III)	−880.9	−3.043
U(IV)	**−794.6**	**−2.059**	Ca(II)	−675.4	−3.500
U(III)	**−698.3**	**−2.412**	Na(I)	−339.3	−3.517
Th(IV)	−957.2	−2.480	Li(I)	−344.8	−3.574
Np(III)	−729	−2.519	Sr(II)	−713.9	−3.700
Eu(III)	−739.7	−2.555	Rb(I)	−361.2	−3.744
Pu(IV)	−739.7	−2.555	K(I)	−362.6	−3.758
Cm(III)	−767	−2.650	Cs(I)	−366.6	−3.800

reduction, this results in some porosity of the fuel. But since uranium is the major component of most types of fuel, the porosity is likely to be insufficient to achieve complete reaction of these active metals. The other chemical process, electrorefining, is thus essential to completely process each batch of fuel. Electrorefining strips away the uranium, which progressively exposes more of the active metals. This progression is illustrated in Figure 6.

Electrorefining is driven by a potential difference applied between the cathode and anode. The fuel basket in this operation is the anode, while the cathode is typically a steel mandrel located adjacent to the anode basket. As in the case of oxide reduction, the process can be controlled either via controlled current or controlled voltage. Use of a Ag/AgCl reference electrode enables the monitoring and control of either the cathode or anode potential. When a constant current is applied, the anode potential rises until uranium oxidation starts to occur, and the cathode potential lowers until uranium reduction starts to occur. These reactions are as follows:

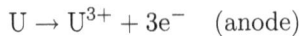

$$U \rightarrow U^{3+} + 3e^- \quad \text{(anode)}$$

$$U^{3+} + 3e^- \rightarrow U \quad \text{(cathode)}$$

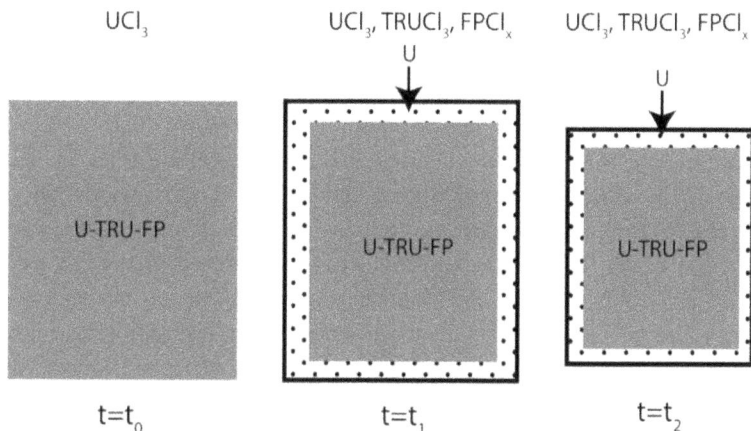

Figure 6. Progression of metallic fuel dissolution during electrorefining in molten salt containing UCl3.

Refer to Table 2 for the standard reduction potentials for uranium and other key metals found in spent fuel ERs. The ions with higher standard reduction potential than U(III)/U are categorized as noble metals, while the metals with lower standard reduction potentials are categorized as active metals. These definitions are relative to U, which oxidizes to U^{3+} at $-2.412\,V$ relative to the Cl^-/Cl_2 reference potential. Note that the standard potentials in this table have been calculated based on the free energies of formation that are given and were calculated using *HSC Chemistry*, version 7.0 [13]. During electrorefining, the noble metals tend to be inert and remain in the fuel basket, while the active metals are converted to chlorides and accumulate in the molten salt electrolyte. Uranium from the spent fuel transports to the cathode. But uranium in the form of UCl_3 in the salt is progressively consumed via oxidation of the active metals. The molten LiCl–KCl has the capacity to dissolve a significant (>10 wt.%) concentration of fission product chlorides. Thus, several batches of fuel can be processed before there is any need to treat or dispose of the salt. In the case of the Mark-IV and Mark-V ERs in the FCF at INL, the mass of molten salt is approximately 500 kg. This provides ample capacity for many batches of fuel to be processed before the salt nears a composition limit. Interestingly, there has yet to be a definitive report on the solubility limit of the fission product chlorides in the eutectic LiCl–KCl. Because of the many potential degrees of freedom (multiple components), it is difficult to design a study to determine at what level of contamination that the salt needs to be kept under. Most of the fission products are considered waste and will need to be eventually stabilized into a waste form for permanent disposal in a deep geologic repository. But the minor actinides should be periodically recovered to be used for making fast reactor fuel.

Uranium metal forms as dendrites during electrorefining in molten eutectic LiCl–KCl [26]. Depending on the design of the cathode, the dendrites either stick to the cathode mandrel or fall off into a product collector. The Mark-IV ER was the first system operated at ANL-West and included a cylindrical cathode mandrel placed parallel to the anode basket [24]. The process was designed to deposit uranium dendrites onto the mandrel, which would be pulled out of the salt at the end of each run. A molten cadmium pool was located below the molten salt to collect any uranium dendrites that fall off of the cathode. The cadmium pool can be anodically polarized to drive dissolution of the uranium and deposition on the cathode mandrel [24]. The Mark-V ER was installed a year later into the FCF for the purpose of processing EBR-II blanket fuel. The mass of blanket fuel was about an order of magnitude higher than the driver fuel that was processed in the Mark-IV ER. The Mark-V ER, thus, needed to achieve much higher processing rates. The physical dimensions of the Mark-IV and Mark-V ERs are identical. But the anode–cathode designs are very different. The Mark-V ER features up to four concentric anode–cathode modules (ACM) [27]. In each ACM, the fuel is loaded into anode baskets, which are surrounded by cathodic tubes. The distance between the anode and cathode is very short, promoting rapid electrotransport. The concentric cylinders in the ACM are rotated in opposite directions to cause continuous scraping of uranium dendrites that form on the cathode cylinders. Those dendrites then fall into product collector baskets attached to the bottom of each ACM. While this design does achieve higher throughput than the Mark-IV design, product morphology can result in poor packing density of uranium in the product collector. The entire ACM must be removed to harvest product, which is now perceived to be an inefficient engineering design. More modern ER designs, such as the PEER system invented by Willit and Williamson from ANL, typically feature means by which to continuously recover uranium product without pulling out fuel baskets until they are completely depleted [28]. Researchers at Korea Atomic Energy Research Institute (KAERI) have developed graphite cathodes for electrorefining [29]. Uranium dendrites readily release from the graphite and can be captured and continuously extracted from the ER [30].

3.2. *Minor actinide recovery*

Refer back to Table 2 and notice that close proximity of the standard reduction potentials for TRU metals compared with uranium. In order to deposit these metals onto the cathode, the potential needs to be about 300 mV lower than the potential needed to recovery uranium. As long as there is UCl_3 in the salt, any deposited TRUs will also spontaneously oxidize back to chlorides. This provides a very real challenge for using electrorefining to recover TRU metals for fast reactor fuel. Low cathode potential is needed to drive TRU deposition faster than it is re-oxidized by UCl_3. Provided there is ample surface area at the cathode and a high concentration of UCl_3 in the salt, U^{3+} reduction readily occurs without much overpotential. This overpotential is defined by the following equation:

$$\eta = E_{\text{cathode}} - E_{\text{eq}}$$

The cathode potential is E_{cathode}, and the equilibrium reduction potential for U^{3+}/U is E_{eq}. By convention, negative potentials are cathodic, and positive potentials are anodic. Thus, the E_{cathode} must decrease in order to induce TRU deposition onto the cathode. It is the overpotential at the cathode that is needed to increase in order to reduce plutonium and other TRU metals. Two basic approaches have been proposed and demonstrated for TRU recovery — employing a liquid metal cathode and increasing the current density at the cathode. The liquid metal that has most extensively been studied is cadmium [31–36]. But bismuth has also garnered some attention [37]. Aluminum has been reported as well as being useful for U/TRU co-deposition, even though it remains in the solid phase during the process [38]. Each of these metals is effective at stabilizing TRU metals by lowering the activity of the reduced state. Activity is defined relative to mole fraction in a given phase by the following equation. Activity coefficient is represented by γ_i, and mole fraction is represented by χ_i:

$$a_i = \gamma_i x_i$$

The electrochemical relevance of activity is found in the Nernst equation:

$$E_{\text{eq}} = E_i^0 + \frac{RT}{zF} \ln \left(\frac{a_{i,\text{ox}}}{a_{i,\text{red}}} \right)$$

where E_i^0 is the standard reduction potential, R is the ideal gas constant, T is absolute temperature, z is the number of electrons transferred per metal ion, F is Faraday's constant, $a_{i,\text{ox}}$ is the activity of the oxidized metal, and

$a_{i,\text{red}}$ is the activity of the reduced metal. In the case of cadmium, it has been hypothesized that intermetallics such as $PuCd_6$ form. This effectively reduced the activity of Pu in the reduced state $(a_{i,\text{red}})$ and increases the equilibrium potential (E_{eq}). Since plutonium's standard reduction potential is less than that of uranium, this results in the equilibrium potentials for U^{3+}/U and Pu^{3+}/Pu getting closer to each other. The activity of reduced metal is also affected by the cadmium but not as much so as Pu.

In addition to providing favorable thermodynamic properties at the cathode, the cadmium also has the benefit of high vapor pressure. This enables its removal from the cathode crucible via high-temperature distillation. This will be further discussed in Section 4. The liquid cadmium cathode (LCC) has been extensively studied at both the labscale and engineering scale by researchers at INL [34–36]. Separation factors have been measured using the labscale experiments to understand how actinides and rare earths can be separated/recovered using the LCC [34, 35]. And kilogram-scale quantities of TRU metals have been collected in engineering-scale LCC experiments [36].

It is very important to understand that the LCC recovers a group actinide product consisting of a mixture of uranium, plutonium, and minor actinides. As long as there is UCl_3 in the salt, uranium reduction cannot be avoided. Increasing the $PuCl_3$ to UCl_3 ratio in the salt will increase the Pu/U ratio in the cathode. In engineering-scale tests, INL reported a Pu/U ratio in the cathode as high as 5.7 [36]. With each successive LCC run, this ratio decreases because of the depletion of $PuCl_3$ from the salt.

In preparing for its engineering-scale LCC tests, researchers at INL added lithium metal to the ER salt to selectively reduce UCl_3 to U metal. This is an effective and quick approach to increasing the $PuCl_3$ to UCl_3 ratio in the salt to promote recovery of a Pu-rich LCC product. The next logical question is whether this can be taken further to achieve a pure Pu metal deposit. The first difficulty with achieving this outcome is that high reduction of UCl_3 is accompanied by significant co-reduction of $PuCl_3$. The relatively close proximity of the free energy of formation of UCl_3 and $PuCl_3$ would likely prevent complete reduction of UCl_3 without some co-reduction of $PuCl_3$. The other issue is that with each deposition of Pu, $PuCl_3$ in the salt would be replaced by UCl_3. This is because there must be an anodic reaction to balance the electrotransfer from each cathodic reaction. Reduction of $PuCl_3$ to Pu must be accompanied by oxidation. An inert anode can be used to oxidize Cl^- to Cl_2 gas. But Cl_2 gas is problematic to handle because of corrosion and can easily be detected. More logically, uranium or spent fuel metal would be used in the anode, and uranium oxidation would be the anodic reaction. The only way to recover Pu from the salt and not increase the uranium concentration is to either generate Cl_2 gas or oxidize a different metal. The metal would have to be more stable as a chloride than $PuCl_3$, otherwise it will start depositing instead of Pu metal. If that metal is uranium, then the U content in the cathode deposits will steadily increase with each Pu recovery run. If it is a more active metal such as lanthanum, it will cause local plutonium reduction rather than on the cathode. Either way, it is exceedingly difficult to use a molten salt ER to recover any substantial quantity of pure Pu metal. This is the basis for the claim that pyroprocessing is proliferation resistant. Granted, proliferation can take other forms besides recovery of pure Pu metal. Thus, we should refrain from calling pyroprocessing proliferation resistant. But it is valid to state that pure Pu recovery via pyroprocessing is very difficult if not entirely unfeasible.

3.3. *Electrowinning*

Electrowinning is a very similar process to U or U/TRU electrorefining. But rather than using spent fuel as the reactive anode, it utilizes an inert anode to generate Cl_2 gas as the oxidation reaction. U, TRU, and/or rare earths can be cathodically deposited while Cl_2 gas is generated at the anode. Effectively, electrowinning extracts metals out of the molten salt rather than out of the fuel. In pyroprocessing system designs, it is usually employed to recover actinides from salt prior to its disposal as waste. Actinides can be recycled to the ER via loading in the anode baskets, and the actinide-free salt can be subjected to fission product separations or direct immobilization in a waste form such as glass-bonded sodalite. Keeping actinides from going to waste has the dual benefits of optimizing resource utilization in the fuel cycle and minimizing impact on the spent fuel repository. Actinides are the major long-lived contributors to dose from a permanent geologic repository. If they can be recycled to fast reactors, the long-term impact of nuclear waste can be greatly diminished. This outcome starts with the electrowinning process, which clearly highlights its importance in the overall pyroprocessing process.

4. Cathode Processing

Metallic actinides recovered from the ER or electrowinning process are coated with molten salt (and possibly cadmium), which freezes upon removal from the heated zone. If the actinides are intended to be used for fuel or will be stored as metal ingots, they must be subsequently processed to remove the salt and consolidate the metal into ingots. This process is commonly referred to as cathode processing using a high-temperature vacuum distillation system. Temperatures need to reach approximately 1200°C in order to fully evaporate the salt and melt the uranium or U/TRU. An excellent overview of the FCF cathode processor (CP) design and operating experience was written by Westphal *et al.* [39]. As explained in this paper, the distillation process is effective based on the following sequence of vapor pressures:

Cadmium metal → chloride salts → actinide metals

The FCF CP is a bottom loading, inductively heated vacuum furnace. It can achieve a maximum temperature of 1200°C and minimum pressure of 10 Pa. Chloride distillation efficiencies have been measured ranging from 98.61% to 99.96% [39]. Cadmium distillation efficiency is even higher with about 35 ppm Cd in the final product. The cathode product is first loaded into a graphite crucible coated with zirconia or a castable zirconia crucible. Crucible coating is important for enabling easy release of the resulting actinide ingot. But the zirconia coating does react with the salt/metal mixture, forming a small amount of actinide dross (oxides). Typically 2–4% of the uranium metal in each batch is converted to uranium oxide dross. Limited processing has been performed using hafnium nitride-coated niobium crucibles [39]. Excellent results were obtained with 6-liter HfN/Nb crucible, but not with an 18-liter version of the crucible. The plasma spray-based coating process is difficult to perform and achieve a full, uniform coating. These special crucibles are also very expensive. Currently, INL is using castable zirconia crucibles to save time by eliminating the slow process of coating the graphite crucibles inside of the FCF hot cell. Dross is still generated from each CP run, so there is room for improvement. In a commercial scale facility equipped to process oxide fuel, the dross can readily be reduced and recycled. Thus, dross formation is not considered to be a show-stopper for commercialization of pyroprocessing.

In addition to dross formation, another problematic side effect of cathode processing is an increase in plutonium contamination in the uranium product. Prior to cathode processing, uranium dendrites can be rinsed to remove salt and analyzed. Virtually no plutonium is found in the dendrite samples after removal of salt. But uranium cathode product samples taken after processing in the CP can contain hundreds of ppm of

Pu. The suspected cause is a shift in the following oxidation–reduction reaction:

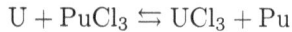

$$U + PuCl_3 \leftrightarrows UCl_3 + Pu$$

This reaction is normally very unfavorable as written with an equilibrium constant several orders of magnitude below one. But during the distillation process, it is expected that UCl_3 evaporates more rapidly than $PuCl_3$ due to differences in vapor pressure. UCl_3 has approximately 10 times higher vapor pressure than $PuCl_3$ [39]. Thus, during distillation, the aforementioned reaction will be shifted to the right as more and more UCl_3 is evaporated. When essentially all of the UCl_3 has evaporated, the equilibrium will shift to form more UCl_3, which will be accompanied by formation of Pu metal. This Pu metal becomes a contaminant in the uranium product ingot. If the uranium will be used for remote fabrication of fast reactor fuel, the plutonium contamination is unimportant. From a proliferation perspective, this plutonium contamination is too low to be a concern. But if the uranium collected needs to be converted into light water reactor fuel produced in commercial fuel fabrication plants, the plutonium contamination can be a serious problem. This has been an impediment to INL selling any of the uranium it has recovered from EBR-II driver fuel to commercial fuel vendors.

5. Metal Waste Processing

After each batch of fuel has been electrorefined, the anode baskets contain undissolved metals — including noble metals (Tc, Ru, Rh, Zr, Pd, Mo), a small fraction of the actinides and cladding hulls. These metals are melted to form a metal waste form. This requires a high-temperature vacuum furnace, similar in design and function as the CP. The metal waste furnace (MWF) performs the following functions [40]:

- Distillation of residual salt on the cladding hulls
- Oxidation of residual sodium metal bonding to sodium chloride (sodium-bonded fuel only)
- Melting the metal alloy to form a consolidated ingot

Cladding hulls comprise the highest mass fraction of metal waste, so their composition is key to designing the metal waste process. For commercial light water reactor fuel, cladding hulls are typically made of Zircaloy with a melting point of about 1855°C. EBR-II metallic U–Zr fuel used stainless steel cladding hulls with a melting point of about 1450°C. For processing metal waste from EBR-II fuel treatment, zirconium is actually added to achieve a nominal 15 wt.% Zr–stainless steel alloy. This is useful for further suppressing the liquidus temperature of the metal alloy. Even with this lowering of the liquidus temperature, induction-heated furnaces are required. One such MWF is currently operating in the Hot Fuel Examination Facility (HFEF) at INL and has been used to produce metal waste forms from the contents of anode baskets used for electrorefining EBR-II fuel.

Achieving high efficiency of actinide dissolution in the ER can significantly lower the noble metal retention efficiency [41]. It is, thus, expected that 1–5% of the uranium from the original spent fuel could be left in the anode baskets at the completion of each run. Of more consequence is that technetium behaves as a noble metal in the ER. Tc-99 is a key fission product that contributes a high dose rate from the spent fuel and is very long lived (half-life = 211,000 years). Immobilization of Tc-99 is essential for management of the spent nuclear fuel waste. Thus, any formulation of the metal waste must adequately stabilize Tc in a phase that has relatively low susceptibility to corrosion. From characterization of metal waste form samples, it has been determined that Tc is sequestered in steel and Fe_2Mo phases [42].

6. Salt Waste Processing

The eutectic LiCl–KCl electrolyte used in the ERs is contaminated by each batch of spent fuel by the active metals, as previously discussed. Those active metals are those that have free energy of formation less than UCl_3 as listed in Table 2. LiCl–KCl has shown to have a high capacity for dissolving these metal chlorides. In fact, the actual solubility limits have not been well established via experimental studies. But there are other factors involved with limiting the concentration of fission products in the salt — including fission heat generation and cathode contamination. Fission heat can especially be problematic if Cs and Sr are contained in the metal feed to the ER. Cathode contamination in the form of ppm levels of rare earths and TRU's can occur as their concentration levels rise in the salt. Fundamentally, there are two options on how to manage the ER salt composition — throw-away option and salt recycling. The throw-away option may also be called *bleed and feed*. More specifically, it involves removing some of the salt from the ER, disposing of it, and replacing it with pure LiCl–KCl. The problem with this approach is that it generates relatively large amounts of waste. The salt may contain 80 wt.% LiCl–KCl, for example, with the remaining salts consisting of actinide and fission product chlorides. Only the fission products need to be disposed. The LiCl–KCl is useful as the electrolyte, and actinides should be recovered to fabricate new fuel. Alternatively, separation processes can be applied to the salt in order to recycle as much of the useful components as deemed to be practical. This serves to minimize generation of waste but adds equipment and complexity to the process.

6.1. *Throw-away option* (*baseline ceramic waste process*)

The throw-away option was initially pursued for the EBR-II Spent Fuel Treatment project, because salt treatment (zeolite ion exchange) was found to be impractical for a limited amount of spent fuel to be processed. ANL (and later INL) developed the ceramic waste process shown in Figure 7 [43–46].

Zeolite-4A is a commercially available aluminosilicate $(Na_{12}(SiO_2)_{12}(AlO_2)_{12})$. It features a three-dimensional (3D) pore network with effective pore diameter of 4 Å. This zeolite is hygroscopic, so a severe drying process is required to remove virtually all absorbed water prior to salt sorption. In the drying process, zeolite is mechanically fluidized while being heated to 500–550°C [44]. The pressure is initially atmospheric. After the majority of the water has been released, applying a rough vacuum was found to be beneficial to further lower the residual moisture concentration. Over about a 12-h heating cycle, the zeolite can be dried to less than 0.5 wt.% water. The zeolite may initially have as high as 20 wt.% water.

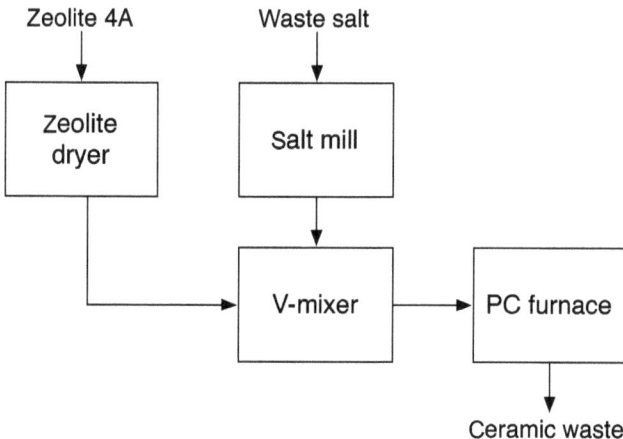

Figure 7. Flowsheet for the ceramic waste process.

ANL-West built and demonstrated operation of a high-temperature V-mixer for the salt sorption step [44, 46]. Both the dried zeolite and salt were size-reduced to a nominal particle diameter range of 45–250 μm. This particle size promotes rapid sorption kinetics while maintaining favorable powder handling characteristics. This process has only been operated in batch mode with multiple transfers of powder between different equipment systems. Thus, good powder flowability is very important. If the zeolite is too fine, it was found to be extremely difficult to induce unloading/loading between containers in the hot cell. The V-mixer design was selected by ANL-West researchers because of its ability to continuously mix the salt and zeolite powders during heating. The V-mixer installed in the HFEF achieved a high temperature of 500°C and held at this level for 12 hours or more to drive near complete sorption of the salt into the zeolite. After cooling, glass frit with a similar particle size range was added to the V-mixer. Glass and salt-loaded zeolite were mixed and unloaded into a canister for final consolidation.

Final consolidation of the ceramic waste form has been demonstrated using two methods — hot isostatic pressing (HIP) and pressureless consolidation (PC). During the EBR-II Spent Fuel Treatment Demonstration Project, a HIP was installed in HFEF that could process canisters containing a maximum of about 1.5 kg of salt-loaded zeolite and glass [44, 47]. For processing in the HIP, these canisters were sealed via welding. The high pressure collapses the canister around the powder. Achieving a temperature of 750°C and pressures up to 25,000 psi resulted in the formation of a glass bonded sodalite waste form. The zeolite-4A encountered a structure change to sodalite, and the glass softened and served as a binder. Further development efforts were successful in eliminating the need for pressure in exchange for higher temperatures. The PC process involved heating the salt-loaded zeolite and glass to 915–950°C without any applied pressure [45, 46]. This also caused the zeolite-4A to convert to sodalite. A very similar ceramic waste form was made using the PC process with only a slightly lower density than that made from the HIP process. Eliminating high pressures from the process effectively eliminated its only safety risk. The PC process is also more adaptable to high throughput consolidation systems that utilize continuous heating of parts. INL demonstrated production of simulated waste forms up to a mass of about 400 kg [48]. But such large waste forms require several days to heat and cool. If the ceramic waste process is eventually developed for a commercial facility, it is more likely that smaller ceramic waste forms would be produced continuously on a conveyer belt-driven furnace.

6.2. *Salt separation methods*

Numerous methods have been envisioned, tested, and benchmarked for separating waste salt from ERs into different streams. This includes ion exchange, chemical reduction, electrochemical reduction, precipitation, and selective crystallization methods. Research in this area is too expansive to completely cover in this chapter. Generally, at least two steps need to be applied in succession — actinide recovery and fission product separation. Actinide recovery is most appropriately achieved using electrowinning, which was previously discussed. The reason for removing actinides first is that they tend to be more reactive than fission products using all methods that have been investigated. In order to make a fission product-rich waste form, the actinides should be removed (and preferably recycled) first. The generic term for actinide removal is *drawdown*. In addition to electrowinning, lithium reduction is a viable method for actinide drawdown [49]. Lithium first reduces UCl_3 to U followed by $PuCl_3$ to Pu and other minor actinides. There is an overlap between minor actinides and rare earths for drawdown, so it is not a simple and complete solution. Ultimately, there usually needs to be a trade-off between actinide drawdown efficiency and carryover of rare earth fission products. After most of the actinides have been recovered, the salt can be treated to concentrate fission products in a waste stream. LiCl–KCl with depleted fission product concentration can then be returned to the ER. Early development by ANL focused on zeolite-A as an ion exchange resin for the salt [50]. When molten salt comes in contact with zeolite-A, two processes occur — ion exchange and occlusion. Ion exchange involves the replacement of framework cations (Na^+ initially) with cations from the molten salt. Occlusion involves absorption of the entire salt molecule into the zeolite pore. This is believed to be a space-filling process, as the capacity actually varies as a function of salt composition [51]. KAERI later developed oxygen sparging [52], zone freezing [53], and phosphate precipitation [54] methods for fission product removal from LiCl–KCl. If the only need is to remove rare earths, electrowinning may also be a viable option. After actinides are removed via electrowinning, rare earths are the next group to be separated from the salt.

References

1. J. P. Ackerman, Chemical basis for pyrochemical reprocessing of nuclear fuel, *Ind. Eng. Chem. Res.* **30**(1) (1991).
2. Y. I. Chang, The integral fast reactor, *Nucl. Technol.* **88**(2), 129–138 (1989).
3. C. E. Till, Y. I. Chang, and W. H. Hannum, The integral fast reactor-an overview, *Prog. Nucl. Energy* **31**(1), 3–11 (1997).
4. R. Benedict, M. Goff, G. Teske, and T. Johnson, Progress in electrometallurgical treatment of spent nuclear fuel, *J. Nucl. Sci. Technol.* **39**(Suppl. 3), 749–752 (2002).
5. H. Lee, G.-I. Park, J.-W. Lee, K.-H. Kang, J.-M. Hur, J.-G. Kim, S. Paek, I.-T. Kim, and I.-J. Cho, Current status of pyroprocessing development at KAERI, *Sci. Technol. Nucl. Ins.* **13**, Article ID 343492 (2013).
6. T. Koyama, Y. Sakamura, M. Iizuka, T. Kato, T. Murakami, and J-P. Glatz, Development of pyro-processing fuel cycle technology for closing actinide cycle, *Procedia Chem.* **7**, 772–778 (2012).
7. S. Vavilov, T. Kobayashi, and M. Myochin, Principle and test experience of the RIAR's oxide pyro-process, *J. Nucl. Sci. Technol.* **41**(10), 1018–1025 (2004).
8. E. J. Karell, K. V. Gourishankar, J. L. Smith, L. S. Chow, and L. Redey, Separation of actinides from LWR spent fuel using molten-salt-based electrochemical processes, *Nucl. Technol.* **136**(3), 342–353 (2001).
9. T. Usami, M. Kurata, T. Inoue, H. E. Simps, S. A. Beetham, and J. A. Jenkins, Pyrochemical reduction of uranium dioxide and plutonium dioxide by lithium metal, *J. Nucl. Mater.* **300**, 15–26 (2002).
10. M. F. Simpson and S. D. Herrmann, Modeling the pyrochemical reduction of spent UO_2 fuel in a pilot-scale reactor, *Nucl. Technol.* **162**, 179–183 (2008).
11. A. Burak and M. F. Simpson, Measurement of solubility of metallic Li in molten $LiCl-Li_2O$, *J. Mater.* **68**(10), 2639–2645 (2016).
12. Y. Kado, T. Goto, and R. Hagiwara, Dissolution behavior of lithium oxide in molten LiCl-KCl systems, *J. Chem. Eng. Data* **53**, 2816–2819 (2008).
13. *HSC Chemistry*, Outotec, Version 7 (2009).
14. O. Levenspiel, *Reactor Engineering*, 2nd edition, Wiley and Sons, New York, pp. 361–366 (1972).
15. L. I. Redey and K. Gourishankar, Direct electrochemical reduction of metaloxides, U.S. Patent 6,540,902 B1, April 1 (2003).
16. C. S. Seo, S. B. Park, B. H. Park, K. J. Jung, S. W. Park, and S. H. Kim, Electrochemical study on the reduction mechanism of uranium oxide in a $LiCl-Li_2O$ molten salt, *J. Nucl. Sci. Technol.* **43**(5), 587–595 (2006).
17. M. Iizuka, Y. Sakamura, and T. Inoue, Electrochemical reduction of (U–40Pu–5Np) O^{2-} in molten LiCl electrolyte, *J. Nucl. Mater.* **359**(1), 102–113 (2006).
18. S. D. Herrmann, S. X. Li, M. F. Simpson, and S. Phongikaroon, Electrolytic reduction of spent nuclear oxide fuel as part of an integral process to separate and recover actinides from fission products, *Sep. Sci. Technol.* **41**(10), 1965–1983 (2006).

19. B. H. Park, I. W. Lee, and C-S. Seo, Electrolytic reduction behavior of U 3 O 8 in a molten LiCl–Li 2 O salt, *Chem. Eng. Sci.* **63**(13), 3485–3492 (2008).

20. S. M. Jeong, H-S. Shin, S.-S. Hong, J-M. Hur, J. B. Do, and H. S. Lee, Electrochemical reduction behavior of U 3 O 8 powder in a LiCl molten salt, *Electrochim. Acta* **55**(5), 1749–1755 (2010).

21. S. D. Herrmann and S. X. Li, Separation and recovery of uranium metal from spent light water reactor fuel via electrolytic reduction and electrorefining, *Nucl. Technol.* **171**, 247–265 (2010).

22. T. Meek, Semi-conductive properties of uranium oxides, In *Proc. Waste Management 2001 Symp.*, Tucson, February 25–March 1 (2001).

23. S. M. Jeong, H. S. Shin, S. H. Cho, J. M. Hur, and H. S. Lee, Electrochemical behavior of a platinum anode for reduction of uranium oxide in a LiCl molten salt, *Electrochem. Acta* **54**, 6335–6340 (2009).

24. S. X. Li, T. Sofu, T. A. Johnson, and D. V. Laug, Experimental observations on electrorefining spent nuclear fuel in molten LiCl-KCl/liquid cadmium system, *J. New Mater. Electrochem. Syst.* **3**, 259–268 (2000).

25. J. L. Willit, W. E. Miller, and J. E. Battles, Electrorefining of uranium and plutonium—A literature review, *J. Nucl. Mater.* **195**, 229–249 (1992).

26. T. C. Totemeier and R. D. Mariani, Morphologies of uranium and uranium–zirconium electrodeposits, *J. Nucl. Mater.* **250**(2), 131–146 (1997).

27. R. K. Ahluwalia, T. Q. Hua, and D. Vaden, Uranium transport in a high-throughput electrorefiner for EBR-II blanket fuel, *Nucl. Technol.* **145**(1), 67–81 (2004).

28. J. L. Willit and M. A. Williamson, U.S. Patent No. 8,097,142, Washington, DC, U.S. Patent and Trademark Office (2012).

29. J. H. Lee, Y. H. Kang, S. C. Hwang, J. B. Shim, E. H. Kim, and S. W. Park, Application of graphite as a cathode material for electrorefining of uranium, *Nucl. Technol.* **162**(2), 135–143 (2008).

30. J. H. Lee, Y. H. Kang, S. C. Hwang, H. S. Lee, E. H. Kim, and S. W. Park, Assessment of a high-throughput electrorefining concept for a spent metallic nuclear fuel-I: Computational fluid dynamics analysis, *Nucl. Technol.* **162**(1), 107–116 (2008).

31. T. Koyama, M. Iizuka, Y. Shoji, R. Fujita, H. Tanaka, T. Kobayashi, and M. Tokiwai, An experimental study of molten salt electrorefining of uranium using solid iron cathode and liquid cadmium cathode for development of pyrometallurgical reprocessing, *J. Nucl. Sci. Technol.* **34**(4), 384–393 (1997).

32. M. Iizuka, T. Koyama, N. Kondo, R. Fujita, and H. Tanaka, Actinides recovery from molten salt/liquid metal system by electrochemical methods, *J. Nucl. Mater.* **247**, 183–190 (1997).

33. T. Kato, T. Inoue, T. Iwai, and Y. Arai, Separation behaviors of actinides from rare-earths in molten salt electrorefining using saturated liquid cadmium cathode, *J. Nucl. Mater.* **357**(1), 105–114 (2006).

34. S. X. Li, S. D. Herrmann, K. M. Goff, M. F. Simpson, and R. W. Benedict, Actinide recovery experiments with bench-scale liquid cadmium cathode in real fission product-laden molten salt, *Nucl. Technol.* **165**(2), 190–199 (2009).

35. S. X. Li, S. D. Herrmann, and M. F. Simpson, Electrochemical analysis of actinides and rare earth constituents in liquid cadmium cathode product from spent fuel electrorefining, *Nucl. Technol.* **171**(3), 292–299 (2010).

36. D. Vaden, S. X. Li, B. R. Westphal, K. B. Davies, T. A. Johnson, and D. M. Pace, Engineering-scale liquid cadmium cathode experiments, *Nucl. Technol.* **162**(2), 124–128 (2008).

37. J. Serp, P. Lefebvre, R. Malmbeck, J. Rebizant, P. Vallet, and J. P. Glatz, Separation of plutonium from lanthanum by electrolysis in LiCl–KCl onto molten bismuth electrode, *J. Nucl. Mater.* **340**(2), 266–270.

38. J. Serp, M. Allibert, A. Le Terrier, R. Malmbeck, M. Ougier, J. Rebizant, and J. P. Glatz, Electroseparation of actinides from lanthanides on solid aluminum electrode in LiCl-KCl eutectic melts, *J. Electrochem. Soc.* **152**(3), C167–C172 (2005).

39. B. R. Westphal, K. C. Marsden, J. C. Price, and D. V. Laug, On the development of a distillation process for the electrometallurgical treatment of irradiated spent nuclear fuel, *Nucl. Eng. Technol.* **40**(3), 163–174 (2008).

40. K. C. Marsden, Production-scale metal waste process qualification, In *Int. Pyroprocessing Research Conf.*, Idaho Falls, August 8–10 (2006).

41. S. X. Li and M. F. Simpson, Anodic process of electrorefining spent nuclear fuel in molten LiCl-KCl-UCl$_3$/Cd System, *J. Miner. Metall. Process.* **22**(4), 192–198 (2005).

42. S. Frank, W. Ebert, B. Riley, H. S. Park, Y. Z. Cho, C. H. Lee, M. K. Jeon, J. H. Yang, and H. C. Eun, Waste stream treatment and waste form fabrication for pyroprocessing of used nuclear fuel, Idaho National Laboratory, INL/EXT-14-34014 (2015).

43. C. Pereira, M. C. Hash, M. A. Lewis, M. K. Richmann, and J. Basco, Incorporation of radionuclides from the electrometallurgical treatment of spent fuel into a ceramic waste form, *MRS Proc.* **556**, 115 (1999).

44. M. F. Simpson, K. M. Goff, S. G. Johnson, K. J. Bateman, T. J. Battisti, K. L. Toews, S. M. Frank, T. L. Moschetti, and T. P. O'Holleran, A description of the ceramic waste form production process from the demonstration phase of the electrometallurgical treatment of EBR-II spent fuel, *Nucl. Technol.* **134**, 263–277 (2001).

45. S. Priebe and K. Bateman, The ceramic waste form process at Idaho National Laboratory, *Nucl. Technol.* **162**(2), 199–207 (2008).

46. M. F. Simpson and P. Sachdev, Development of electrorefiner waste salt disposal process for the EBR-II spent fuel treatment project, *Nucl. Eng. Technol.* **40**(3), 175–182 (2008).

47. C. Pereira, M. Hash, M. Lewis, and M. Richmann, Ceramic-composite waste forms from the electrometallurgical treatment of spent nuclear fuel, *J. Mater.* **49**(7), 34–40 (1997).

48. M. C. Morrison, K. J. Bateman, and M. F. Simpson, Scale-up of ceramic waste forms for the EBR-II spent fuel treatment process, In *Int. Pyroprocessing Research Conf.*, Idaho National Laboratory, INL/CON-10-19439 (2010).

49. M. F. Simpson, T. S. Yoo, D. LaBrier, M. Lineberry, M. Shaltry, and S. Phongikaroon, Selective reduction of active metal chlorides from molten LiCl-KCl using lithium drawdown, *Nucl. Eng. Technol.* **44**(7), 767–772 (2012).

50. R. K. Ahluwalia, H. K. Geyer, C. Pereira, and J. P. Ackerman, Modeling of a zeolite column for the removal of fission products from molten salt, *Ind. Eng. Chem. Res.* **37**(1), 145–153 (1998).

51. T. S. Yoo, S. M. Frank, M. F. Simpson, P. A. Hahn, T. J. Battisti, and S. Phongikaroon, Salt-zeolite ion exchange equilibrium studies for complete set of fission products in molten LiCl-KCl, *Nucl. Technol.* **171**(3) (September 2010).

52. Y. J. Cho, H. C. Yang, H. C. Eun, E. H. Kim, and I. T. Kim, Characteristics of oxidation reaction of rare-earth chlorides for precipitation in LiCl-KCl molten salt by oxygen sparging, *J. Nucl. Sci. Technol.* **43**(10), 1280–1286 (2006).

53. Y. Z. Cho, T. K. Lee, J. H. Choi, H. C. Eun, H. S. Park, and G. I. Park, Eutectic (LiCl-KCl) waste salt treatment by sequential separation process, *Nucl. Eng. Technol.*, **45**(5), 675–682 (2013).

54. V. A. Volkovich, T. R. Griffiths, and R. C. Thied, Treatment of molten salt wastes by phosphate precipitation: Removal of fission product elements after pyrochemical reprocessing of spent nuclear fuels in chloride melts, *J. Nucl. Mater.* **323**(1), 49–56 (2003).

Chapter 3

Electrochemical Studies of Uranium in LiCl–KCl Eutectic Salt for an Application of Pyroprocessing Technology

Dalsung Yoon* and Supathorn Phongikaroon

Department of Mechanical & Nuclear Engineering
Virginia Commonwealth University, 401 West Main St.
Richmond, VA 23284, USA
*yoond2@vcu.edu

1. Introduction

As the power production by nuclear power plant has increased around the world, the management of the used nuclear fuel (UNF) will become an important issue due to political, economic, and societal concerns in the nuclear industry [1]. The total amount of UNF cumulatively generated worldwide by 2014 was 204,421 tHM, and it has continued increasing from year to year [2]. The used fuel storage capacity in 2014 was 201,722 tHM, but the global reprocessing capacity was only 3800 tHM/year [2]. Therefore, in terms of saving fuel resources and solving the issue of storage capacity, the significance of recovering components from the used fuel will continue to grow in the future [1]. Reprocessing UNF can be more invaluable when it is considered that 96% of uranium remains after the fuel is permanently removed from reactor, which can be re-used after suitable retreatments [1, 3]. Two methods have been widely investigated and implemented for the reprocessing of the nuclear fuel, which are referred to as aqueous reprocessing and pyroprocessing [3]. The aqueous reprocessing utilizes a method known as plutonium uranium reduction extraction (PUREX), which is the most common and well-developed technique. In the PUREX process, pure U and Pu are separated through chemical adjustments and several cycles of solvent extractions by using highly concentrated nitric acid [3–5]. By experiencing the renaissance of nuclear energy worldwide, pure U and Pu productions gave rise to several concerns about proliferation of nuclear materials. From the safeguarding aspect of reprocessing UNF, the pyroprocessing technology has been considered as an alternative method for future reprocessing [1, 3, 5, 6], which accomplishes recovery of uranium, plutonium, and other minor actinide (MA) elements by the way of high-temperature electrorefining. Specially, the fact that plutonium is recovered as a mixture with the MA materials gives strong proliferation resistance to pyroprocessing compared with PUREX system. The main objective of the technology is to use the recovered uranium, plutonium, and actinides into fast reactor fuel fabrication, and minimize the generation of the nuclear waste. While the pyroprocessing technology is developed for treating the metal waste from fast reactors, the oxide form of nuclear waste from light water reactor (LWR) can also be processed by developing an additional stage, which is referred to as the oxide reduction process. Although this technology has not yet reached to the commercialization stage like the PUREX technology, it has been extensively studied and developed by many countries including USA, South Korea, China, Japan, Russia, and several European countries. In this chapter, the pyroprocessing technology will be generally reviewed and an overview of electrochemical and thermodynamics of uranium chloride, which is a major component in this process, will be followed.

2. Pyroprocessing

Pyroprocessing (also known as electrochemical process, electrometallurgical reprocessing, or pyrochemical technology) was originally developed by the Argonne National Laboratory (ANL) and currently is being operated at the Idaho National Laboratory (INL) to treat used metallic fuels from Experimental Breeder Reactor-II (EBR-II) [7–9]. This technology uses molten salt electrolytes as media instead of using acid solutions or organic solvents [3, 8]. In these electrolytes, U, Pu, and MAs can be recovered by electrochemical reduction on cathode electrodes. In comparison with the conventional aqueous methods, which have superior maturity to date, the pyroprocessing technology provides the following unique benefits: (1) intrinsic proliferation-resistant features by Pu recovery as a mixture, (2) compact facilities for fuel recovery and fabrication, (3) critically safe condition for processing high enriched fuel, and (4) rapid on-site support for fast reactor fuel cycle [5, 10]. In addition, a low radiation sensitivity of the salt electrolyte allows an early reprocessing of UNF after its discharge [1], which can help minimizing the chance of a loss of coolant accidents in the spent fuel pool [11]. With these noble features, the main purpose of the pyro-processing technology is not only to treat the irradiated nuclear fuels but to reduce volume of the nuclear waste, recycle actinides, and close the fast reactor fuel cycle [12]. Therefore, there are considerable ongoing research and development on pyroprocessing technologies in many countries [3].

Figure 1 illustrates schematically the flow sheet of the process [4, 5]. Despite of its original development purpose, the oxide form of used fuels from LWR can also be treated through the head-end process, known as electrolytic oxide reduction. In the electrolytic oxide reduction process, the used oxide fuel is loaded into a cathode basket in a molten LiCl–Li$_2$O salt at 923 K and chemically reduced into metal form by the reaction with Li metal, which is electrochemically deposited on the cathode [13]. In 2006, the INL successfully converted 50 g of the used LWR fuel into the metal form in their hot cell facility [14]. Then, the metal fuels are transferred into the anode basket of an electrorefiner (ER). While U, Pu, MAs, and rare earth (RE) materials are anodically dissolved into the LiCl–KCl eutectic salt from the anode basket, only U is recovered on the solid cathode by controlling the voltage applied on the cathode. After that, the residual U, Pu, and MAs are simultaneously collected using a liquid cadmium cathode (LCC) because the activity of the elements gets very small in liquid metal [15, 16].

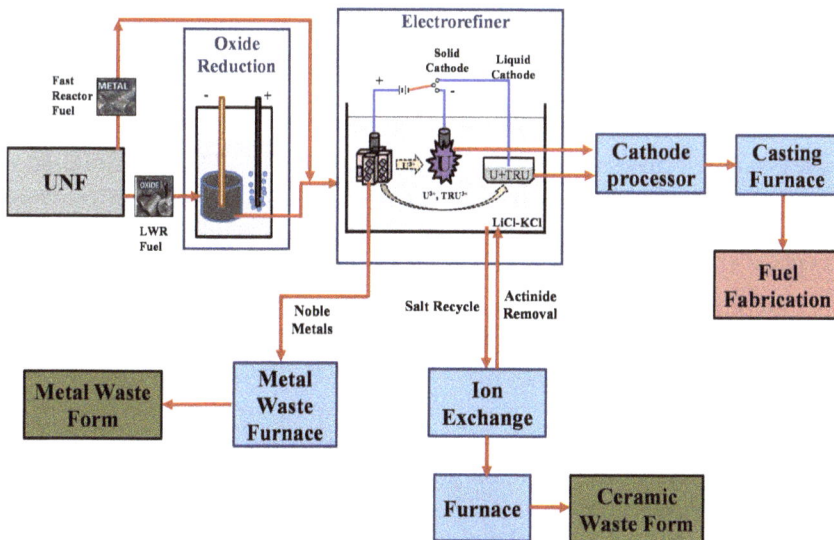

Figure 1. Schematic flowchart of pyroprocessing based on used nuclear fuel treatment.

The deposits (U and U–Pu–MA mixtures) from the cathode electrodes go to the cathode processor. The process is basically in a high-temperature vacuum furnace where the adhering salt or cadmium can be evaporated and pure metal products are left behind [9]. The ingot products from the cathode processor are fabricated into new metal fuel in an injection-casting furnace [7], which can be used in fast reactors. The electrolyte salt from the ER system is recycled by removing actinides through the ion exchange method, and the separated fission products (Cs, Rb, Sr, Ba, Br, I, Y, Sm, Eu, etc.) are immobilized into ceramic waste forms [9]. After operating the ER system, zirconium and noble metal fission products are left in the anode basket, which are processed into alloy metal waste form for disposal [9].

3. Electrorefiner

The ER is the key component in the pyroprocessing technology, in which the primary separation of U and MAs from fission products is being performed [16, 17]. The system contains a LiCl–KCl eutectic salt (44.2 wt.% LiCl and 55.8 wt.% KCl), and the operating temperature ranges from 723 to 773 K [3]. There are two engineering scale ERs (Mark IV and Mark V) at INL (USA), which are currently operated to treat the metallic used driver and blanket fuels, respectively, from EBR-II [17]. Figure 2 shows the drawings of both INL ERs, which have similar size but Mark V has an upgraded electrode design [17]. The Mark IV ER has been used to treat the driver fuels (highly enriched uranium at about 63% of U-235), whereas Mark V ER treats the blanket fuels (which contains depleted uranium) [18]. To date, 830 kg of heavy metals have been processed via Mark IV ER and numerous studies have been performed and reported in the literature [17]. The technical issues that are currently of concern include current efficiency, uranium recovery efficiency, zirconium recovery, and understanding of cadmium effects [17].

Here, UNFs are loaded in the anode basket and lowered into the LiCl–KCl molten salt. As current is applied through the cell, U metal is oxidized into U^{3+} ions from the anode basket [17]. Fission products, which have more negative redox potential than that of U, are being oxidized together (these products are referred to as active products), whereas noble metals that have more positive redox potential than U stay in the anode basket [17]. Therefore, the main goal is to dissolve as much U as possible from the used fuels at the anodic side with minimal oxidation of the noble metals into the salt [17] and collect uranium metal at the cathode side. Li and Simpson in 2005 [18] reported that 99.7% of U from fission products could be dissolved into the salt over several runs, but

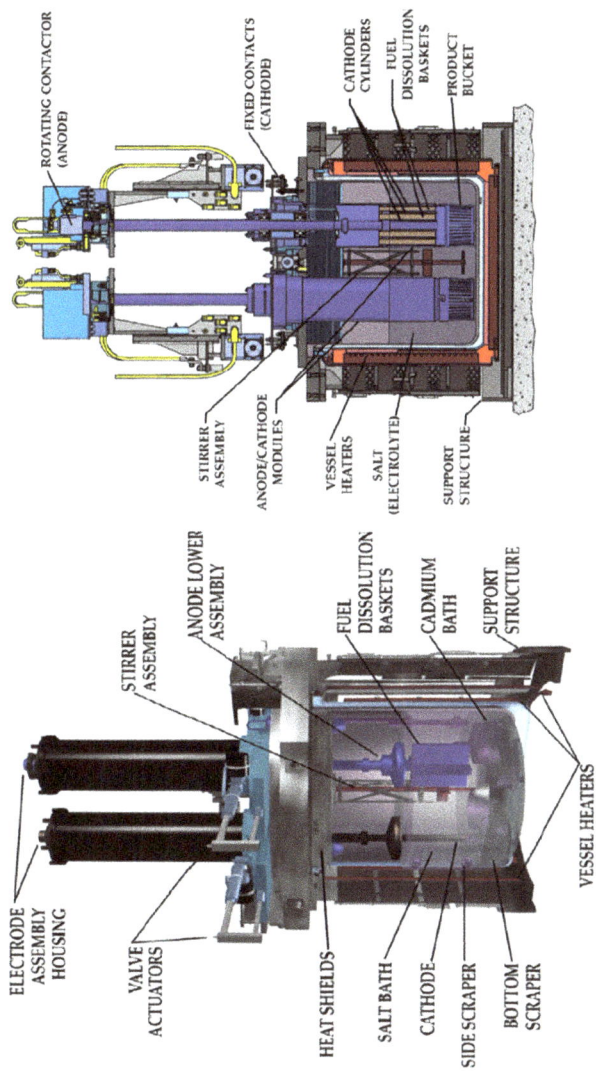

Figure 2. Engineering scale electrorefiner at INL Fuel Conditioning Facility, Mark IV (left) and Mark V (right) [18].

it caused Zr and noble metal co-dissolution into the salt. For the cathode electrodes, there are two electrode types: (1) solid electrode (typically stainless steel) and (2) the LCC. By using the solid electrode, selective deposition of pure U is possible by controlling the applied potential. On the other hand, reduction potentials of actinide elements including Pu and U become very close on the LCC, so that MAs are inevitably recovered along with U and Pu into the LCC process [16]. In general, U has a tendency to be deposited in form of a dendrite [16, 17], which prevents the co-deposition of MA on LCC. Therefore, several studies have been performed to develop the LCC structures for avoiding dendrite formation on LCC [16, 19, 20]. Overall, the main reaction schemes in the ER can be described as follows [21]:

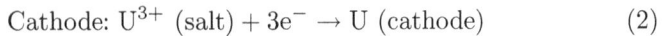

$$\text{Anode: U (anode)} \rightarrow \text{U}^{3+} \text{ (salt)} + 3\text{e}^- \tag{1}$$

$$\text{Cathode: U}^{3+} \text{ (salt)} + 3\text{e}^- \rightarrow \text{U (cathode)} \tag{2}$$

$$\text{Net reaction: U (anode)} \rightarrow \text{U (cathode)} \tag{3}$$

4. Reviews of Electrochemical, Thermodynamic, and Kinetic Properties of U in LiCl–KCl Eutectic

4.1. *Experimental preparation*

The eutectic salt, LiCl–KCl, actinide chloride, and RE chloride salt have hygroscopic characteristic (absorbing water molecules from the surrounding environment and hold them), and the absorbed moistures are likely to produce soluble hydroxides during the heating of the salt. Therefore, extreme care should be taken in order to prevent the extra chemical reactions with water molecules. Even though high-purity anhydrous salts are commercially available, they still have moisture contents below 50 ppm. Thus, some studies purged dry chlorine gas [22] or HCl [23] through the fused salt in order to remove the hydroxides. In addition, several researchers [24–26] retained the solid salts below melting temperature at about 573 K for several hours to dry out the water contents from the salt.

The actinide and RE metals (including their chloride forms) are reactive toward oxide-based ceramics (crucibles or instrumentation sheath) [27–29]; thus, new species, most probably oxychlorides of the salt, container cations, and/or an electrically conductive film can be formed on ceramic surfaces [30]. As a consequence, when in contact with oxide ceramic, the concentration of the chloride salt may change as a function of both time and temperature. Oxygen-free containment such as glassy carbon or metals helps

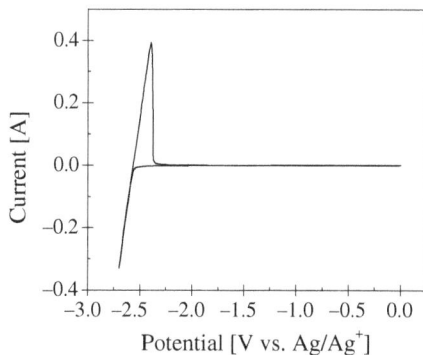

Figure 3. Cyclic voltammogram of pure LiCl–KCl eutectic salt at 773 K at the scan rate of $0.1\,V\,s^{-1}$. Tungsten rod (2 mm in diameter) was used as cathode electrode, and the surface area was $0.471\,cm^2$.

avoiding this issue; however, oxide crucibles can be utilized with careful pre-purification, short contact duration at that temperature. In addition, three-electrode system has been used to understand the fundamental properties of the nuclear materials in the salt system. Here, the common experimental reference electrode is Ag^+/Ag reference electrode, which mostly uses the glass-based membranes (thin-layered glass tubing) for ionic conductivity. This glass material is very vulnerable to the RE metals at high temperature, so that the black oxide film can be seen on the Ag^+/Ag reference, which possibly have an influence on the data measurements during the experiments.

Therefore, it is reasonable to prepare pure LiCl–KCl eutectic salt in advance and verify the absence of impurity peaks and background current densities via cyclic voltammetry (CV) test. Figure 3 shows an example of CV curve in pure LiCl–KCl at 773 K. Potential window is between -2.5 and $+1.2\,V$ based on Ag^+/Ag reference electrode, and no additional reaction should be observed. Then, the target salt elements can be added to the LiCl–KCl eutectic. During the experiments, it is good practice to check CV curves between experiments for oxide formations or concentration change in the electrolyte.

4.2. *Molten salts in electrorefining*

Molten salts have several advantages over aqueous solvents when serving as a medium for UNF reprocessing. The advantages include high radiation

Table 1. Physical properties of the salt mixtures [32–34].

Properties	LiCl–KCl	NaCl–KCl	BeF$_2$–LiF	CaF$_2$–LiF
Composition (A–B, mol.% B)	40	50	67	77
Eutectic melting temperature (K)	623	931	733	1042
Molar mass (g mol^{-1})	55.25	66.50	32.89	37.93
Density at melting point (kg m^{-3})	1644.6	1604.1	2055.7	2066.8
Viscosity at melting point (Pa s)	0.2798	0.2128	0.1947	—
Electrical conductance (S m^{-1})	63.72	220.07	128.37	624.68

resistance, low criticality concern, low vapor pressure, low secondary wastes, and high stability [31]. Thus, physical, chemical, and electrochemical properties of the salts and its interactions with actinides and fission products must be considered. Currently, two chloride salt mixtures (LiCl–KCl and NaCl–KCl) and two fluoride salt mixtures (BeF$_2$–LiF and CaF$_2$–LiF) have been recognized as good candidates for pyroprocessing [32–34]. The compositions of these salts and their melting temperatures are given in Table 1. LiCl–KCl molten salt is widely selected as a candidate for electrorefining system because of its lower melting eutectic temperature. Typically, electrorefining system is operated at a temperature ranging between 723 and 773 K.

4.3. *CV of UCl$_3$ in LiCl–KCl*

CV is the most common electrochemical measurement where the current response is obtained according to the potential scan. If the potential exceeds certain element's reduction potential toward negative direction, the reduction reaction begins on the working electrode and corresponding reduction peak current can be observed at that potential. Figure 4 is an example of CV curves in LiCl–KCl–1 wt.% UCl$_3$ salt at 773 K between potentials of 0 and −2.35 V (vs. Ag$^+$/Ag). Reduction and oxidation peaks for U^{4+}/U^{3+} is shown at around −0.5 V (vs. Ag$^+$/Ag, I_a/I_c^-), and U^{3+}/U reaction can be observed at III$_a$/III$_c$ (around −1.5 V vs. Ag$^+$/Ag). In addition, small pre-peaks have been observed on the reduction and oxidation processes at II$_a$ and II$_c$, respectively. These pre-peaks have been reported by many researchers [35–37]. Reddy *et al.* [35] and Serrano and Taxil [37] further studied the pre-peaks by observing the peak current at different scan rates and concentrations, and concluded that the peaks were being attributed by an adsorption and desorption of uranium monolayer on the working electrode.

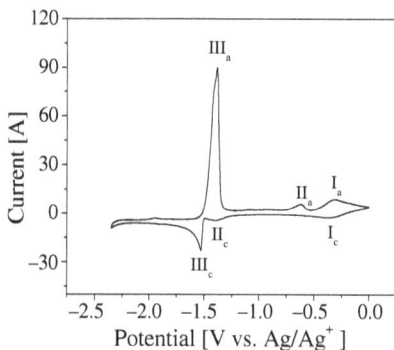

Figure 4. Cyclic voltammograms in LiCl–KCl–1.0 wt.% UCl_3 at 773 K.

4.4. *Redox process and apparent standard potential of* U^{3+}/U

The electrochemical process of U^{3+}/U reduction–oxidation (which is called "redox") reaction is governed by the Nernst equation exhibiting the equilibrium potential for the reaction with respect to temperature and concentrations of oxidant/reductant [38]:

$$E^{eq}_{U^{3+}/U} = E^{0}_{U^{3+}/U} + \frac{RT}{nF} \ln \left(\frac{a^{n+}_{U}}{a_{U}} \right) \tag{4}$$

where $E^{eq}_{U^{3+}/U}$ is the equilibrium potential between U metal and U^{3+} ions, E^{3+}_{U}/U^{0} is the standard reduction potential, which is the theoretical potential when the cell is ideally reversible at equilibrium states with a solution concentration of 1 mol L^{-1} at 1 atm and 298 K, R is the universal gas constant (8.314 J mol^{-1} K^{-1}), F is the Faraday's constant ($96{,}485$ C mol^{-1}), and T is the absolute temperature (K). The a^{3+}_{U} is the activity of UCl_3, which can be expressed by $a_{U^{3+}} = \gamma^{3+}_{U} \times \chi^{3+}_{U}$ where χ^{3+}_{U} is the activity coefficient and χ^{3+}_{U} is the mole fraction of U. Thus, the equation can be expressed using the mole fraction [38]:

$$E^{eq}_{U^{3+}/U} = E^{0*}_{U^{3+}/U} + \frac{RT}{nF} \ln \left(\frac{\chi^{3+}_{U}}{\chi_{U}} \right) \tag{5}$$

where $E^{0*}_{U^{3+}/U}$ is the apparent standard potential (this is often referred to as a "formal potential"). The apparent standard potential is the function of temperature and activity coefficient; therefore, it can be determined by plotting $E^{eq}_{U^{3+}/U}$ against $\ln(\chi_{U^{3+}})$. This property gives an insight into

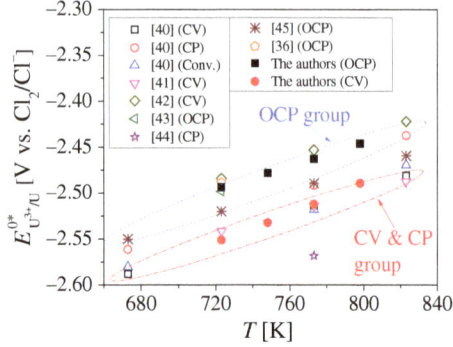

Figure 5. The averaged values of $E^{0*}_{U^{3+}/U}$ via CV and OCP technologies, compared with literature values.

the reduction potential that the operator will be able to apply for the U recovery in the ER. Therefore, several researchers have reported the values of $E^{0*}_{U^{3+}/U}$, in which they used different temperatures, concentrations, reference electrodes, and experimental methods [24, 39–44], as shown in Figure 5. The reported data values show good agreement, with 50 mV deviation from the average values, which are linearly dependent with temperature in general. However, there are two separated trends depending on measurement methods, which can be seen in the literature studies as well as the resulting data sets by us (Yoon and Phongikaroon). The data measured by CV and chronopotentiometry (CP) are shown in more negative range compared with that from open circuit potential (OCP). This may be because the responses of CV and CP methods are affected by diffusion of the elements in the salt and bring overpotential on the electrode. Therefore, in general, the OCP method is more suitable to understand the equilibrium potential and apparent standard potential between the anion and cation. Typically, the property of E^{0*} is considered as independent of concentration; however, this must be experimentally evaluated.

4.5. Activity coefficient of UCl₃

Once $E^{0*}_{U^{3+}/U}$ values were determined, thermodynamic properties of UCl_3 in LiCl–KCl eutectic salt can be further estimated using following equations:

$$E^{0*}_{U^{3+}/U} = E^{0}_{U^{3+}/U} + \frac{RT}{nF} \ln(\gamma_{U^{3+}}) \tag{6}$$

where $\gamma_{U^{3+}}$ is the activity coefficient of UCl_3 in LiCl–KCl. Since $E^{0}_{U^{3+}/U}$ is a theoretical value at the aforementioned ideal state, the value does not exist in reality. Therefore, super-cooled liquid state was commonly considered as ideal as possible in the real system [39, 40, 42], and Eq. (6) can be re-written using the basic thermodynamic equation as following:

$$\Delta G^{SC}_{UCl_3} - \Delta G^{0*}_{UCl_3} = RT \ln(\gamma_{U^{3+}}) \tag{7}$$

where $\Delta G^{SC}_{UCl_3}$ is the Gibbs free energy at super-cooled liquid state. The data for the pure substance can be found in several references [3, 45]. If $\Delta G^{SC}_{UCl_3}$ is known, $\gamma_{U^{3+}}$ can be calculated. Although the melting temperature (T_m) of CeCl$_3$ salt is 1100 K in nature, UCl$_3$ is dissolved in LiCl–KCl eutectic salt under a liquid phase; therefore, some researchers used the fusion energy between solid and liquid phases. Thus, ΔG^{SC} can be re-written by considering the fusion energy between liquid and solid states through the following relationship:

$$\Delta G^{SC} = \Delta G^{Formation} + \Delta G^{Fusion} \tag{8}$$

where $\Delta G^{Formation}$ is the Gibbs free energy for the formation (kJ mol^{-1}) and ΔG^{Fusion} is the Gibbs energy for the fusion between liquid and solid phases. The second term can further be expressed as

$$\Delta G^{Fusion} = \Delta H^{Fusion} - T\Delta S^{Fusion} + \int_{T_m}^{T} \Delta C_p dT - T \int_{T_m}^{T} \frac{\Delta C_p}{T} dT \tag{9}$$

where ΔH^{Fusion} is the enthalpy of fusion (kJ mol^{-1}), ΔS^{Fusion} is the entropy of the fusion at T_m (kJ mol^{-1} K^{-1}), and ΔC_p is the heat capacity between T_m and T (kJ mol^{-1} K^{-1}).

We found that the pure thermodynamic data for UCl$_3$ at the super-cooled liquid state from a thermodynamic text book [45] and calculated thermodynamic values (ΔG^{Fusion} and ΔG^{SC}), which are summarized in Table 2. The resulting values are compared with the reported values from

Table 2. Thermodynamic information from the literature [45] and calculated values.

	723 K	748 K	773 K	798 K	T_m (1100 K)
$\Delta G^{Formation}_{UCl_3}$ (kJ mol^{-1})	-703.03	-697.83	-692.64	-687.45	
$\Delta H^{Fusion}_{UCl_3}$ (kJ mol^{-1})					46.44
$\Delta S^{Fusion}_{UCl_3}$ (J mol^{-1} K^{-1})					41.84
C_p (J mol^{-1} K^{-1})	111.22	111.94	112.66	113.38	129.70
$\Delta G^{Fusion}_{UCl_3}$ (kJ mol^{-1})	14.76	13.96	13.12	12.26	
$\Delta G^{SC}_{UCl_3}$ (kJ mol^{-1})	-688.26	-683.87	-679.52	-675.19	

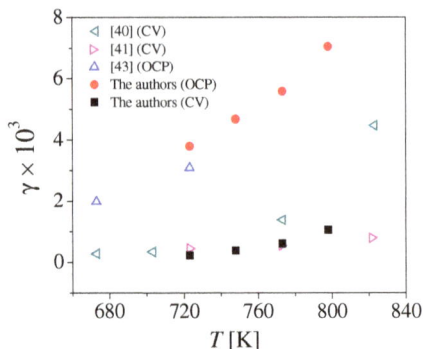

Figure 6. The values of $\gamma_{U^{3+}}$ measured via OCP and CV methods compared with literature data.

literature studies as shown in Figure 6. Here, a similar trend can be observed as seen in the results of the apparent standard potentials (Figure 5). The values of $\gamma_{U^{3+}}$ measured using CV and OCP are divided into two different trends according to the temperature. Between CV and OCP methods, significant discrepancies were found up to 10^2 order of magnitude, and the OCP method generally provides higher values and shows a steep increase rate when the temperature rises. The discrepancy arises because the calculation is based on the results of the apparent standard potential. Also, there is a challenge on obtaining $\Delta G_{UCl_3}^{SC}$ from literatures, and the exponential term $(\exp(\Delta G_{UCl_3}^{SC} - \Delta G_{UCl_3}^{0*})/RT)$ causes even bigger distinction among the resulting data. Therefore, sufficient data sets need to be collected $\gamma_{U^{3+}}$ in order to demonstrate thermodynamic trends of U in the salt system.

4.6. *Diffusion coefficient of U^{3+} in LiCl–KCl salt*

Mass transfer from and to the electrodes in an electrochemical cell can be expressed using the diffusivity or diffusion coefficient, which is constant between molar flux and concentration gradient [46]. The value of the diffusion coefficient affects the mass transport of the species in the molten salt, the efficiency of the system, and the maximum current that the system can support. Thus, knowledge on the diffusion coefficient of UCl_3 in LiCl–KCl eutectic melt provides essential data for optimization of U electrorefining. The diffusion coefficients of U^{3+} have been electrochemically measured in the salt by using CV and CP techniques. From the peak current height in the CV method, a model for the reversible soluble/insoluble diffusion system (electrodeposition on the electrode) was developed by Berzins and

Delahay [47]. The model used the assumption that the deposited species (in metal form) has a constant activity at unity. The equation can be expressed as

$$i_p = 0.611 nFSC_0 \left(\frac{nFvD}{RT} \right)^{1/2} \tag{10}$$

Therefore, Eq. (10) allows one to calculate the diffusion coefficient of UCl_3 by plotting i_p versus the square root of the scan rate. CP is applying a constant current to the working electrode and observing potential changes as a function of time. When current is applied on the electrode, the reduction potential of U^{3+}/U can be measured at a constant value; however, this potential will rapidly drop toward a more negative potential for seeking another reduction process when the interfacial concentration of UCl_3 is depleted. This potential transition time is referred to as a transition time, τ. Once the transition time is measured using the CP experiments, the diffusion coefficient of UCl_3 (or other interested elements) can be calculated using the Sand equation:

$$I_d\sqrt{\tau} = \frac{nFSC\sqrt{\pi D}}{2} \tag{11}$$

where I_d is the applied current (A). The diffusivity generally follows Arrhenius temperature relationship, which can be expressed as

$$D = D_0 \exp\left(\frac{-E_a}{RT} \right) \tag{12}$$

where D_0 is the pre-exponential factor and E_a is an activation energy (kJ mol^{-1}) for the diffusion. Therefore, the activation energy for the diffusion can be calculated from the slope when $\ln(D)$ is plotted versus $1/T$.

Figure 7 shows reported diffusion values of UCl_3 in LiCl–KCl from several studies [24, 35, 36, 39, 40, 48]. It appears that the diffusion coefficient is linearly dependent to temperature; however, data are scattered specially at high temperature (for example, a standard deviation of $D_{U^{3+}}$ is 1.5×10^{-5} cm^2 s^{-1} from the mean value of 3.02×10^{-5} cm^2 s^{-1} at 823 K). The CP method generally provides higher values than those obtained using the CV method. Figure 8 shows the diffusion coefficients measured by us, showing that the values of diffusion coefficients are weakly affected by concentration change up to 2 wt.% of UCl_3. However, the values seem to be affected by concentration at 4 wt.% UCl_3. Several studies have reported the decrease in the diffusion coefficient with increasing concentration of UCl_3. The most

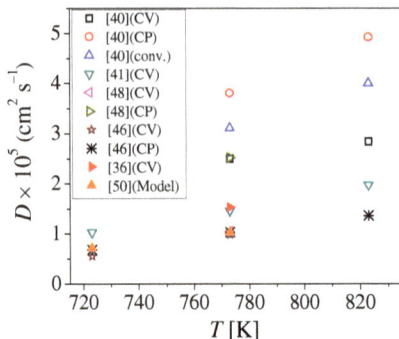

Figure 7. Plot of the diffusion coefficients of UCl₃ in LiCl–KCl salt reported by various researchers.

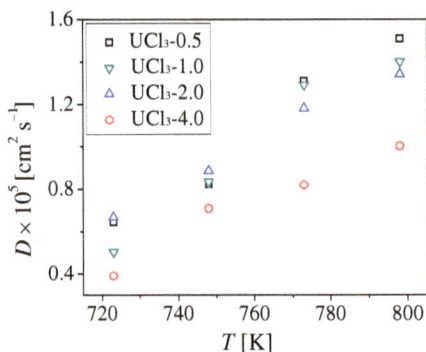

Figure 8. Diffusion coefficients of UCl₃ at concentrations ranging from 0.5 wt.% to 4 wt.% in LiCl–KCl as a function of temperature, measured by us (Yoon and Phongikaroon, 2017).

supporting explanation is the ohmic (IR) drop effect at higher concentrations. When current is flowing through the electrochemical cell, there is a potential drop between the reference electrode and the working electrode, which is caused by electrolyte resistance (conductivity). The ohmic drop can be compensated by obtaining solution resistance by using the electrochemical impedance spectroscopy (EIS) technique, and the true working electrode potential can be adjusted. Figure 9 compared the CV curves with and without IR drop compensation. The compensated CV curve shows a larger peak current according to the adjusted true potential. The IR drop is more significant with larger distance between the reference and working

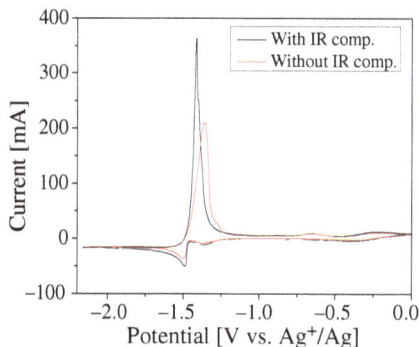

Figure 9. Comparison of CV curves with IR compensation and without IR compensation in LiCl–KCl–1 wt.% UCl_3 at 773 K.

electrodes, larger electrode area, faster scan rate, and higher concentration. Therefore, it is important to consider the solution resistance and compensate this prior to the CV measurements. In general, commercial potentiostat/galvanostats provide the function of IR compensation; therefore, the solution resistance can be automatically compensated in advance to the CV measurements with the advanced features in instruments.

4.7. *Exchange current density* (i_0) *of* U^{3+}/U

Exchange current density is an important parameter to understand the kinetics of electrochemical reactions (U^{3+}/U) on the electrode surface. Additionally, the values are essential to the physics-based model used in Butler–Volmer equation, which shows fundamental relationship between current and overpotential applied on the electrode. The equation can be expressed as

$$i = i_0 \left\{ \exp\left[\frac{\alpha_a n F \eta}{RT}\right] - \exp\left[-\frac{\alpha_c n F \eta}{RT}\right] \right\} \tag{13}$$

where i_0 is the exchange current density, i is the current, α_a is the anodic charge transfer coefficient, α_c is the cathodic charge transfer coefficient, and η is the activation overpotential. There are several methods to measure the value of i_0: Tafel plot, linear polarization (LP), and EIS. However, the i_0 of U^{3+}/U couple has not been well measured and understood because of the challenges in the measurement. In 2009, Choi *et al.* [49] performed LP experiments in 3.3 wt.% UCl_3–LiCl–KCl salt where Cd co-existed at the bottom of the vessel. The researchers conducted the experiments by using different materials for the electrode and concluded that the different materials of electrodes result in different values of i_0, ranging from 0.0202 to 0.0584 A cm^{-2}. Ghosh et al. [50] reported that the value is $8 \pm 2 \times 10^{-3}$ A cm^{-2}, using the measurements of Tafel plot, which stands far from the trend of other literature studies. In 2015, Rose et al. [51] measured the i_0 of U^{3+}/U in 5 wt.% UCl$_3$–LiCl–KCl salt by using Tafel plot method. Tafel plots were analyzed using the Oldham–Mansfeld model in a very small overpotential region, which is theoretically not a Tafel region. The value of i_0 from that study was from 0.0695 to 0.220 A cm^{-2}. Recently, Lim et al. [52] reported the i_0 values of U by using the Tafel and LP methods; this published study is the only source with reported i_0 data at different temperatures and concentrations. All reported values of i_0 for U^{3+}/U are plotted and compared in Figure 10. From these literature results, there are still missing data sets for i_0 of U in order to fully understand the general trend. Also, a meaningful comparison among the reported data is not possible due to the data discrepancy resulting from the differences in the experimental conditions, the size of the system, and the data acquisition methods. Furthermore, the challenges and uncertainties in measuring the electrode surface area during the Tafel and LP measurements have never been clearly reported. Therefore, further studies must be conducted in the same experimental conditions and system sizes. The measurements at different concentrations and temperatures under the same experimental environments would be necessary in order to build up a meaningful data base for i_0 of U^{3+}/U in LiCl–KCl salt.

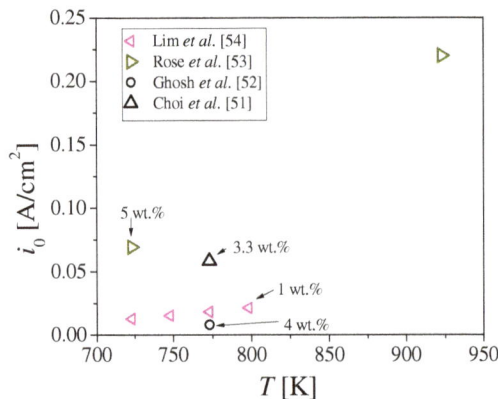

Figure 10. Plot of i_0 from literature studies (measuring with a tungsten electrode).

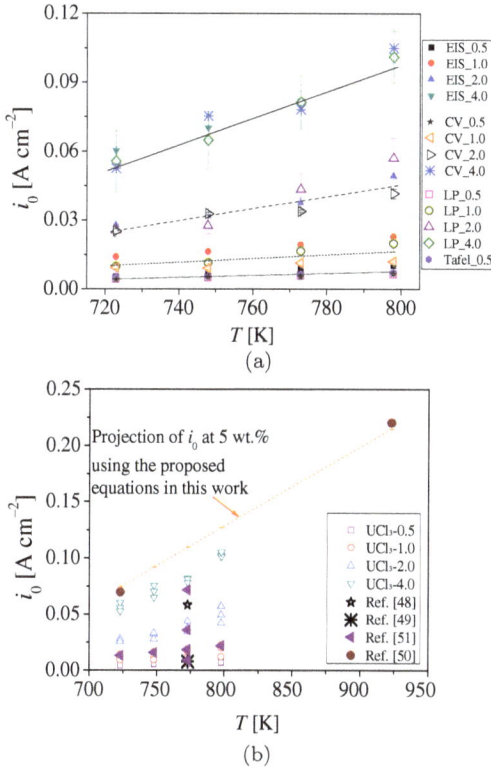

Figure 11. (a) The i_0 values measured using the four different methods in this study [53] and (b) the reported values of i_0 for U/U^{3+} from different literature studies are being superimposed.

We [53] employed four techniques including LP, Tafel, EIS, and CV methods to investigate the i_0 kinetics of U^{3+}/U couple at UCl_3 concentrations ranging from 0.5 to 4 wt.% and temperature from 723 to 798 K. Figure 11(a) shows the results obtained using the four techniques. The i_0 values from EIS measurements show the most linear trends regarding to the concentration and temperature, whereas the values measured using CV method are more scattered and do not have a strong linear correlation with the concentration and temperature. Overall, the results from the four methods are in a moderate agreement, with 33% relative difference from their average values. Particularly, the EIS measurements for 1 wt.% UCl_3 in this study agree moderately well with those from Lim *et al.*'s [52] study. However, the LP results in this study are significantly influenced by the scan

Table 3. Linear models for concentration (mole fraction) and temperature (K) dependence [53].

		Equations	R^2
Temperature (T) dependence of i_0	UCl$_3$–0.5:	$6.752 \times 10^{-5}\ T - 0.0436$	0.973
	UCl$_3$–1.0:	$1.221 \times 10^{-4}\ T - 0.0750$	0.987
	UCl$_3$–2.0:	$2.682 \times 10^{-4}\ T - 0.1678$	0.932
	UCl$_3$–4.0:	$5.494 \times 10^{-4}\ T - 0.3393$	0.956
Concentration (C)* relationship of i_0	723 K:	$8.823\ C$	0.993
	748 K:	$10.274\ C$	0.993
	773 K:	$12.047\ C$	0.995
	798 K:	$15.012\ C$	0.996

rate, which reputes the conclusion from the study of Lim *et al.* that the scan rate (5–$100\,\text{mV s}^{-1}$) has no significant effect on the i_0 value.

Because the EIS method provides the most linear trends and agreements with other literature data, the results from the EIS measurements have been selected for further analysis without considering the deviations (error bars) of the data. The linear relationships of i_0 were modeled using a linear equation, $y = ax + b$ (these fitted values are summarized in Table 3). Here, the equations in Table 3 can be used to determine any desired i_0. These equations were being extrapolated to estimate the i_0 of U^{3+}/U at the $5\,\text{wt.\%}$ concentration under 723 and 923 K reported by Rose *et al.* [51]. The resulting calculations reveal that the values of the i_0 are $0.0739\,\text{A cm}^{-2}$ ($C = 5\,\text{wt.\%}$ (0.00848 mole fraction) and $T = 723\,\text{K}$) and $0.214\,\text{A cm}^{-2}$ ($C = 5\,\text{wt.\%}$ (0.00848 mole fraction) and $T = 923\,\text{K}$) agreeing well with the values of i_0 reported by Rose *et al.* [51]. The projection for i_0 value at $5\,\text{wt.\%}$ UCl$_3$ can be checked in Figure 11(b) (see a line of orange dots). Although literature i_0 values were being measured under different electrochemical scales and cell configurations, the results are in a good agreement (see Figure 11(b)). Thus, it can be postulated that the effect of system scale and configuration has a weak influence on the i_0 kinetics. In order to confirm this argument, a large number of data collections will be required in different electrochemical environments.

4.8. *Effects of RE elements co-existing in ER*

Although the ER salt system contains a number of fission products including actinide (An) and RE components [18], the uranium fundamental behaviors in multi-elements have not been well studied. Therefore, we investigated a quaternary salt (LiCl–KCl–UCl$_3$–GdCl$_3$) to gain an insight on the effect of other elements on U properties, which should be happening in a real ER system. The effects of $GdCl_3$ on UCl_3 behaviors were explored at 773 K by adding certain amounts of $GdCl_3$ to LiCl–KCl–UCl_3 salt (Sample Nos. 1–6 in Table 4). The concentration ratio of $UCl_3/GdCl_3$ varies from 1 to 4 through the samples, and the diffusion coefficients, equilibrium potentials, and exchange current densities were measured for the salt samples. Figure 12 shows the measured CV data in LiCl–KCl–1 wt.% UCl_3–1 wt.% $GdCl_3$ at scan rate ranging from 50 to 200 mV s^{-1}. Redox peaks for Gd^{3+}/Gd couple (IV_c and IV_a, respectively) were observed at potentials more negative than U^{3+}/U redox peaks; therefore, it can be considered that the U^{3+}/U reaction occurs independently whereas Gd^{3+}/Gd reaction is happening along with U deposition on the electrode. Figure 13 shows the diffusion coefficients of UCl_3 measured in Sample Nos. 1–6 along with the results reported in Figure 8 at 773 K. By adding $GdCl_3$ to the LiCl–KCl–UCl_3 system, the diffusion coefficient of UCl_3 becomes smaller

Table 4. Electrolyte concentration and the working electrode surface area for each experimental run.

No.	Samples	UCl_3 (wt.%)	$GdCl_3$ (wt.%)	Electrode surface area (cm^2)
1	U0.5–Gd0.5	0.59	0.49	0.565
2	U1.0–Gd0.25	1.04	0.29	0.584
3	U1.0–Gd0.5	1.45	0.47	0.503
4	U1.0–Gd1.0	1.09	0.95	0.522
5	U2.0–Gd1.0	1.71	0.96	0.471
6	U2.0–Gd2.0	1.94	2.09	0.452

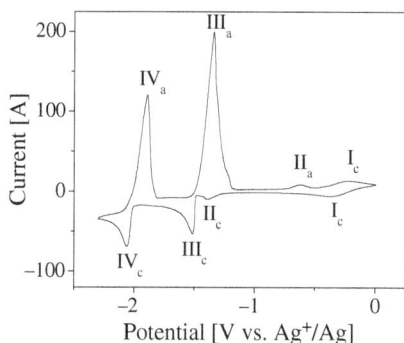

Figure 12. Cyclic voltammogram in LiCl–KCl–1 wt.% UCl_3–1 wt.% $GdCl_3$ at 773 K, measured at a scan rate of 200 mV s^{-1}.

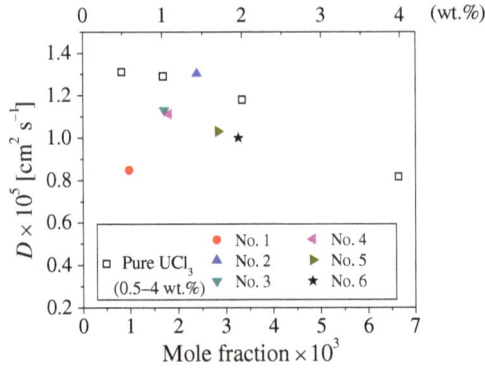

Figure 13. Plots of diffusion coefficients measured in LiCl–KCl–UCl$_3$–GdCl$_3$ mixtures (No. 5–10), which were compared with original data sets (Section 3.6) at 773 K [54].

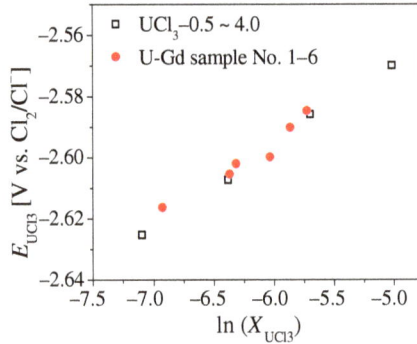

Figure 14. The equilibrium potentials of U^{3+}/U measured with presence of GdCl$_3$, which were superimposed in the original trend of the equilibrium potential at 773 K, measured by [54].

by $0.2 \times 10^{-5} \sim 0.3 \times 10^{-5}\,\mathrm{cm^2\,s^{-1}}$. The values decreases slightly with increasing concentrations of UCl$_3$ and GdCl$_3$, but UCl$_3$–GdCl$_3$ mixtures need to be examined at concentrations higher than 4 wt.%. At lower concentration of UCl$_3$ (Sample No. 5), GdCl$_3$ co-existence significantly affects the diffusion behavior of UCl$_3$, which reveals that diffusivity of UCl$_3$ in the ER system may be considerably affected by the concentration of other elements (e.g., actinide and lanthanide elements). However, further studies need to be done with different elements in order to understand the evidence of decreasing diffusivity. This may be affected by the physical and chemical interactions among the particles. Figure 14 shows the equilibrium potentials

Table 5. Thermodynamic properties of UCl_3 in $LiCl–KCl–UCl_3–GdCl_3$ mixture salts [54].

No.	Sample	$E^{0*}_{U^{3+}/U}$ (V vs. Cl_2/Cl^-)	$\Delta G^{0*}_{UCl_3}$ (kJ mol^{-1})	$\gamma^{3+}_U \times 10^3$
1	U0.5–Gd0.5	$-2.46 \pm 6.0 \times 10^{-3}$	-712.8 ± 17.4	$5.64 \pm 1.53 \times 10^{-1}$
2	U1.0–Gd0.25	$-2.46 \pm 4.2 \times 10^{-4}$	-713.2 ± 1.2	$5.26 \pm 9.8 \times 10^{-2}$
3	U1.0–Gd0.5	$-2.47 \pm 4.5 \times 10^{-5}$	-713.8 ± 1.3	$4.80 \pm 9.7 \times 10^{-3}$
4	U1.0–Gd1.0	$-2.46 \pm 4.0 \times 10^{-4}$	-712.6 ± 11.6	$5.80 \pm 1.1 \times 10^{-1}$
5	U2.0–Gd1.0	$-2.46 \pm 2.7 \times 10^{-4}$	-712.1 ± 7.6	$6.30 \pm 7.5 \times 10^{-2}$
6	U2.0–Gd2.0	$-2.46 \pm 1.2 \times 10^{-4}$	-711.4 ± 3.5	$6.98 \pm 3.8 \times 10^{-2}$

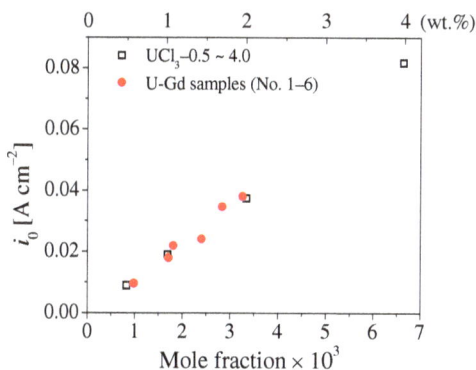

Figure 15. The exchange current densities measured in U–Gd samples, compared with the values measured with pure UCl_3 salts at 773 K [54].

as function of concentration in Sample Nos. 1–6 at 773 K, which are superimposed on the pure UCl_3 data as seen in Figure 5. The resulting values follow the linear dependence line, with 0.2% relative difference indicating that the results of $E^{0*}_{U^{3+}/U}$ and $\gamma_{U^{3+}}$ also show good agreement with the original data with pure UCl_3; therefore, it could be postulated that thermodynamic properties of UCl_3 is scarcely dependent on the presence of $GdCl_3$ in $LiCl–KCl–UCl_3$ salt system. The calculations of $E^{0*}_{U^{3+}/U}$ and $\gamma_{U^{3+}}$ are summarized in Table 5, and show good consensus with the data sets measured with pure UCl_3 salts (reported in Sections 3.4 and 3.5). Furthermore, the exchange current densities of U^{3+}/U were reported with the $GdCl_3$ additions and the resulting data are plotted in Figure 15. The values are laid along with the linear model, indicating that the kinetic parameters of

U^{3+}/U are also independent on $GdCl_3$. The resulting data and trends from the experiments with the mixture salts provide a useful insight into behaviors of U such as electrochemical, thermodynamic, and kinetic behaviors with $GdCl_3$ existence in LiCl–KCl; however, further experimental database should be accumulated to understand U behavior among the various fission products, which is occurring in the real ER system. Unfortunately, the thermodynamic and kinetic properties of U in multi-components have not been well explored in previous literature studies.

5. Summary

The pyroprocessing technology has been examined in several countries including the USA, Russia, France, Japan, South Korea, and India. The purpose of this technology is not only to treat the irradiated nuclear fuel but also to reduce volume of the nuclear waste, recycle actinides, and close the fast reactor fuel cycle. The heart of this technology is the ER where pure U and An elements are being recovered in LiCl–KCl eutectic salt at 773 K. This recovery can be accomplished using electrolysis; therefore, many studies have been done to understand the electrochemical, thermodynamic, and kinetic behaviors of those products. Yet, several data sets reported in the past have been scattered due to challenges in the experimental measurements and systematic designs. In addition, the fundamental understanding on the special nuclear materials in multi-components, which is similar to the ER salt system has been insufficient. This chapter has provided the literature review including an on-going study on UCl_3 in LiCl–KCl showing significant progress in this research area. Definitely, further studies should be performed to establish the reliable data pool for U, other actinides, and lanthanide elements in LiC–KCl salt system, which will give fundamental understanding of the nuclear waste management and signatures for material accountability and safeguards application in ER process for the pyroprocessing technology.

References

1. IAEA, *Spent Fuel Reprocessing Options*, IAEA-TECDOC-1587, Vienna (2008).
2. *Nuclear Energy Data*, NEA No. 7246, OECD (2015).
3. M. F. Simpson and J. D. Law, *Nuclear Fuel Reprocessing*, INL/EXT-10-17753, February (2010).
4. T. Todd, *Nuclear Regulatory Commission Seminar*, March 25 (2008).
5. S. Phongikaroon, *EGMN 691 — Nuclear Fuel Cycle*, Lecture #1, Fall (2015).

6. M. Iizuka, Diffusion coefficients of cerium and gadolinium in molten LiCl–KCl, *J. Electrochem. Soc.* **145**, 84–88 (1998).
7. Y. I. Chang, The integral fast reactor, *Nucl. Technol.* **88**, 129–138 (1989).
8. J. P. Ackerman, Chemical basis for pyrochemical reprocessing of nuclear fuel, *Ind. Eng. Chem. Res.* **30**, 141–145 (1991).
9. C. E. Till and Y. I. Chang, *Plentiful Energy — The Story of the Integral Fast Reactor* (2011).
10. Z. Wang, D. Rappleye, and M. F. Simpson, Voltammetric analysis of mixtures of molten eutectic LiCl-KCl containing $LaCl_3$ and $ThCl_4$ for concentration and diffusion coefficient measurement, *Electrochim. Acta* **191**, 29–43 (2016).
11. Y. Amano, The Fukushima Daiichi accident, Report GC(59)/14, IAEA, Vienna (2015).
12. OECD NEA, Pyrochemical separations in nuclear applications, A Status Report, NEA No. 5427, Paris, ISBN: 92-64-02071-3 (2004).
13. E. Choi and S. Jeong, Electrochemical processing of spent nuclear fuels: An overview of oxide reduction in pyroprocessing technology, *Prog. Nat. Sci. Mater. Int.* **25**(6), 572–582 (2015).
14. S. D. Herrmann, S. X. Li, and M. F. Simpson, Electrolytic reduction of spent light water reactor fuel bench-scale experiment results, *J. Nucl. Sci. Technol.* **44**, 361–367 (2007).
15. K. Uozumi, M. Iizuka, T. Kato, T. Inoue, O. Shirai, T. Iwai, and Y. Arai, Electrochemical behaviors of uranium and plutonium at simultaneous recoveries into liquid cadmium cathodes, *J. Nucl. Mater.* **325**, 34–43 (2004).
16. T. Koyama, M. Iizuka, Y. Shoji, R. Fujita, H. Tanaka, T. Konayashi, and M. Tokiwai, An experimental study of molten salt electrorefining of uranium using solid iron cathode and liquid cadmium cathode for development of pyrometallurgical reprocessing, *J. Nucl. Sci. Technol.* **34**, 384 (1997).
17. M. F. Simpson, Developments of spent nuclear fuel pyroprocessing technology at Idaho National Laboratory, (2012).
18. S. X. Li and M. F. Simpson, Anodic process of electrorefining spent driver fuel in molten LiCl-KCl-UCl_3/Cd system, *Minerals and Metallurgical Processing*, **22**, 192–198 (2005).
19. S. Kim, D. Yoon, Y. You, S. Paek, J. Shim, S. Kwon, K. Kim, H. Chung, D. Ahn, and H. Lee, In-situ observation of a dendrite growth in an aqueous condition and a uranium deposition into a liquid cadmium cathode in an electrowinning system, *J. Nucl. Mater.* **385**, 196–199 (2009).
20. D. Vaden, S. X. Li, B. R. Westphal, K. B. Davies, T. A. Johnson, and D. M. Pace, Engineering-Scale Liquid Cadmium Cathode Experiments, In *Int. Pyroprocessing Research Conf.*, 1, INL/CON-06-11544 (2006).
21. R. Taylor, *Reprocessing and Recycling of Spent Nuclear Fuel*, 1st edition, Woodhead Publishing, Elsevier, Oxford, UK (2005).
22. D. L. Maricle and D. N. Hume, A new method for preparing hydroxide-free alkali chloride melts, *J. Electrochem. Soc.* **107**(4), 354–356 (1960).
23. H. A. Laitinen, W. S. Ferguson, and R. A. Osteryoung, Preparation of pure fused lithium-chloride-potassium chloride eutectic solvent, *J. Electrochem. Soc.* **104**(8), 516–520 (1957).

24. R. O. Hoover, M. R. Shaltry, S. Martin, K. Sridharan, and S. Phongikaroon, Electrochemical studies and analysis of 1–10 wt.% UCl$_3$ concentrations in molten LiCl–KCl eutectic, *J. Nucl. Mater.* **452**, 389–396 (2014).

25. D. Yoon and S. Phongikaroon, Electrochemical properties and analysis of CeCl$_3$ in LiCl-KCl eutectic salt, *J. Electrochem. Soc.* **162**, E237–E243 (2015).

26. D. Yoon, S. Phongikaroon, and J. Zhang, Electrochemical and thermodynamic properties of CeCl$_3$ on liquid cadmium cathode (LCC) in LiCl-KCl eutectic salt, *J. Electrochem. Soc.* **163**, E97–E103 (2016).

27. K. E. Johnson and J. R. Mackenzie, Samarium, europium, and ytterbium electrode potentials in LiCl-KCl eutectic melt, *J. Electrochem. Soc.* **116**(12), 1697–1703 (1969).

28. S. A. Kuznetsov, H. Hayashi, K. Minato, and M. Gaune-Escard, The influence of oxide ions on the electrochemical behavior of UCl$_4$ and UCl$_3$ in a LiCl-KCl eutectic melt, *Electrochem.* **73**(8), 630–632 (2005).

29. K. R. Kvam, D. Bratland, and H. A. Øye, The solubility of neodymium in the systems NdCl$_3$-LiCl and NdCl$_3$-LiCl-KCl, *J. Mol. Liquids* **83**(1–3), 111–118 (1999).

30. K. C. Marsden and B. Pesic, Evaluation of the electrochemical behavior of CeCl$_3$ in molten LiCl-KCl eutectic utilizing metallic Ce as an anode, *J. Electrochem. Soc.* **158**(6), F111–F120 (2011).

31. J. Zhang, Electrochemistry of actinides and fission products in molten salts — data, *J. Nucl. Mater.* **447**, 271–284 (2014).

32. G. Janz, *Molten Salt Handbooks*, Academic Press, New York and London (1967).

33. G. J. Janz, G. L. Gardner, U. Krebs, and R. P. T. Tomkins, *Molten Salts: Volume 4, Part 1, Fluorides and Mixtures Electrical Conductance, Density, Viscosity, and Surface Tension Data*, New York (1974).

34. G. J. Janz, R. P. T. Tomkins, C. B. Allen, J. R. Downey, Jr, G. L. Gardner, U. Krebs, and S. K. Singer, *Molten Salts: Volume 4, Part 2, Chlorides and Mixtures — Electrical Conductance, Density, Viscosity, and Surface Tension Data*, New York (1975).

35. B. P. Reddy, S. Vandarkuzhali, T. Subramanian, and P. Venkatesh, Electrochemical studies on the redox mechanism of uranium chloride in molten LiCl–KCl eutectic, *Electrochim. Acta* **49**, 2471–2478 (2004).

36. M. M. Tylka, J. L. Willit, J. Prakash, and M. A. Williamson, Application of voltammetry for quantitative analysis of actinides in molten salts, *J. Electrochem. Soc.* **162**, H852–H859 (2015).

37. K. Serrano and P. Taxil, Electrochemical reduction of trivalent uranium ions in molten chlorides, *J. Appl. Electrochem.* **29**, 497 (1999).

38. C. M. A. Brett and A. M. O. Brett, *Electrochemistry Principles, Methods, and Applications*, Oxford University Press Inc., New York (1993).

39. P. Masset, D. Bottomley, R. Konings, R. Malmbeck, A. Rodrigues, J. Serp, and J. Glatz, Electrochemistry of uranium in molten LiCl-KCl eutectic, *J. Electrochem. Soc.* **152**, A1109–A1115 (2005).

40. S. A. Kuznetsov, H. Hayashi, K. Minato, and M. Gaune-Escard, Electrochemical transient techniques for determination of uranium and rare-earth

metal separation coefficients in molten salts, *Electrochim. Acta* **51**, 2463–2470 (2006).

41. O. Shirai, H. Yamana, and Y. Arai, Electrochemical behavior of actinides and actinide nitrides in LiCl–KCl eutectic melts, *J. Alloys Compounds* **408–412**, 1267–1273 (2006).

42. J. J. Roy, L. F. Grantham, D. L. Grimmett, S. P. Fusselman, C. L. Krueger, T. S. Storvick, T. Inoue, Y. Sakamura, and N. Takahashi, Thermodynamic properties of U, Np, Pu, and Am in molten LiCl-KCl eutectic and liquid cadmium, *J. Electrochem. Soc.* **143**, 2487–2492 (1996).

43. L. Martinot, *Gmelin Handbuch der Anorganischen Chemie*, Springer-Verlag, New York (1984).

44. Y. Sakamura, T. Hijikataa, K. Kinoshitaa, T. Inouea, T. S. Storvickb, C. L. Kruegerb, J. J. Royc, D. L. Grimmett, S. P. Fusselman, and R. L. Gay, Measurement of standard potentials of actinides (U,Np,Pu,Am) in LiCl–KCl eutectic salt and separation of actinides from rare earths by electrorefining, *J. Alloys Compounds* **271–273**, 592–596 (1998).

45. Barin, *Thermochemical Data of Pure Substances*, Third edition, Wiley-VCH Verlag GmbH (1995).

46. R. O. Hoover, Uranium and zirconium electrochemical studies in LiCl-KCl eutectic for fundamental applications in used nuclear fuel reprocessing, Ph.D. Thesis, University of Idaho, Idaho Falls, ID (2014).

47. T. Berzins and P. Delahay, Oscillographic polarographic waves for the reversible deposition of metals on solid electrodes, *J. Am. Chem. Soc.* **75**, 555–559 (1953).

48. W. Zhou and J. Zhang, Chemical diffusion coefficient calculation of U^{3+} in LiCl-KCl molten salt, *Prog. Nucl Energy* **91**, 170–174 (2016).

49. I. Choi, B. E. Serrano, S. X. Li, S. Hermann, and S. Phongikaroon, Determination of exchange current density of U^{3+}/U couple in LiCl-KCl eutectic mixture, In *Proc. GLOBAL 2009*, Paris, France, September 6–11 (2009).

50. S. Ghosh, S. Vandarkuzhali, N. Gogoi, P. Venkatesh, and G. Seenivasan, Anodic dissolution of U, Zr and U–Zr alloy and convolution voltammetry of $Zr^{4+}|Zr^{2+}$ couple in molten LiCl–KCl eutectic, *Electrochim. Acta* **56**, 8204–8218 (2011).

51. M. A. Rose, M. A. Williamson, and J. Willit, Determining the exchange current density and Tafel constant for uranium in LiCl/KCl eutectic, *ECS Electrochem. Lett.* **4**, C5–C7 (2015).

52. K. H. Lim, S. Park, and J. Yun, Study on exchange current density and transfer coefficient of uranium in LiCl-KCl molten salt, *J. Electrochem. Soc.*, **162**, E334–E337 (2015).

53. D. Yoon and S. Phongikaroon, Measurement and analysis of exchange current density for U/U^{3+} reaction in LiCl-KCl eutectic salt via various electrochemical techniques, *Electrochim. Acta* **222**, 170–179 (2017).

54. D. Yoon and S. Phongikaroon, Electrochemical and thermodynamic properties of UCl₃ in LiCl-KCl eutectic salt system and LiCl-KCl-GdCl₃ System, *J. Electrochem. Soc.*, **164**(9), E217–E225 (2017).

Chapter 4

Safeguards for Pyroprocessing

Wentao Zhou[*,†] and Jinsuo Zhang[‡,§,¶]

*School of Nuclear Science and Engineering,
Shanghai Jiao Tong University, Shanghai, China

†Nuclear Engineering Program, Ohio State University, OH, USA

‡Nuclear Science and Engineering Program, Virginia Tech, VA, USA
§zjinsuo5@vt.edu

1. Introduction

The dawn of usage of nuclear energy can be dated back to 1938 when German scientists Hahn and Strassman [1, 2] found that uranium (U) was split into lighter elements when struck by neutrons. Later, Meitner and Frisch [3] attributed the loss of atomic mass to the generation of energy by Einstein's theory. Scientists throughout the world began to think and discuss the possibility of a self-sustaining chain reaction to continuously release the energy inside atoms, among whom Fermi proposed the model of using graphite to moderate the neutrons in a cube-like frame [4, 5]. This was the theoretical basis of Chicago Pile-1, which achieved criticality on December 2, 1942 and was the world's first artificial nuclear reactor [6].

There were 441 commercial reactors in operation worldwide at the end of 2015, with a total electric capacity of 382.9 GW(e), and another 68 are under construction [7]. Nuclear power is playing an increasingly important role in energy usage and its use could be increased up to 70% by 2030 [7]. The spent nuclear fuel (SNF) accumulation in 2015 was around 266,000 tHM worldwide. The number is increasing at a rate of about 7,000 tHM per year. However, the reprocessing capacity is only around 4,200 tHM per year [8].

¶Corresponding author.

In USA, 99 nuclear reactors are in operation generating about 20% of the nation's total electricity consumption [9]. The SNF storage is about 75,000 tHM [8], and increasing by around 2,000 tMH each year [10]. This SNF is stored at 75 sites in 33 states in pools or dry casks [10]. Managing SNF properly is inevitable to develop nuclear energy sustainably.

Reprocessing could be a promising method considering that around 96% of its original U remains in the SNF [11], recycling of which significantly improves the efficiency of fuel resource and decreases the volume of waste. Also, returning the actinides back to the reactors dramatically decreases the time needed for the waste to decay to the normal radiation level and minimizes its radiotoxicity, as shown in Figure 1 [12]. The time when the radioactivity decays to the level of U ore decreases from several million to several hundred years after the removal of actinides. It is going to be easier to handle the decay heat and will benefit the repository management [13]. However, reprocessing initially was not for these purposes but essentially used in World War II, mainly focusing on how to strip the plutonium (Pu) from the SNF for the purpose of being used in weapons [14–16]. The first method selected was the bismuth phosphate (BiO_4P) process [17], which was used to produce all the Pu of the Fat Man atomic bomb detonated over Nagasaki [18]. Later the reduction and oxidation (REDOX) [19] process, which is safer and more efficient, replaced the BiO_4P process in 1952 [20]. But currently, the standard method widely used is the Plutonium–URanium Extraction (PUREX) process developed at Knolls Atomic Power Laboratory and Oak Ridge National Laboratory (ORNL) [21]. It was applied in

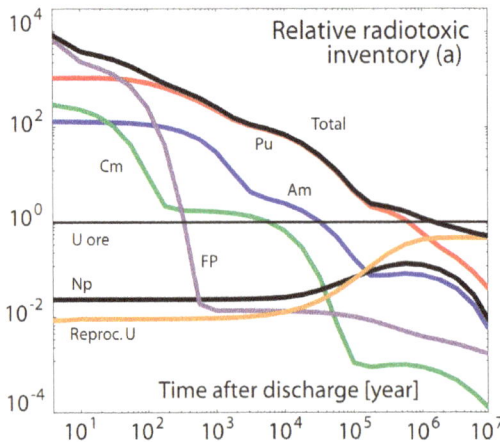

Figure 1. The relative radiotoxic inventory of spent uranium oxide (UOX) [12].

the Savannah River site in 1954 and the Hanford site in 1956 and also all the commercial reprocessing plants after that [21–23].

After the war, the reprocessing of SNF was considered necessary due to a perceived scarcity of U resource with the development of commercial nuclear powers to harness the nuclear power for peaceful use [24]. The Atomic Energy Commission (AEC) issued licenses to several companies, such as Nuclear Fuel Services, General Electric Company, and Allied-General Nuclear Services, to construct and operate the reprocessing plants [24]. However, the Carter administration banned the commercial reprocessing of SNF on April 7, 1977 announcing "We will defer indefinitely the commercial reprocessing and recycling of the plutonium produced in the U.S. nuclear power program" [25] mainly due to the threat of nuclear proliferation by the diversion of Pu and the consideration to encourage other countries to follow its lead [26]. For the last four decades, all the SNF in USA is stored in either wet pools or dry casks for interim storage [27]. Yucca Mountain was designated as the solo repository to permanently store the SNF in 2002 by The Bush Administration [28, 29]; however, this was terminated by the Obama administration in 2010 [30]. On the other timeline, Experimental Breeder Reactor-II (EBR-II) started to operate in 1964 after the success of Experimental Breeder Reactor-I (EBR-I) [31]. The main objective was to demonstrate the operation of liquid metal-cooled fast breeder reactor with on-site integral fuel cycle [31]. The Fuel Cycle Facility (FCF) stopped operating temporarily in 1969 after adequate demonstration has been done while the reactor did not shut down until 1994 [32]. Melt refining technology used in the fuel recycle has the disadvantage of losses of actinides in slag phase and increase of impurities over time. It was replaced by electrochemical pyroprocessing later [33], which can generally produce products with higher purity [34]. A 3-year project to demonstrate the feasibility of electrometallurgical techniques to treat the spent fuel in Department of Energy (DOE) complex was conducted in Argonne National Laboratory west (ANL west) from 1996 to 1999 [35]. And the technology has been incorporated to DOE's Advanced Fuel Cycle Initiative program since 2002 [36]. From 1996 to 2012, more than 4 metric tons of heavy metal (MTHM) spent fuel from EBR-II has been reprocessed based on pyroprocessing method [37]. Now the technology is being actively investigated in USA, Japan, Korea, and Europe [37–40].

2. Pyroprocessing Facility

Pyroprocessing is an electrochemical method to recycle the actinides contained in SNF based on the molten salt electrolyte. It was originally

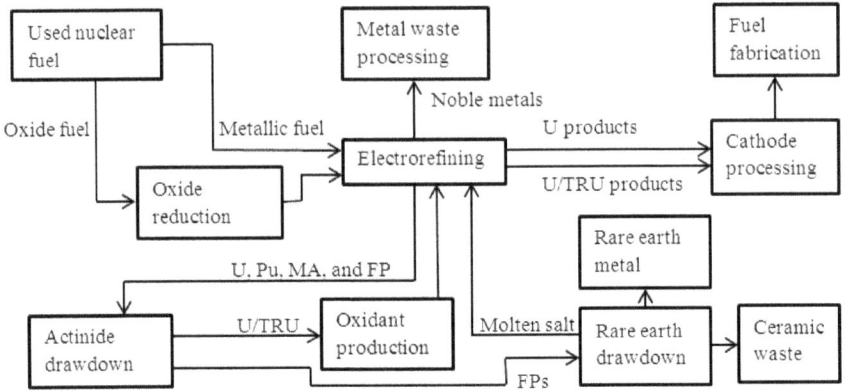

Figure 2. Conceptual flowsheet of pyroprocessing to process oxide and metallic fuel.

developed by ANL [41, 42] and used to process the metallic fuel discharged from EBR-II as a part of the Integral Fast Reactor (IFR) program initiated in 1984 to demonstrate the fast reactor on-site fuel cycle closure [43]. The conceptual flowsheet of the method is illustrated in Figure 2.

Even though it was initially designed for the treatment of metallic used fuels from fast reactors, pyroprocessing can also be extended to process oxide fuel from light water reactors (LWRs) with an additional step of oxide reduction based on a molten Li_2O–LiCl salt, which has been well reviewed by Choi $et\ al.$ [44]. The key element of the pyroprocessing is the electrorefining process, where U and transuranium (TRU) elements are electrochemically separated from noble metals (NMs) and fission products (FPs), as shown in Figure 3 [45].

Metallic SNF or reduced oxide fuel is charged in a perforated basket as the anode, inside which active elements, such as actinides and active FPs, are oxidized into LiCl–KCl electrolyte. Table 1 shows the representative used driver fuel composition for EBR-II and their standard potentials [47–50]. The elements whose potentials are lower than that of U are dissolved into LiCl–KCl electrolyte with U and others are left in the anode basket. At the cathode side, U is deposited on the solid cathode selectively as the dendrite form by controlling the applied current. Residual U, Pu, and minor actinides are then deposited into liquid cadmium cathod (LCC) or on a solid electrode based on electrorefiner design. Both products go through cathode processing to clean the salt or Cd by distillation before injection casting and fuel fabrication [51, 52]. More active FPs such as lanthanide metals and alkaline earth metals remain in electrolyte with some actinides. Actinide drawdown is applied to clean the actinides, which are

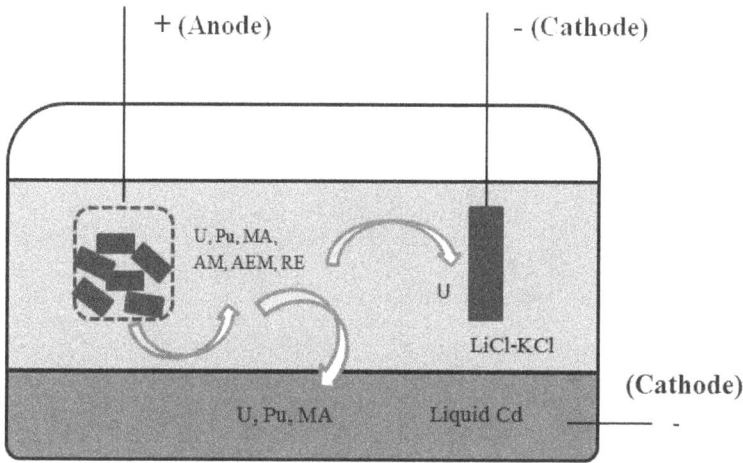

Figure 3. Schematic figure of typical design of electrorefiner.

Note: AM: Alkali metal FPs; AEM: Alkaline earth metal FPs; RE: Rare earth FPs; MA: Minor actinides.

oxidized and returned to the electrorefiner for the current support [53]. The remaining rare earth FPs are cleaned by rare earth drawdown. Other waste, for example Cs and Sr, is prepared in ceramic form for the final disposal [53]. Cladding materials and other NMs residing in anode basket or at the bottom of electrorefiner are consolidated into metallic ingot in metal waste furnace for the final disposal [37].

The reactions in electrorefiner can be expressed as

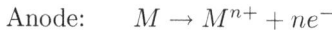

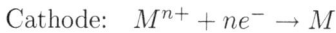

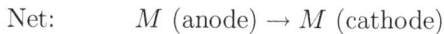

$$\text{Anode:} \quad M \rightarrow M^{n+} + ne^-$$

$$\text{Cathode:} \quad M^{n+} + ne^- \rightarrow M$$

$$\text{Net:} \quad M \text{ (anode)} \rightarrow M \text{ (cathode)}$$

Pyroprocessing is a dry process without involvement of water. The LiCl–KCl molten salt has high radiation resistance. Therefore, it has the potential capability to process large amount of hot fissile materials that even only have been cooled for several months through remote control with little worry about criticality risks and electrolyte degradation [54, 55]. Economic advantage is also expected due to the usage of a more compact site with fewer steps, less equipment and footprint [56, 60]. Additionally, Pu is co-deposited with U and other minor actinides as U/TRU products, which provides a barrier to the proliferation. Therefore, pyroprocessing is being considered as a promising alternative to the traditional PUREX process to recycle the SNF.

Table 1. Representative used driver fuel compositions and standard potential.

Element	Wt.% in the used fuel [46]	$E°$ (V vs. Ag/AgCl) at 723 K [47, 48]
Noble elements		
Br	0.007	0.920
Te	0.112	0.64
Ru	0.407	0.615
Rh	0.111	0.526
Pd	0.090	0.513
I	0.048	0.473
As	0.005	0.283
Mo	0.771	0.119
Sb	0.004	0.087
Ag	0.004	0.000
Cu	0.003	0.295
Sn	0.015	−0.355
Nb	0.002	−0.41
Se	0.019	−0.459
Cd	0.007	−0.589
V	0.003	−0.806
Ti	0.077	−1.010
Zr	10.81	−1.088
Eu	0.011	−1.471[a]
Active elements		
U	80.60	−1.496
Np	0.041	−1.519
Pu	0.413	−1.570
Am	—	−1.642 [49]
Gd	0.005	−2.066
Nd	0.930	−2.097
Y	0.126	−2.109
La	0.284	−2.126
Pm	0.011	−2.147[a]
Ce	0.542	−2.183
Pr	0.269	−2.316[a]
Sr	0.217	−2.429[a] [50]
Na	2.160	−2.5
Sm	0.177	<−2.5
Eu	0.011	<−2.5

Note: [a]Values at 773 K.

3. Safeguards of Special Nuclear Materials

According to the International Atomic Energy Agency (IAEA), the objective of safeguards is the timely detection of the diversion of special nuclear
materials (SNMs) with significant quantities from peaceful to non-peaceful

use [57], which is imperative for the civil use of nuclear energy. Timely detection indicates that the time from diversion to detection should be less than the time needed to convert the nuclear materials to the component of a nuclear explosive device (conversion time) in order to allow for evaluation and response. And significant quantities are the quantities minimally required to manufacture a nuclear explosive device for different kinds of SNMs [57]. The conversion time and significant quantity provided by IAEA are listed in Tables 2 and 3 [58]. Thus, when applying the criteria for pyroprocessing, the over-arching objective is the detection of the diversion of 8 kg Pu within one month of diversion. The U also needs to be safeguarded but to a less extent. The IAEA basically needs to verify the material flows and inventories of Pu and U to ensure that the operation is the same as declared and to detect any possible diversion scenarios [59]. Once the materials unaccounted for exceed the significant quantity in a conversion time, alert should be given to stop the nuclear facility and close the material balances.

Table 2. Significant quantities for different kinds of materials [58].

Material	Significant quantities
Direct use nuclear material	
Pu (^{238}Pu<80%)	8 kg Pu
^{233}U	8 kg ^{233}U
High enriched U (^{235}U$\geq$ 20%)	25 kg ^{235}U
Indirect use nuclear material	
Low enriched U (^{235}U<20%)	75 kg ^{235}U
Natural U	10 t natural U
Depleted U	20 t depleted U
Th	20 t Th

Table 3. Conversion time for different kinds of materials [58].

Material	Significant quantities
Pu, HEU or ^{233}U metal	Order of days (7–10)
PuO$_2$, Pu(NO$_3$)$_4$ or other pure Pu compounds; a HEU or ^{233}U oxide or other pure U compounds; MOX or other non-irradiated pure mixtures containing Pu, U (^{233}U + ^{235}U $\geq$ 20%); Pu, HEU, and/or ^{233}U in scrap or other miscellaneous impure compounds	Order of weeks (1–3)
Pu, HEU or ^{233}U in irradiated fuel	Order of months (1–3)
U containing <20% ^{235}U and ^{233}U; Th	Order of months (3–12)

4. Safeguards Issues in Pyroprocessing

Considering a commercial facility with the reprocessing capacity of 100 tHM per year, Pu involved in the process could be up to 1 Mt, which is hundreds of times its significant quantity. Because all the reprocessing plants nowadays are based on the PUREX process, unique safeguarding methods for it have been well developed. However, key differences between PUREX process and pyroprocessing [60] challenge the application of these developed safeguards approaches to the pyroprocessing facilities. These key differences include the following sections.

4.1. *No accountability tank*

The mostly traditional method for safeguards is the material control and accountability by destructive assay (DA) or nondestructive assay (NDA), which, however, is more suitable for the PUREX process. In the PUREX process, all the SNF is dissolved into accountability tank to generate a homogenous solution. By sampling the solution, U and Pu inventories can be determined easily by DA or NDA methods. Additionally, due to the continuity of the process, Pu and U can be readily tracked by analyzing the flow and separation conditions [61]. However, the method is hard to apply in pyroprocessing, where the SNF is dissolved into LiCl–KCl electrolyte electrochemically instead of chemically. The anode dissolution happens at the same time as the cathode deposition. It is a dynamic process, and at no time during the process will all the SNF distribute only in electrolyte. It is thus impossible to determine inventories of Pu and U by sampling the electrolyte only. Other NDA methods to directly measure the Pu composition in SNF assembly, such as neutron counting, could have an uncertainty of several percent since it needs to be based on the DA to determine the Cm/Pu ratio first, which introduces uncertainty because of the heterogeneity of the fuel [62]. A significant quantity of Pu can be accumulated quickly even for a throughput of 100 tHM/year.

4.2. *Inability to flush out*

In the aqueous process, a flush-out can be conducted to close the material balance with low uncertainty to determine whether diversion of SNM happens. In pyroprocessing, however, it may not be feasible to apply it. With the main purpose to separate the U from TRU and FPs by electrochemical method, U should be kept above a certain concentration to support the applied current and guarantee the purity of the deposition [63].

Removing all the salt and actinides would interrupt the process and affect its efficiency [64].

4.3. *Electrorefiner*

The electrorefiner is where all the separations happen. With more and more batches reprocessed, Pu will be accumulated to a high concentration. A measurement with low uncertainty could result in a discrepancy of 1 SQ Pu. Also, the large processing volume with multiple cathodes makes the measurement more complicated [60]. The dendritic solid deposition also requires new approaches for composition assay considering the difficulty to obtain a homogeneous sample.

4.4. *Harsh environment*

The electrorefining is generally run at the elevated temperature ranging from 450°C to 550°C. The molten salt and metal solutions are highly corrosive [65]. Such an environment will be very challenging for the equipment and instruments used for safeguards.

4.5. *Little experience*

Unlike the aqueous processes that have been commercialized for more than half a century [22], the pyroprocessing now is still in development and no commercial facility is running. The design of such facilities is not well defined yet. Not much experience or literature thus has been accumulated for reference. All these features of pyroprocessing set barriers for the application of existing safeguards approaches and require more effective and proper material tracking and assay technologies.

5. Safeguards Approaches

Taking into account all these features, Los Alamos [66] proposed four prospective safeguards approaches for a pyroprocessing facility.

5.1. *Neutron balance method by Cm accounting*

The principle is the assumption that Pu and Cm have very similar separation behaviors resulting from the similar thermodynamic properties. They are thus never separated from each other during all the separation processes. In essence, gross neutron is measured on each material stream and inventory, such as pins entering the system, the electrorefiner, waste, and

U/TRU product. The method detects diversion by checking the conservation of gross neutron rather than by measuring the amount of Pu. Therefore, the amount of Pu is not known during the operation but can be calculated after determining Pu/Cm ratio in U/TRU product by NDA or DA at the end of the operation.

5.2. *Assay of Pu in spent fuel via Pu/Cm ratio and DA*

This method is a modification of the neutron balance method mentioned in Section 5.1. In addition to the neutron measurement and DA or NDA on the U/TRU in Section 5.1, it also includes DA on a number of rod pieces in head-end process to determine the Pu/Cm ratio. Therefore, by detecting the gross neutron for each material stream, the amount of Pu can be calculated with the assumption of constant Pu/Cm ratio. However, techniques with low detection error are required for this method to measure the Pu amount in SNF.

5.3. *Electrorefiner assay*

This method mainly focuses on closing the material balance in electrorefiner. It is achieved by sets of assays on Pu in the electrorefiner salt before removal, in the removed and recharged salt, in all cathode products, in the metal waste, and in the recovered salt. It requires homogeneous mixing of electrolyte solution and techniques to assay various materials. Also, the operation process may be impacted due to massive and complex assays.

5.4. *Homogenized input*

The method involves a homogenization step before electrorefining to have representative samples for Pu assay. Pu/Cm ratio can thus be obtained for the material accountancy in the downstream. The method is straightforward and most accurate; however, it may modify the current conceptual design of the pyroprocessing [65]. All these approaches will need process monitoring, integrated video, and neutron monitoring to monitor the separation process, and material entry and removal.

In the following sections, emphasis is placed on the thermodynamic properties of U and Pu in molten salt and the techniques being developed to assay the material inventories in electrorefiner due to the presence of the most concentrated actinides inside.

6. Fundamental Data of U and Pu in LiCl–KCl Eutectic Salt

As discussed above, tons of Pu and U are involved in a commercial pyro-processing facility, mainly in the electrorefiner and also in the molten salt cleanup steps. Safeguards issue has to be addressed before it can be used safely. Due to limited operation and safeguards experience, the method to safeguard the pyroprocessing facility is still unclear to us. Unfortunately, traditional methods developed for aqueous process are either incapable or introduce unacceptable uncertainties. Considering the special design and operation pattern of pyroprocessing, new technologies and safeguards approaches targeting it have to be developed [60], especially for the electrorefiner.

Fundamental data of U and Pu in LiCl–KCl molten salt determines their separation behaviors in electrorefining, which affects the operation efficiency and material accountancy. These data mainly include solubility, apparent potential, activity coefficient, diffusion coefficient, and exchange current. They have been widely studied using various electrochemical techniques [49, 67–89].

Solubility determines the maximum concentration gradient a species can reach in anode diffusion layer because the surface concentration cannot exceed its solubility limit. This value can be assessed from the phase diagram calculation. Phase diagram of LiCl–KCl–UCl$_3$ ternary system has been assessed by Ghosh *et al.* [74], which yields a solubility of

$$\log S_{UCl_3}(\text{mole fraction}) = 0.1980 - \frac{466.14}{T} \qquad (1)$$

Recently, we calculated the phase diagram of LiCl–KCl–PuCl$_3$ system, as shown in Figure 4. It indicates two eutectics and one quasi-peritectic. One eutectic (E1) involving LiCl, PuCl$_3$, and K$_2$PuCl$_5$ occurs at 616 K and another one (E2) involving LiCl, KCl, and K$_3$PuCl$_6$ at 589 K. The quasi-peritectic involving K$_2$PuCl$_5$, K$_3$PuCl$_6$, and LiCl appears at 658 K. There is also a monovariant eutectic involving LiCl and K$_2$PuCl$_5$ at 690 K. Accordingly, the solubility of PuCl$_3$ was reported to be

$$\log S_{PuCl_3}(\text{mole fraction}) = 0.2415 - \frac{478.37}{T} \qquad (2)$$

Potential is the deciding factor for separating one element from another. When the applied potential is more negative than the redox potential of an element, the element can be reduced and deposited out. For a reduction

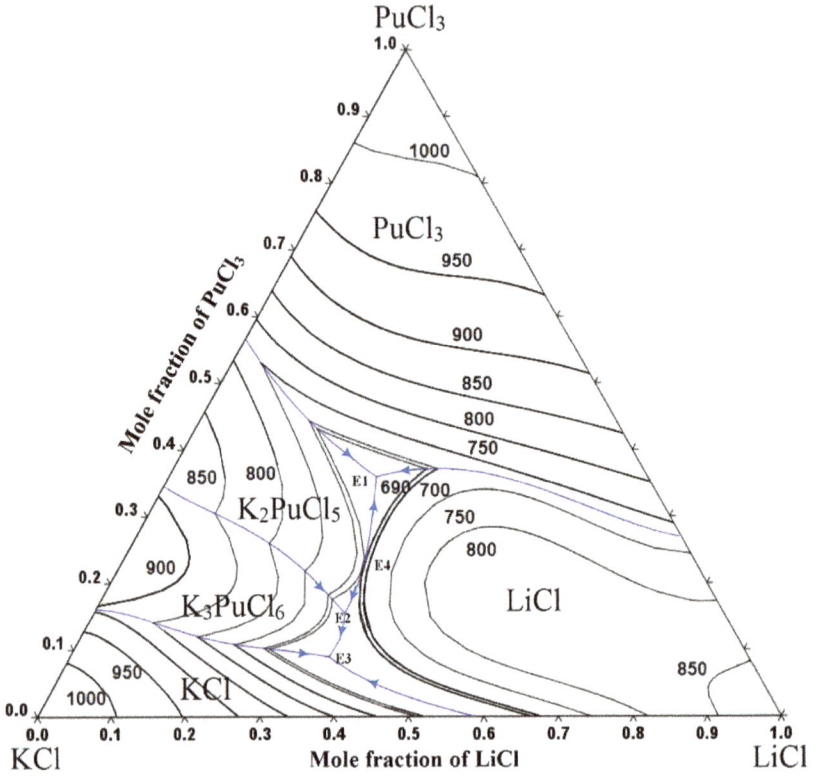

Figure 4. Calculated liquidus projection for LiCl–KCl–PuCl$_3$ system.

reaction shown as

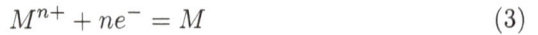

$$M^{n+} + ne^- = M \tag{3}$$

The equilibrium potential can be expressed by the Nernst equation

$$E_{eq} = E^0_{MCl_n} + \frac{RT}{nF} \ln \left(\frac{a_{MCl_n}}{a_M} \right) \tag{4}$$

where E^0 is the standard potential, R is the universal gas constant, T is the temperature in Kelvin, n is the number of electrons involved, F is the Faraday constant, and a is the activity coefficient. It can be written as

$$a = \gamma x \tag{5}$$

where x is the mole fraction, and γ is the activity coefficient, which is a measurement of the deviation from the ideal solution. For a metal deposition,

a is reduced to unity. Generally, the activity coefficient of metal chloride is lumped into the standard potential, which gives

$$E_{eq} = \left[E^0_{MCl_n} + \frac{RT}{nF} \ln(\gamma_{MCl_n}) \right] + \frac{RT}{nF} \ln(x_{MCl_n})$$

$$= E^{ap} + \frac{RT}{nF} \ln(x_{MCl_n}) \tag{6}$$

where E^{ap} is called the apparent potential. Figures 5 and 6 plot the apparent potentials and activity coefficients, respectively, of UCl$_3$ and PuCl$_3$ in LiCl–KCl eutectic molten salt. The apparent potential agrees with other values within a range of 60 mV and shows a linear relationship with the temperature. The activity coefficient also increases with the temperature, but presents a discrepancy up to one order of magnitude.

Diffusion coefficient is crucial to understand the transport of materials in LiCl–KCl electrolyte by Fick's law. Generally, the diffusion coefficient could be expressed by the Arrhenius law with the temperature

$$D = D_0 \exp\left(\frac{-E_a}{RT} \right) \tag{7}$$

where D_0 is pre-exponential factor, and E_a is the activation energy. Figure 7 shows the diffusion coefficients of UCl$_3$ and PuCl$_3$ in LiCl–KCl eutectic molten salt. It shows an approximately linear relationship with the temperature but scatters widely.

Also, some studies investigated diffusion coefficient of UCl$_3$ under different concentrations, as shown in Figure 8. However, no obvious conclusion about the concentration dependence can be reached due to large discrepancy in the data.

Exchange current is an important measurement about how fast electrons can be transferred in a reaction. Its value represents the half-reaction current when a reaction reaches equilibrium. For a soluble–insoluble transition, exchange current can be expressed by

$$j_0 = nFk_0 C_O^{b\,(1-\alpha)} \tag{8}$$

where k_0 is the rate constant, C_O is the concentration of oxidant, and α is the electron transfer coefficient. By different methods, namely CV, Tafel plot, electrochemical impedance spectroscopy (EIS), and linear polarization (LP), exchange current for the UCl$_3$ in LiCl–KCl with various concentrations at different temperatures has been studied [90–93]. Figure 9 plots the exchange currents at different concentrations and temperatures. Basically, exchange current increases with the temperature. Also, it shows a

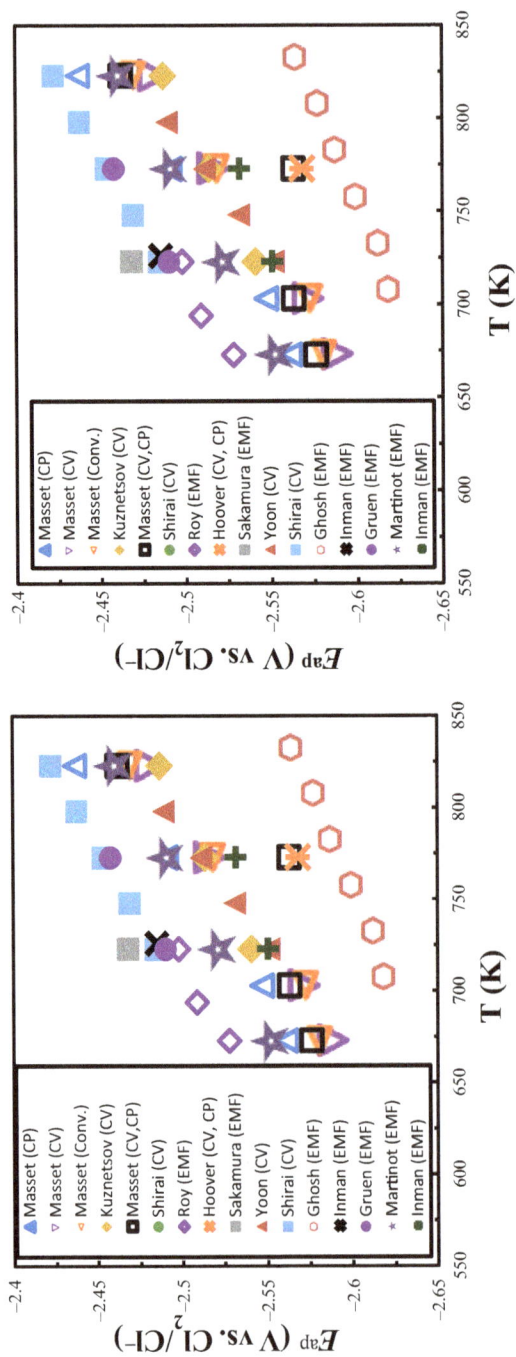

Figure 5. The apparent potentials of UCl$_3$ and PuCl$_3$ in LiCl–KCl eutectic.

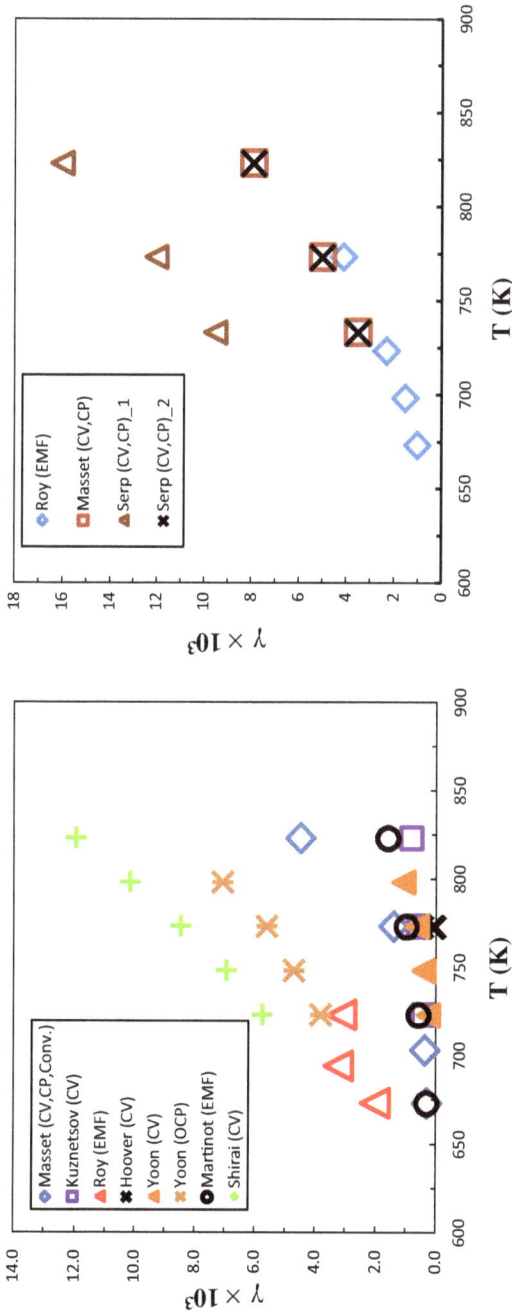

Figure 6. Activity coefficients of UCl$_3$ and PuCl$_3$ in LiCl-KCl eutectic.

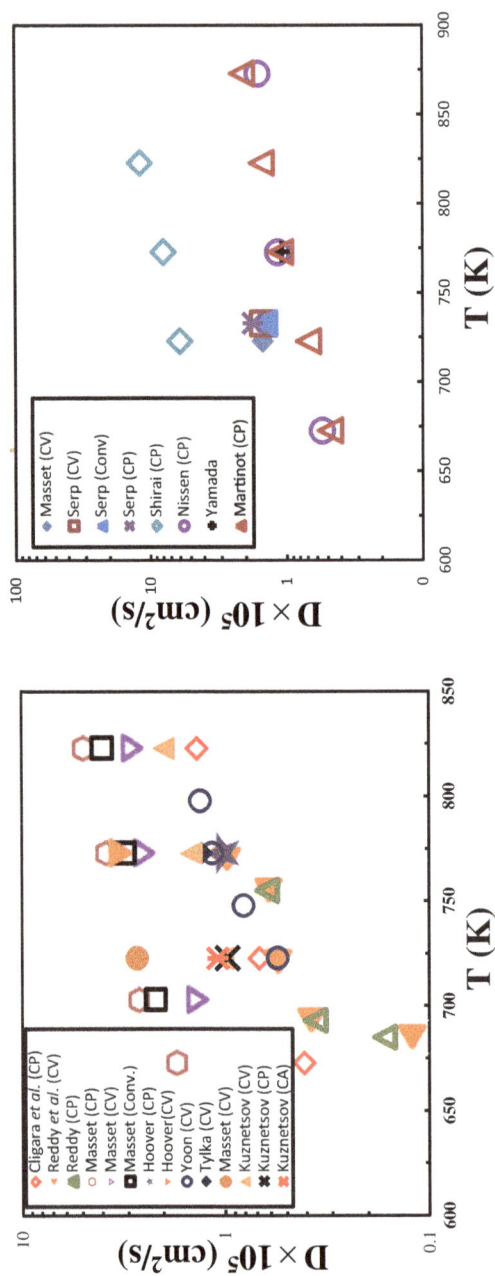

Figure 7. Diffusion coefficients of UCl_3 and $PuCl_3$ in $LiCl$-KCl eutectic at different temperatures.

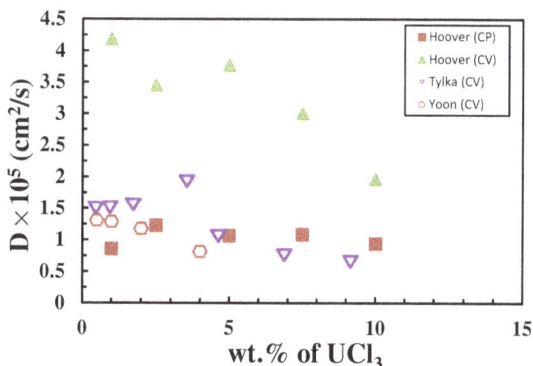

Figure 8. Diffusion coefficient of UCl₃ in LiCl–KCl eutectic at 773 K with different concentrations.

rough straight line with concentration, which indicates a value of zero for α according to Eq. (8). However, no references found report such a low value. There are no investigations for the exchange current of Pu.

Obviously, extensive studies have been conducted on the electrochemical behaviors of U and Pu in eutectic LiCl–KCl salt. However, almost all of them only focused on the dilute solution, when the thermodynamic properties depend little on the solute's concentration. But in pyroprocessing, actinide concentration can be up to 10 wt.% when the concentration dependence of these properties has to be considered and studied for a reliable prediction of the separation process and safeguards implementation [94]. Very recently, some studies started to focus on the concentration dependence of these fundamental data. By molecular dynamics simulation, we [95, 96] studied the activity coefficient, apparent potential, and diffusion coefficient of UCl₃ in LiCl–KCl molten salt up to a concentration of around 3 mol.% (about 16 wt.%). The results are shown in Figures 10–12. The results show that activity coefficient and apparent potential depend strongly on the concentration. With the mole fraction increasing from 0.5% to 3.0%, the activity coefficient of U^{3+} in LiCl–KCl molten salt increases from 6.08×10^{-3} to 3.62×10^{-2} and 7.46×10^{-3} to 4.36×10^{-2} at 723 K and 773 K, respectively. The study shows that diffusion coefficient varies little at concentration below 7.5 wt.%, beyond which it shows an increase followed by a decrease. However, the variation of these values is not significant enough to be seen clearly when compared with the literature, as plotted in Figure 13. Using the method, Wang [97] calculated the properties of other elements, namely La, Y, Sc, and Tb. Their results show the

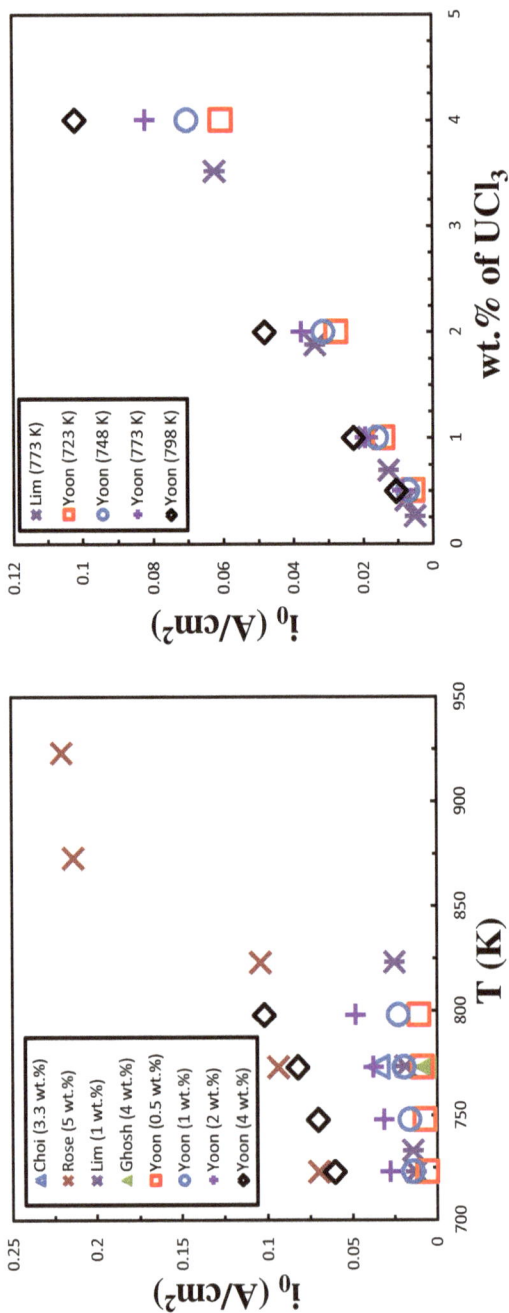

Figure 9. Exchange current of UCl$_3$ in LiCl–KCl eutectic at different temperatures and concentrations.

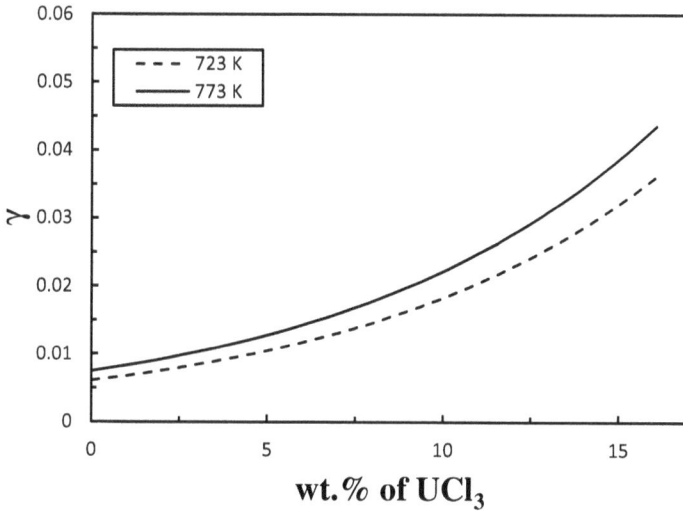

Figure 10. Activity coefficient of UCl₃ in LiCl–KCl molten salt.

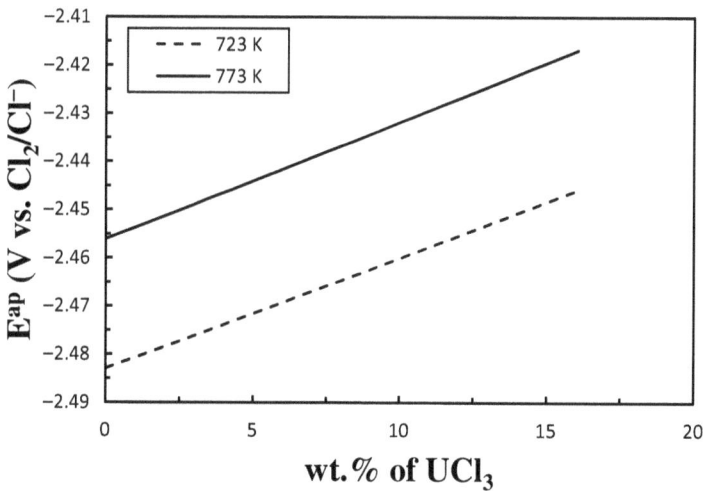

Figure 11. Apparent potential of UCl₃ in LiCl–KCl molten salt.

same pattern. Bagri and Simpson [94] studied the concentration dependence of the activity coefficient of La up to a concentration of around 3 mol.%. Figure 14 shows comparison of the results from Wang [97] and Bagri and Simpson [94]. Even though there is difference in the values, they

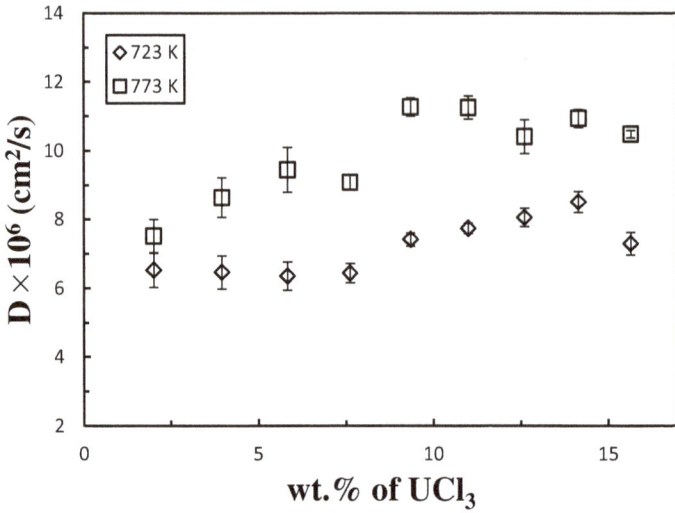

Figure 12. Chemical diffusion coefficient of U^{3+} in LiCl–KCl.

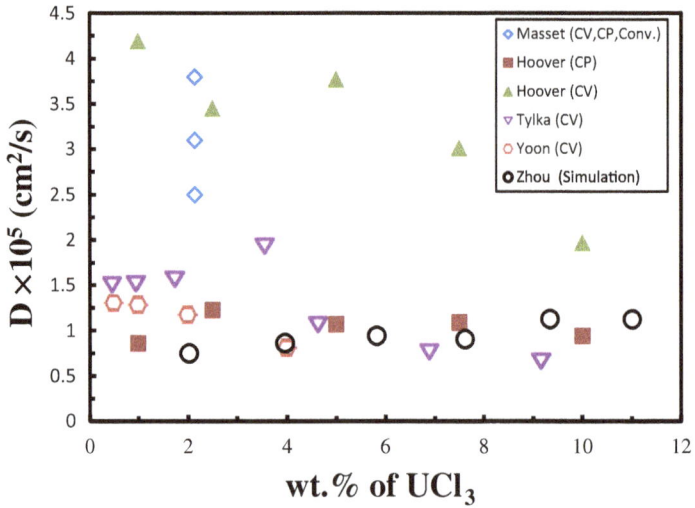

Figure 13. Comparison of chemical diffusion coefficient of U^{3+} in LiCl–KCl at 773 K.

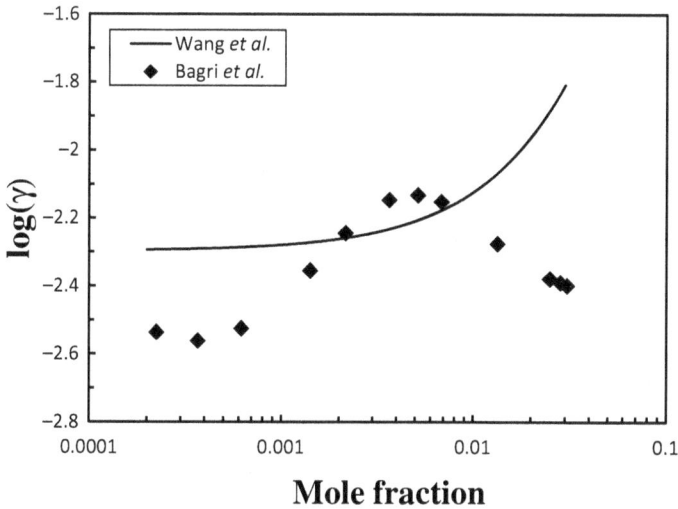

Figure 14. Activity coefficient of La in LiCl–KCl molten salt reported by Wang *et al.* [97] and Bagri *et al.* [94] at 773 K.

both reported an increasing activity coefficient of up to about 0.5 mol.% of La. However, Wang's result shows a continuous increase after that, but the value from Bagri proceeds to decrease. Bagri [98] later performed similar experiments on $GdCl_3$. However, the activity coefficient shows a pattern of decrease and then almost constant increase against the increase–decrease of $LaCl_3$. Without more data, it is hard to claim which one is closer to reality. But it is worth noting that all the studies show a strong dependence of activity coefficient on the concentration. They provide a new perspective to understand the behaviors of nuclear materials in LiCl–KCl eutectic salt.

7. Safeguards Methods

Considering that the separation of actinides from FPs mainly occurs in the electrorefiner, most of the safeguards methods currently are focusing on it. Compared with traditional aqueous process, pyroprocessing has the features of discrete batches, high temperature, and electrochemical process. Safeguards approaches being developed now are either dealing with these features, for example in the remote control method, or using them, as in the methods based on electrochemical techniques. Safeguards methods attracting most of the researches in this area are reviewed below.

7.1. Goals-driven method

Proposed by Wigeland *et al.* [62], goals-driven method is based on the nature of discrete batches of pyroprocessing. From the chopped fuel to final electrorefining, it is a step-by-step process. Basically, each item has a unique path. Therefore by monitoring its motions and positions, safeguards can be achieved by considering any abnormal movement conflicting with normal route has the potential for diversion. The weakness of approach is that it is only capable of detecting non-standard movements, and so if one wants to know the material composition for the balance closure, other DA or NDA methods are still necessary.

7.2. Laser-induced breakdown spectroscopy (LIBS)

The schematic of laser-induced breakdown spectroscopy (LIBS) is shown in Figure 15 [99]. The main components include LASER, reflective optics and lenses, fiber optics, and spectrometer to do the measurement and analysis.

This technology uses laser to ablate the surface of a sample to create plasma. These excited plasmas will emit characteristic photons when returning to the ground states. By spectrally resolving the light, types of species in sample can be differentiated based on the spectrum. The concentration can be revealed according to the intensity of the emission line [100–102]. According to the thermodynamic theory, when an excited atom j returns to its ground state i, the energy or intensity of its spectral line

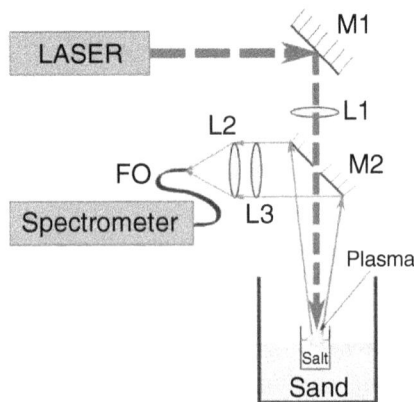

Figure 15. Schematic diagram of LIBS [99].

can be expressed by [103]

$$I = h\upsilon_{ji} A_{ji} N g_j Q^{-1} \exp\left(-\frac{E_j}{kT}\right) \tag{9}$$

where h is the Planck constant, υ_{ji} is the frequency of the transition j to i, A_{ji} is the Einstein coefficient, N is the population of species j in the sample, Q is the partition function of relevant species, g_j and E_j are, respectively, the statistical weight and energy of j, k is the Boltzmann constant, and T is the plasma temperature. However, because of the unknown composition of the sample, the relation between N and concentration may be difficult to express. Generally, a calibration curve will be generated in advance with known concentration to interpolate or extrapolate to other values in experiments [103].

Thanks to the use of fiber optics, collection and analysis of signal *in situ* is not required. Therefore, it can achieve a remote control and operate in stand-off distances with even 100 m between remote location and experimental site [104], which is particularly significant in an electrorefining system with hostile environment, such as high radiation and elevated temperature, so as to avoid direct contact.

The method is considered essentially as non-destructive because only a small quantity of material is consumed during the test. The short duration of laser pulse, typically 0.1–10 μs [105], and fast data collection in less than 1 min provide the potential of near-real-time online detection.

Witehouse *et al.* [106] analyzed the compositions of spent fuel residues through LIBS spectroscopy. The spectra recorded matched considerably with the contaminant material of zirconium molybdate (ZM). The feasibility of the method for quantitative measurements was demonstrated by calibration curve generation for different concentrations of calcine and ZM. Nicholas *et al.* [107] tried to apply LIBS to electrochemical process for safeguards purpose. A preliminary test was carried out on solid LiCl–KCl eutectic salt with up to 1.5 wt.% actinide chlorides (U, Np, and Pu). However, even though significant signals were captured attributing to the Li and K, almost no response recorded to any of the actinides. Weisberg *et al.* [108] reported a further study to measure the Eu and Pr concentration in molten eutectic LiCl–KCl. For a better prediction, the multivariate calibration method of partial least squares regression was applied to analyze the relationship between signal lines and concentration. This study presented very promising results that both Eu and Pr can reach an absolute accuracy of 0.13 wt.% with concentrations ranging from 0 to 3.01 wt.% and from 0 to 1.04 wt.%, respectively. When considering improving the beam alignment, removing the floating film and limiting the salt spray, the accuracy could be even better.

The feasibility of LIBS for safeguards purpose has been demonstrated preliminarily, but more work is merited mainly in three aspects to minimize the uncertainties and obtain accurate results. First, how to guarantee that the salt being ionized can represent the salt in electrorefiner? The signal from the floating film with unknown compositions or the salt spray may distort the results of concentration. Some preliminary laser pulses to ablate the floating film can be a method to ionize the representative salt. And the salt spray can be mitigated by decreasing the pulse energies by using fiber optic and optical probe [108].

Second, how to optimize the parameters of a LIBS system? The parameters in a LIBS system include laser pulse energy, time delay to collect the signal, lens position and distance, fiber material, pulse number before useful signal can be collected, and saturation of spectrometers. These parameters should be optimized to get high signal intensity and provide proportional relationship between signal and concentration [103].

Finally, how to analyze the signal to reveal the solute concentration? The most common way is to generate the calibration curves between signal intensity and concentration [103, 106, 108]. However, previous studies [108] indicated large uncertainties when the relationship between signal intensity and concentration were directly plotted. Multivariate analysis methods, such as partial least squares may be an alternative to deconvolute the signal to derive concentration information, but more demonstrations are needed.

7.3. *Neutron balance method*

The SNF emits neutrons spontaneously, which is mainly attributed to the Cm. A typical neutron emission from a pressurized water reactor (PWR) is shown in Figure 16 [109]. As discussed in Section 5.1, the method is based on the assumption that Pu and Cm have very similar thermodynamic properties and they are likely to track each other within the separation processes. Therefore, the gross neutron emission at different batches reflects the distribution of Pu and can be used to detect its diversion. Table 4 shows the electrochemical properties of Pu and Cm in LiCl–KCl eutectic at 773 K. For the aqueous process, neutron balance method has been successfully applied to the high-level liquid waste, spent fuel assemblies, and leached hulls of

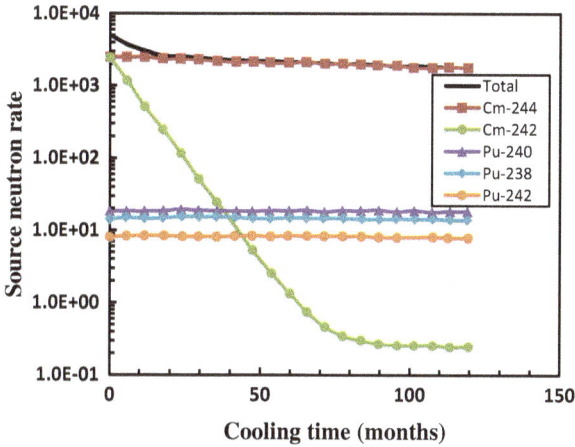

Figure 16. The relative neutron emission rates from the major isotopes in a PWR spent fuel assembly after exposure of 30 GWd/tU.

Table 4. Electrochemical properties of Pu and Cm in LiCl–KCl eutectic at 773 K [111].

	Activity coefficient	Apparent potential E^{ap} (V vs. Cl_2/Cl^-)	Diffusion coefficient D (cm^2/s)
Pu	$9.793e^{-3}$	-2.755	$1.122e^{-5}$
Cm	—	-2.861	$1.259e^{-4}$

a reprocessing plant [109, 110], which revealed the potential to apply it to the pyroprocessing system. Quantitative analysis of Pu could be achieved if we know the initial Pu/Cm ratio in SNF and the neutron emission at key points in the reprocessing. At least, the balance Pu can be detected based on the neutron balance in different batches of the pyroprocessing facility.

But before the method can be applied, the assumption that Pu tracks Cm during separation process should be validated. Experiments were conducted in Idaho National Laboratory (INL) [112] to investigate it. However, the Cm concentration in ERB-II spent fuel was below the detection limit. Even though the Pu/Cm ratio was determined as 1.4×10^4 in LWR BR-3 spent fuel, the Cm still cannot be detected in electrorefiner with this fuel. Due to the great difficulty encountered in assaying the Cm concentration, the method was not verified in this experiments. Recently, Gonzalez [113] studied the co-deposition properties of Pu and Cm by a 1D transient electrorefiner model Enhanced REFIN with Anodic Deposition (ERAD) [114].

An expected concentration ratio of Pu/Cm, which is around 100 in practice, however, deposited out none of Cm on the solid cathode while Pu deposited with U at high applied current. The Pu/Cm ratio in molten salt electrolyte decreased with time. Cm could only have a positive deposition when increasing its initial concentration in molten salt to 1.0 wt.%. But the Pu/Cm ratio on the cathode was three orders of magnitude higher than that in salt. This preliminary study indicated that neutron balance method may not be a reliable approach for safeguards purpose but more experiments are needed for the final validation.

7.4. *Electrochemical technique*

Because pyroprocessing itself is an electrochemical process, electrochemical techniques are readily used. Available methods mainly include cyclic voltammetry (CV), chronopotentiometry (CP), chronoamperometry (CA), anodic stripping voltammetry (ASV), square wave voltammetry (SWV), and normal pulse voltammetry (NPV).

Basically, electrochemical techniques are those methods measuring the responses of current or potential when applying a potential or current with certain patterns, respectively, to reveal the properties and behaviors of reactants in a system.

7.4.1. *Cyclic voltammetry*

For the CV method, the potential is linearly scanned to a direction followed by an opposite direction with the same vertex potentials, as shown in Figure 17(a). Correspondingly, reduction and oxidation processes happen

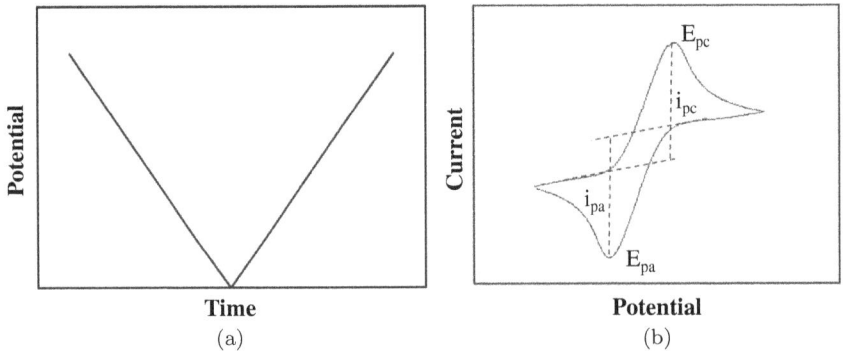

Figure 17. Pattern of (a) applied potential and (b) typical current response in CV.

alternately to generate cathodic and anodic currents. Peak currents appear due to the depletion of species on the electrode surface, as shown in Figure 17(b). These values can be related with species concentration. For the soluble/soluble transition, Delahay [115] derived the expressions of

$$i_p = 0.446nFAC \left(\frac{nFDv}{RT}\right)^{1/2} \tag{10}$$

$$i_p = 0.499nFAC \left(\frac{\alpha nFDv}{RT}\right)^{1/2} \tag{11}$$

where A is the electrode surface area and v is the scan rate. These two equations are applicable to reversible and irreversible systems, respectively. For the soluble/insoluble transition, Berzins and Delahay [116] gave the expression below to describe a reversible system:

$$i_p = 0.611nFAC \left(\frac{nFDv}{RT}\right)^{1/2} \tag{12}$$

When the experiment is carried out under different scan rates, the concentration of certain species can be derived from the slope of i_p vs. $v^{1/2}$ plot.

7.4.2. *Chronoamperometry*

CA is a potential control method to monitor the variation of the current. A driving potential, E_d, is applied to the working electrode from an initial potential, E_i, as shown in Figure 18(a). The species concerned can be reduced on the electrode by setting the driving potential to be lower than

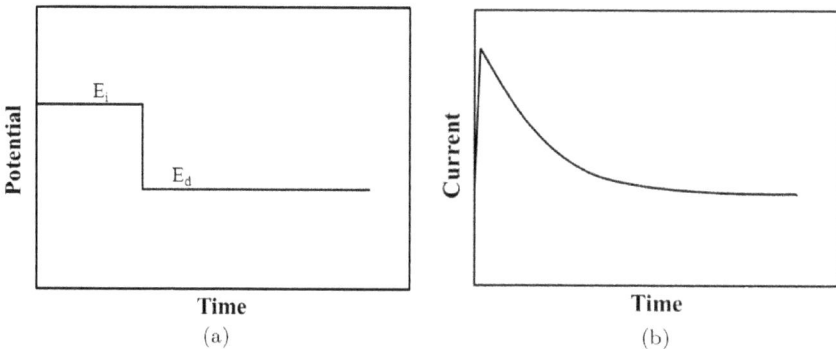

Figure 18. Pattern of (a) applied potential and (b) typical current response in CA.

its redox potential. With the depletion of species on the electrode surface and progress of the diffusion layer, the current drops with time due to the diffusion limit, as shown in Figure 18(b).

If a reaction is diffusion-controlled, the current drop with time can be expressed by Cottrell equation [117].

$$i(t) = \frac{nFAD^{1/2}C}{\pi^{1/2}t^{1/2}} \tag{13}$$

7.4.3. Chronopotentiometry

CP is a method similar to CA, but it controls the current to monitor the potential. A driving current, i_d, is applied to the working electrode after the pre-step current i_i, as shown in Figure 19(a). The noblest element would be reduced first on the electrode to support the current. When it is not enough to carry the current, potential would drop rapidly to reduce the second noblest element. The process repeats until current stops or all elements are depleted. The lasting time for the plateau is referred as transition time, τ. It can be related with the concentration by Sand's equation [118].

$$i_d\tau^{1/2} = \frac{nFAC(\pi D)^{1/2}}{2} \tag{14}$$

7.4.4. Anodic square voltammetry

ASV has two steps. First, a reducing potential is applied to the working electrode to electroplate the analyte on the electrode. The potential is then linearly increased to an anodic potential to oxidize the analyte, which

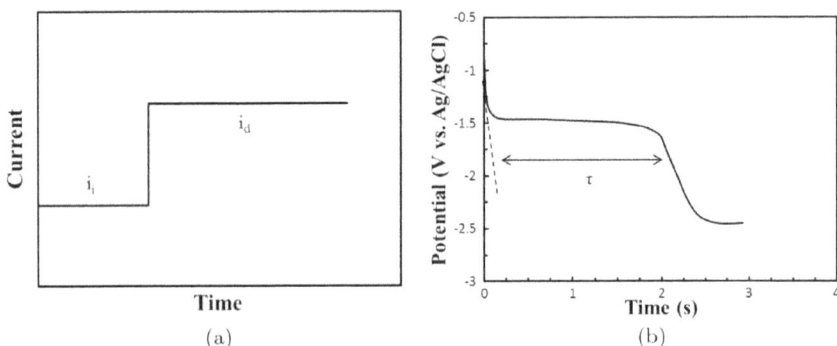

Figure 19. Pattern of (a) applied current and (b) typical potential response in CP [119].

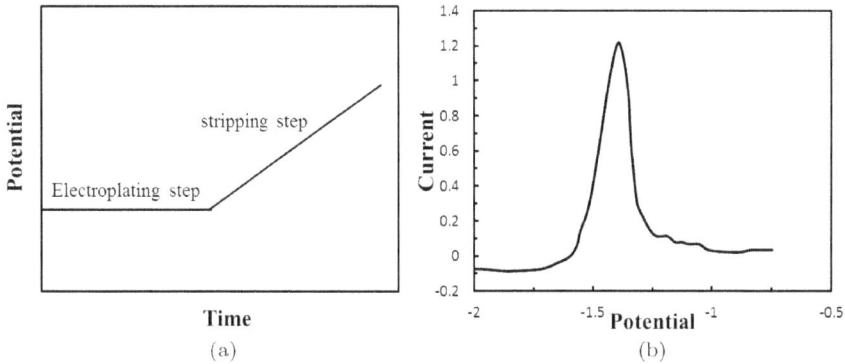

Figure 20. Pattern of (a) applied potential and (b) typical current response in ASW [72].

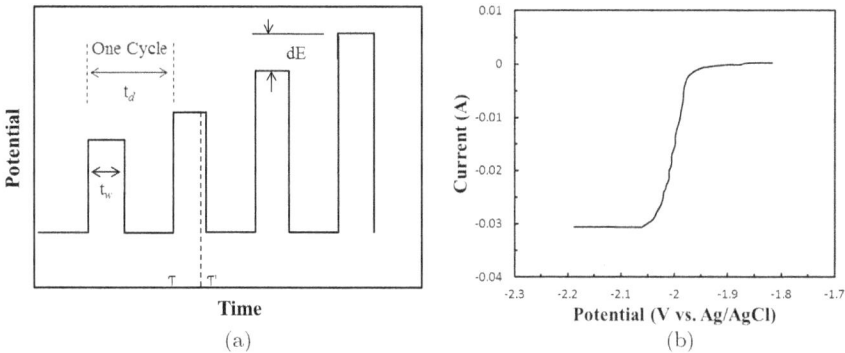

Figure 21. Pattern of (a) applied potential and (b) typical current response in NPV [120].

generates a peak potential, as shown in Figure 20. Sometimes, a cleaning step with oxidizing potential is added before the electroplating step to fully remove the analyte from the electrode. ASV is similar to the anodic sweep in CV. Therefore, the current peak should also have linear relationship with the concentration under the same scan rate.

7.4.5. *Normal pulse voltammetry*

NPV is a potential control method. It is similar to the square wave method but with the potential height increased for each pulse, as shown in Figure 21(a). The wave can be described by the pulse width, t_w, distance between two pulse t_d, and pulse increment dE. After a time τ, the current is sampled at a time τ' near the end of the pulse. The typical response of current is shown in Figure 21(b). The current approaches zero with large potential and is limited by the diffusion process when the potential goes negative. Basically, the process is similar to the CA method, and the current at τ' can be expressed using Cottrell equation [117]

$$i(t) = \frac{nFAD^{1/2}C}{\pi^{1/2}(\tau' - \tau)^{1/2}} \tag{15}$$

7.4.6. Square wave voltammetry

SWV is similar to the NPV method but has a different potential pattern. The applied potential can be treated as the superposition of regular square wave and staircase wave, as shown in Figure 22(a). Parameters used to describe the wave include the period t_p, pulse height E_h, and potential increment dE of each cycle. Two current samples can be taken from each cycle, the forward sample at the end of the first half cycle and the reverse sample at the end of the second half cycle. Both have a peak at certain potential, as shown in Figure 22(b). The difference between forward and reverse samples is referred to as Δi. Its peak has the expression of [121]

$$i_p = \frac{nFACD^{1/2}}{\pi^{1/2}t_p^{1/2}}\Delta\psi_p \tag{16}$$

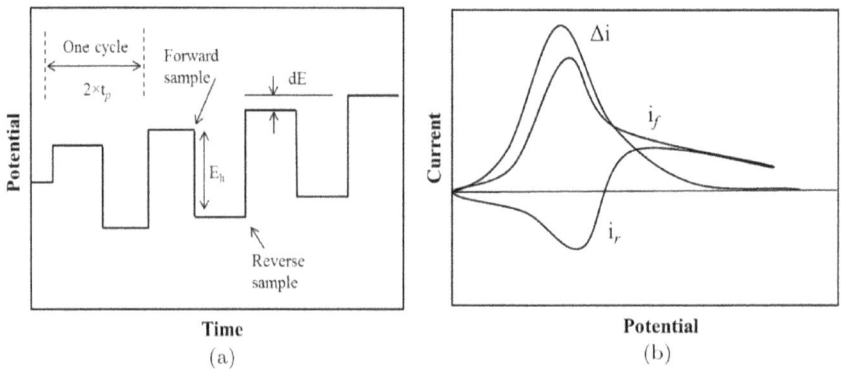

Figure 22. Pattern of (a) applied potential and (b) typical current response in SWV [121].

where the $\Delta\psi_p$ gauges the peak height in SWV relative to the limiting response in NPV with the same parameters. The peak of i_f and i_r can be also related to the concentration in similar expressions but by replacing $\Delta\psi_p$ with ψ_f and ψ_r, respectively.

These methods are chosen because their peak information, coulomb pass, steady potential, or steady current can be related with the solute concentration [47, 48]. Therefore an online real-time detection of the actinide concentrations may be achieved.

A bunch of studies have been carried out based on these methods. Hoover *et al.* [72] studied the LiCl–KCl–UCl₃ with 1–10 wt.% UCl₃ concentrations by using CV, CP, and ASV techniques. ASV peak height, CV anodic peak height and cathodic peak height were particularly studied with various concentrations. The results showed that CV cathodic peak gave best linear trend with concentration and could be a promising way for concentration detection. Kim *et al.* [122] tested CV and CA using 9 wt.% Nd. CV showed pretty good linearity, but CA only had a linear current–concentration curve when the current was taken at the very small time of 0.9 s. They reported that the CA method could be nested to electrorefining system without interfering with the main operation system by the alternate applications of cathodic and anodic currents. Both researches gave exciting information that electrochemical method may have the capability for the actinide quantification. However, they only studied the concentration detection for unique species. Multiple species are needed when considering the complexity of the electrorefiner. Iizuka *et al.* [120] reported the studies of multicomponent system with SWV and NPV methods. In this study, SWV showed a high sensitivity and more separated peaks; however, a reasonable linear relationship between peak current and concentration only presented at the concentration below 0.1 wt.% Pu, which is much lower than the practical concentration of few weight percent in electrorefining [123]. The NPV seemed to be more promising since the reduction current of U and Pu were proportional to their concentrations up to about 1.7 wt.%. Also, their signals were pretty stable and separated with the presence of FPs. However, the concentration of minor actinides, such as neptunium and americium, could be hard to determine with the appearance of Pu due to their close standard potentials, which resulted in highly overlapped signal.

Considering the deconvolution of the overlap signal, several methods have been proposed. One idea is to keep the potential at constant somewhere before the appearance of the next wave, and then repeat the experiment with further potential to plot the next peak after reinitiating the conditions [121]. A similar method involves the stop of potential before the foot of next peak and then waiting for the decay of current to zero before continuing the potential scan. Therefore, basically there are no baselines needed to determine for every peak [121]. Another method using semi-differential transformation was reported by Palys *et al.* [124]. This method is based on the additivity of semi-differential operator, i.e.,

$$\frac{d^{1/2}}{dt^{1/2}}i(t) = \frac{d^{1/2}}{dt^{1/2}}i_1(t) + \frac{d^{1/2}}{dt^{1/2}}i_2(t) \tag{17}$$

After removing undesired peaks from the semi-derivative curve, the one-peak voltammetric curve could be semi-integrated back through

$$i(t) = \frac{d^{-1/2}}{dt^{-1/2}}\left(\frac{d^{1/2}}{dt^{1/2}}i(t)\right) \tag{18}$$

Generally, the semi-derivative peaks would be much more separated than the original ones. However, all these three methods mentioned require that the peaks in voltammetric curve are separated to some degree. And the first two methods can be too complicated for the near-real-time detection when applied to the electrorefining where more than 30 types of species coexist [120].

Recently, Rappleye *et al.* [125] tried to analyze the signal from NPV to determine the concentrations of actinides by multivariate analysis techniques. Two similar methods of principle component regression and partial least squares were investigated. The key point is to figure out the principle components that have a significant variance to decrease the dimension of variables to address the multicollinearity but still keep most of the information. Therefore, these methods can deal with the signals even almost completely overlapped. As expected, these two methods predicted more accurate U and Pu concentrations in LiCl–KCl than the differentiated NPV method based on the data from Iizuka *et al.* [120]. However, analysis on experimental data with multiple-analytes is still required to determine its accuracy with more complex melt environments.

Basically, there are three issues that still need to be resolved before electrochemical techniques can be used fully to safeguard the pyroprocessing facilities. They are:

1. How to extend the reliable linear relationship to high concentration up to 10 wt.%.
2. How to deconvolute the overlapped signals involved in multi-component systems to extract reliable information.

3. How the basic parameters are affected by the concentration since all the fittings currently assume that thermodynamic parameters keep constant at all concentrations which may not be right.

8. Safeguards Model

Kinetic model development is another promising method for the safeguards. It integrates the basic data and electrochemical theories to predict the material transport in the electrorefiner. The signals from a model include anode and cathode potentials, partial current of each element, and composition changes in anode, electrolyte, and cathode. Therefore, besides safeguarding the pyroprocessing by tracking the materials via monitoring the composition changes in different batches, it can also be used to optimize the facility design, for example the Pu/U ratio and applied current, and predict the processing performance, like the efficiency and purity of the deposited materials.

Early models were only based on the thermodynamic equilibriums between pairs of elements and their oxidants existing at the anode–electrolyte and cathode–electrolyte interfaces. Main parameters used in these models are the Gibbs energy and activity coefficient. Take the transport of U and Pu from liquid cadmium anode (LCA) to LCC as an example. They have a reaction of

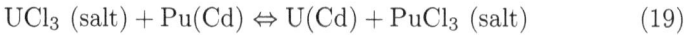

$$\text{UCl}_3 \text{ (salt)} + \text{Pu(Cd)} \Leftrightarrow \text{U(Cd)} + \text{PuCl}_3 \text{ (salt)} \tag{19}$$

The distribution of metals in liquid Cd and molten salt electrolyte is determined by the equilibrium constant, K_{eq}

$$K_{eq} = \exp\left(-\frac{\Delta G_f}{RT}\right) = \frac{N_{\text{U,Cd}} N_{\text{PuCl}_3,\text{ms}}}{N_{\text{Pu,Cd}} N_{\text{UCl}_3,\text{ms}}} \times \frac{\gamma_{\text{U,Cd}} \gamma_{\text{PuCl}_3,\text{ms}}}{\gamma_{\text{Pu,Cd}} \gamma_{\text{UCl}_3,\text{ms}}} \tag{20}$$

We define K_x as

$$K_x = K_{eq} \times \frac{\gamma_{\text{Pu,Cd}} \gamma_{\text{UCl}_3,\text{ms}}}{\gamma_{\text{U,Cd}} \gamma_{\text{PuCl}_3,\text{ms}}} = \frac{N_{\text{U,Cd}} N_{\text{PuCl}_3,\text{ms}}}{N_{\text{Pu,Cd}} N_{\text{UCl}_3,\text{ms}}} \tag{21}$$

where ΔG_f is the Gibbs energy of the reaction, N is the mole number of a species in each phase, respectively. At the beginning, the mole numbers of U and Pu in LAC, molten salt, and LCC are set to be $N_{\text{U,Cd,a}}$, $N_{\text{Pu,Cd,a}}$, $N_{\text{UCl}_3,\text{ms}}$, $N_{\text{PuCl}_3,\text{ms}}$, $N_{\text{U,Cd,c}}$, $N_{\text{Pu,Cd,c}}$, respectively. After a duration of electrorefining, t, some U and Pu will be dissolved into molten salt ($\Delta N_{\text{U,Cd,a}}$, $\Delta N_{\text{Pu,Cd,a}}$) and deposited into LCC ($\Delta N_{\text{U,Cd,c}}$, $\Delta N_{\text{Pu,Cd,c}}$).

According to the thermodynamic equilibrium and mass balance, the following equations are established:

$$\frac{(N_{U,Cd,a} - \Delta N_{U,Cd,a})(N_{PuCl_3,ms} + \Delta N_{Pu,Cd,a} - \Delta N_{Pu,Cd,c})}{(N_{Pu,Cd,a} - \Delta N_{Pu,Cd,a})(N_{UCl_3,ms} + \Delta N_{U,Cd,a} - \Delta N_{U,Cd,c})} = K_{x,a}$$

(22)

$$\frac{(N_{U,Cd,c} - \Delta N_{U,Cd,c})(N_{PuCl_3,ms} + \Delta N_{Pu,Cd,a} - \Delta N_{Pu,Cd,c})}{(N_{Pu,Cd,c} - \Delta N_{Pu,Cd,c})(N_{UCl_3,ms} + \Delta N_{U,Cd,a} - \Delta N_{U,Cd,c})} = K_{x,c}$$

(23)

$$n_U \Delta N_{U,Cd,a} + n_{Pu} \Delta N_{Pu,Cd,a} = n_U \Delta N_{U,Cd,c} + n_{Pu} \Delta N_{Pu,Cd,c} \qquad (24)$$

where n is the oxidation state. If the percentage transported (P_t), or current profile, $i(t)$, is given, one has

$$\frac{\Delta N_{U,Cd,a} + \Delta N_{Pu,Cd,a}}{N_{U,Cd,a} + N_{Pu,Cd,a}} = P_t \qquad (25)$$

or

$$n_U \Delta N_{U,Cd,a} + n_{Pu} \Delta N_{Pu,Cd,a} = \int_0^t i(t)dt \qquad (26)$$

Then these equations can be solved and composition changes of each phase can be derived. They were proposed by Johnson *et al.* [126] and Ackerman *et al.* [127, 128]. Nawada *et al.* [129] introduced the method to simulate the U and Pu transport in molten salt under 16 conditions, which was enhanced later by Ghosh *et al.* [130] with a more robust code called PRAGAMAN. These models did not consider the transport or electron transfer process, which may prevent the reaction from reaching equilibrium.

Another bunch of models are the diffusion control models. These models assume that all the material diffusing to or out of the electrode react instantly. Therefore, the current for each species, for example, at the cathode, is

$$i = nFD_{ms}A_c \frac{C_b - C_s}{\delta} \qquad (27)$$

or

$$i = nFK_{ms}A_c(C_b - C_s) \qquad (28)$$

where δ is the thickness of Nernst diffusion layer, and K_{ms} is the mass transfer coefficient in molten salt. C_b is bulk concentration in molten salt and C_s is the surface concentration on the molten salt/electrode interface. The electrode potential is expressed as

$$E = E_0 + \frac{RT}{nF} \ln \left(\frac{\gamma_{ms} x_{ms}}{\gamma_{ed} x_{ed}} \right) \tag{29}$$

where the subscript *ed* denotes the parameters for the electrode. Based on this theory, Kobayashi *et al.* [131] reported the TRAIL model to study the multicomponent transport in molten salt. In their model, the Nernst's diffusion layer was assumed, whose thickness at different interfaces was determined by polarization experiments. Iizuka *et al.* [132, 133] conducted a similar simulation to explain the electrorefining of U–Pu–Zr alloy but added diffusion layers in solid anode considering the formation of Zr porous layer during electrorefining. Zhang [134, 135] constructed another model based on the diffusion control where the mass transfer coefficient rather than diffusion layer thickness was used as one of the important factors. Therefore, the flow conditions can be considered in this model by relating it with the mass transfer coefficient. One of the important assumptions in these models is the reaction equilibrium on the electrode surface so the Nernst equation can be applied at the interface to calculate the electrode potential. However, all the models discussed above only considered the thermodynamic properties and diffusion at the interface but not the kinetic process on the electrode surface. Improved models took into account both the diffusion process in diffusion layer and electron transfer process on the electrode by applying the Butler–Volmer equation [113, 114, 136–139]. The 1D transport of materials taking into account the diffusion, convection, and migration in electrolyte can be expressed by

$$\frac{\partial C}{\partial t} = D \frac{\partial^2 C}{\partial x^2} - v \frac{\partial C}{\partial x} + \frac{nFD}{RT} \frac{\partial}{\partial x} \left(C \frac{\partial \Phi}{\partial x} \right) \tag{30}$$

where v is the flow velocity toward the electrode surface, Φ is the electric potential. Current density due to mass transport is

$$j_t = D \frac{\partial C}{\partial x} - vC + \frac{nFD}{RT} C \frac{\partial \Phi}{\partial x} \tag{31}$$

The Faraday process is described by the Butler–Volmer equation. The current density is

$$j_f = j_0 \left\{ \frac{C_{O,s}}{C_{O,b}} \exp[-(n\alpha F/RT)\eta] - \frac{C_{R,s}}{C_{R,b}} \exp[(1-\alpha)(nF/RT)\eta] \right\} \tag{32}$$

where j_0 is the exchange current expressed as

$$j_0 = nFk_0 C_{O,b}^{(1-\alpha)} C_{R,b}^{\alpha} \tag{33}$$

where $C_{R,b}$ reduces to unit when the reductant is insoluble. η is the over-potential

$$\eta = E - E_{eq} \qquad (34)$$

Some studies [136] simplified the $\frac{C_{O,s}}{C_{O,b}}$ and $\frac{C_{R,s}}{C_{R,b}}$ to 1 considering close values of surface and bulk concentrations. It is questionable because of the conflicts with the transport equation, which is based on the concentration gradient. By considering both the diffusion and electron transfer processes, Hoover *et al.* [136] studied the current and polarization properties of U and Zr under various conditions in Mark-IV. They also reported the anode potential and system resistance with the additional element Pu [137]. Bae *et al.* [140] developed the model of REFIN. The model not only included diffusion but also electro-migration. Cumberland and Yim [114] developed a comprehensive model of ERAD based on previous REFIN code with a series of improvements, such as an anode passivation layer. The code was then applied to analyze the cyclic voltammogram, diffusion coefficient, and exchange current of UCl_3 in LiCl–KCl electrolyte [139]. Another similar model was reported by Ghosh *et al.* [139] to investigate the cyclic voltammetry, and cathodic and anodic polarization with inert electrodes and liquid Cd yielding different kinds of elements. Generally, steady state was assumed and there was no accumulation process on the electrode–electrolyte surface in these models, which means the current due to the Butler–Volmer equation is equal to the current due to diffusion. The electrode potential constitutes of both the equilibrium potential from the Nernst equation and the overpotential needed to overcome the energy barrier. Recently, commercial computational fluid dynamics (CFD) codes were applied either individually or with the 1D code as another method to assess the electrorefining performance [141–143]. Basically, the system was divided into discrete cells, and then the finite element method was applied to solve the governing equations in each cell. They broke the traditional homogeneous setting in 1D model and could plot the distribution of diffusion layers, current density, electrode potential, and streamline in three dimensions. The CFD models have the advantage of taking into account the flow conditions but generally are computationally expensive or only applicable to a specific design [114].

8.1. *Model construction*

Recently, Zhou *et al.*, developed an integrated model by combining the transport and electron transfer theory [144, 145]. Instead of solving the diffusion equation, they used Eq. (28) to describe the transport process.

The reason is that in Eq. (30), v denotes the flow velocity toward the electrode surface. However, for an electrorefiner, where generally the electrodes are rotated and the solution is constantly being stirred, the velocity toward the electrode surface is not well defined. Therefore, it is hard to take into account the flow condition when solving the mass transport equation. When using the mass transfer coefficient, its value can be related to the electrode rotation by developed correlations. For example, the most common relation used for rotation cylinder electrode is the Eisenburg equation [146]

$$K = 0.0791U^{0.7}d^{-0.3}v^{-0.344}D^{0.644} \tag{35}$$

where U is the rotational velocity ($U = \pi dw$), d is the electrode diameter, v is the kinematic viscosity of the molten salt ($v = \mu/\rho$), and μ is the dynamic viscosity. It can be concluded that for the same flow conditions and electrode, mass transfer coefficient of a species is proportional to $D^{0.644}$. The solubility of species in liquid Cd is also considered. Other factors taken into account include porous Zr layer in solid anode and fundamental data dependence on the concentration.

8.2. *Model validation*

Because of the scarce data regarding the experiments between LCA and LCC, the model was only validated in three cases, including LCA to solid cathode, solid anode to LCC and solid anode to solid cathode.

8.2.1. *Liquid anode to solid cathode*

Transport from liquid anode to solid cathode was validated by the experiments carried out by Tomczuk [147] in ANL to study the behaviors of U and Pu. Figures 23–26 show the comparison of simulation results with experimental data. The results show that the model predicts the material distribution in electrorefining cell very well. The small mismatch can be due to the 1.5 A current applied in the simulation. In the experiment, the author mentioned the current was 1–2 A but did not present the exact current pattern.

8.2.2. *Solid used fuel anode to solid cathode*

The experiment was taken from the work by Koyama *et al.* [148]. In the run #4, they used unirradiated metal alloy fuel of U–20Pu–10Zr as the anode and solid steel as the cathode.

Figure 27 shows the comparison of anode potential, which indicates that the simulation results agree well with the experimental data. There are two

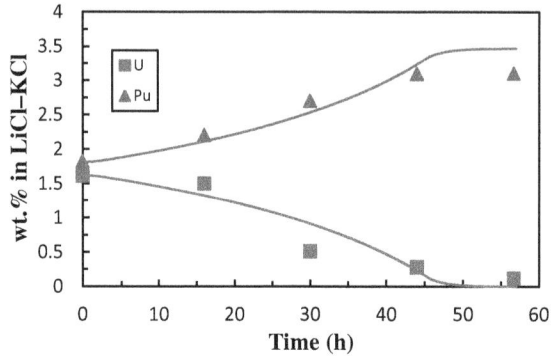

Figure 23. Wt.% of U and Pu in LiCl–KCl molten salt electrolyte.

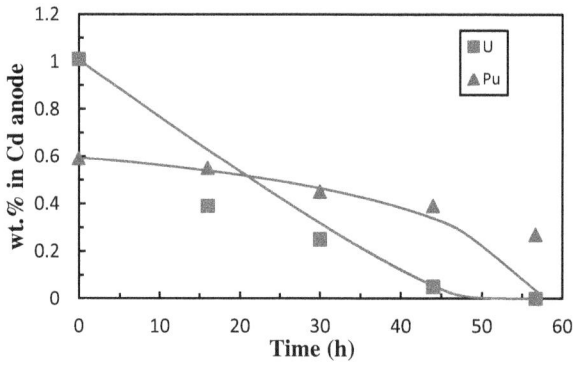

Figure 24. Wt.% of U and Pu in LCC.

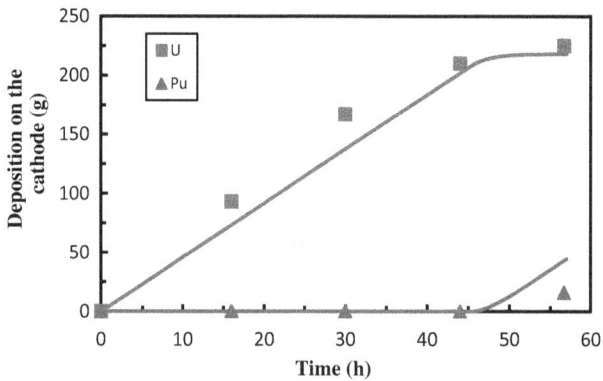

Figure 25. Amount of U and Pu deposited on the solid cathode.

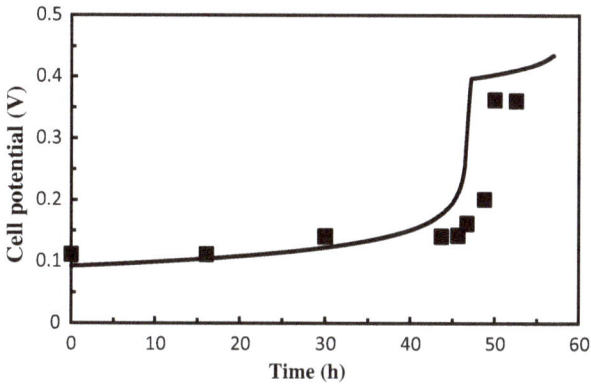

Figure 26. Cell potential of the electrorefining cell.

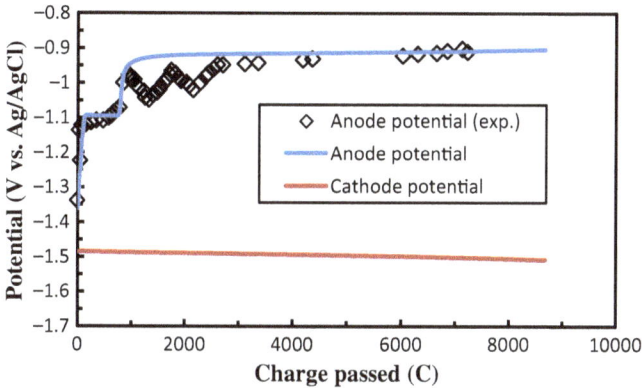

Figure 27. Anode and cathode potentials.

plateaus for the anode potential. The fluctuation between them in experiments can be caused by the non-uniform porous layer during electrorefining. The first plateau corresponds to the dissolution of U and Pu. The sudden increase represents the beginning of the dissolution of Zr.

Figures 28 and 29 plot the comparison of the amount of U, Pu, and Zr in electrolyte and solid cathode at the end of the experiments and simulation. It shows that the data of U and Pu agree quite well with the experimental results, while Zr in the electrolyte was overestimated. This makes sense when considering that low concentrations are hard to be detected accurately. Figure 28 shows increase of Pu and decrease of U in electrolyte because U and Pu both dissolved from the anode, but U dominated the deposition on the cathode. Figure 29 shows that Zr co-deposited with U

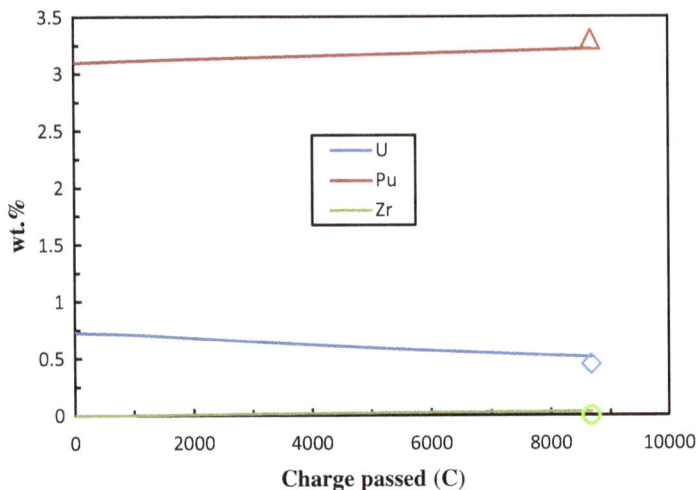

Figure 28. Concentration in the molten salt, markers represent the final concentrations of the species in experiments denoted by the line with the same color.

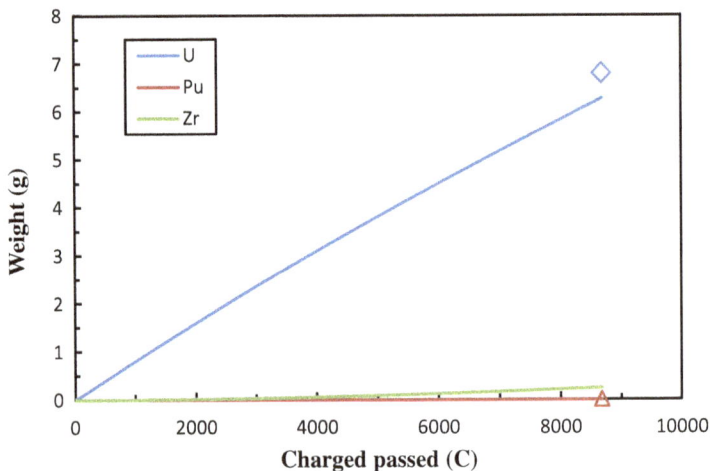

Figure 29. Weight deposited on the cathode, markers represent the final concentrations of the species in experiments denoted by the line with the same color.

due to its high redox potential but low concentration in electrolyte and no Pu was found on the cathode, which corresponds to the flat cathode potential shown in Figure 27. Considering the material balance, some Zr should be reduced on the cathode in experiments even though it was not determined because Zr dissolved from the anode while its final concentration in electrolyte decreased.

8.2.3. *Solid used fuel anode to LCC*

The transport from solid anode to liquid cathode was validated by run #2 carried out by Koyama *et al.* [148].

Figures 30–32 plot the results from the simulation. Figure 30 indicates that the potential was predicted very well. There is a sudden increase–decrease–increase pattern for the anode potential. That is due to the competition between dissolution of Zr and actinides. With the dissolution of Zr, it would deposit on the cathode simultaneously, which resulted in an increase in the cathode potential, as shown in the trailer of the plot in Figure 30. Figures 31 and 32 show the composition of electrolyte and liquid Cd at the end of the experiment and the simulation. The concentration of UCl_3 in LiCl–KCl molten salt is higher, but overall, the agreement is pretty good. What should be noted is that 6.0 wt.% deposition of Pu in liquid Cd has exceeded its solubility, as shown in Figure 33. Concentration of Pu is kept constant after its saturation, which should be taken into account in model development.

8.3. *Integrated model construction and case study*

Figure 34 shows the schematic flow of the integrated model that can be used to predict the material transport during pyroprocessing. It mainly includes

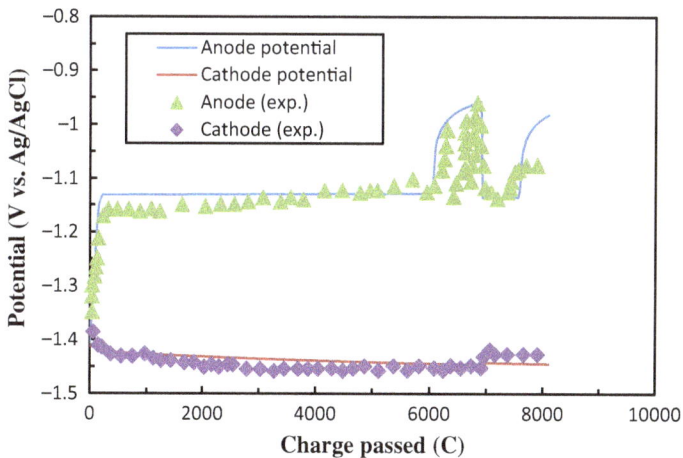

Figure 30. Anode and cathode potentials.

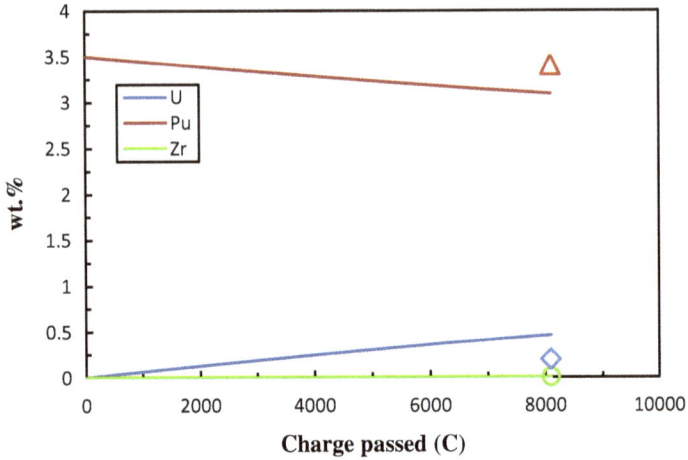

Figure 31. Concentration in the molten salt, markers represent the final concentrations of the species in experiments denoted by the line with the same color.

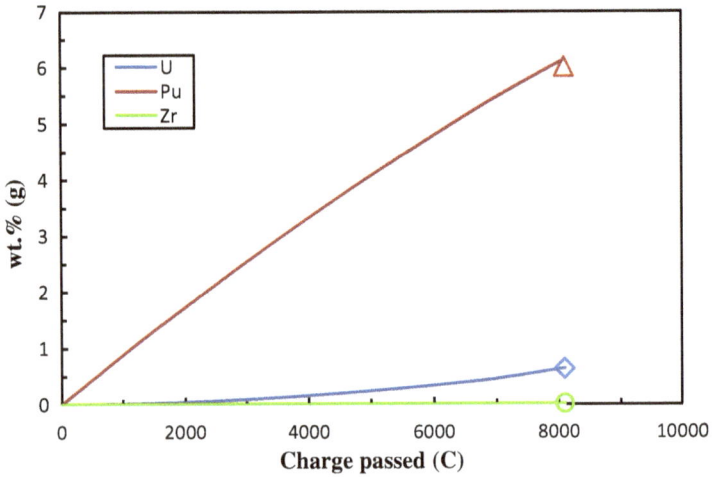

Figure 32. Wt.% in the cathode, markers represent the final concentrations of the species in experiments denoted by the line with the same color.

U separation on a solid cathode, U/TRU separation into liquid cathode, and actinide and rare earth drawdown processes.

Initially, the electrolyte composition, anode and cathode properties, current applied, flow conditions, and fundamental data are provided to conduct

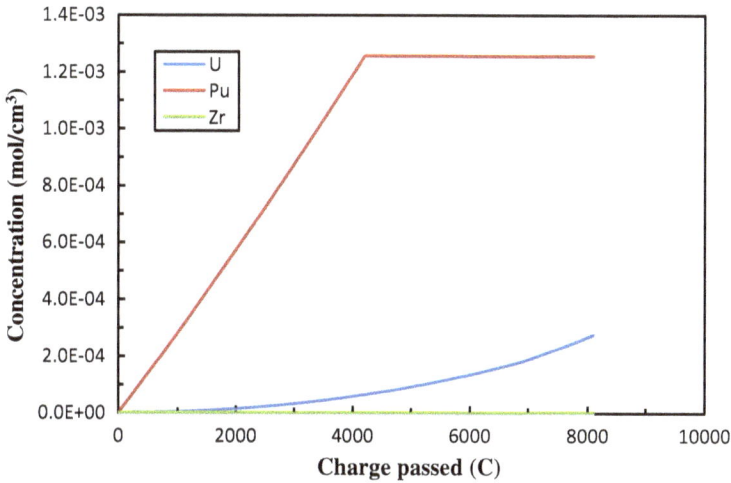

Figure 33. Concentrations of elements in liquid Cd cathode.

the electrorefining to deposit U on a solid electrode. Different criteria can be used to switch the solid electrode to liquid Cd electrode for the U/TRU separation, i.e., cycle number and $PuCl_3$ concentration. One cycle number stands for the exhaustion of U in one anode basket. After the decontamination of majority of the U and Pu, the molten salt is electrolyzed to drawdown the remaining actinides. An applied potential or required separation factors need to be provided in this step due to incomplete separation of Am without deposition of rare earth elements. Finally, the rare earth elements are collected by applying a negative enough potential after the removal of actinides. Basically, the output from the last step is used as the input of the next one. During all the simulations, essentially, dissolution fraction of anode, composition changes in electrolyte and liquid Cd, deposition amount in cathode, anode and cathode potentials, partial current of each element, and separation factors are recorded to be used as the safeguards signatures. When the measurements and signals from the processing do not follow the same route as the simulation, an alarm is triggered to alert the inspectors to stop the processing and close the material balance to detect possible diversions of U and Pu.

An example was studied as below. All the anodes were solid used fuel anodes because they are the ready-to-use form after the fuel is discharged from fast reactor or obtained by reducing the oxide fuel from LWR. Seven species, namely Zr, U, Pu, Am, Gd, Ce, and La were considered. Potential

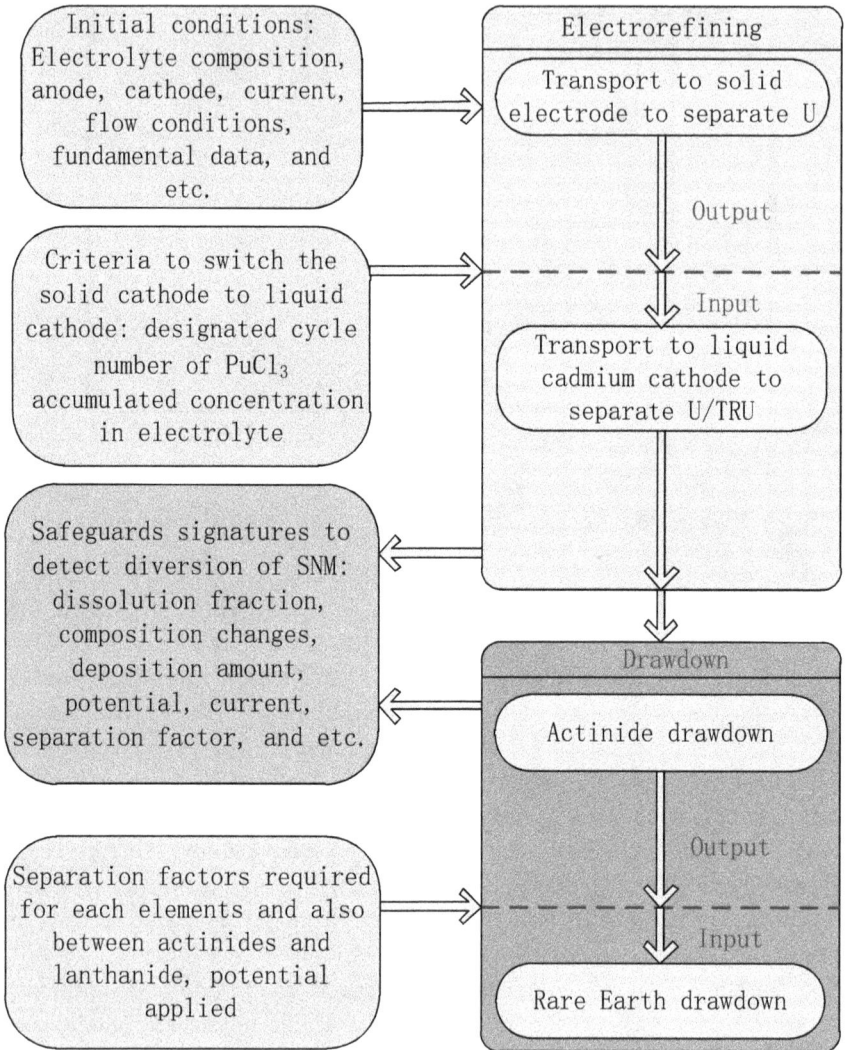

Figure 34. Schematic flow of the integrated model.

of Zr is next to the actinides. Potentials of Am and Gd are next to the lanthanides and actinides, respectively. Therefore, their preformation can represent how the NMs, actinides, and lanthanides can be separated from each other. Their wt.% in used fuel is shown in Table 5 according to the composition of EBR-II used driver fuel shown in Table 1. The Am was set to be 0.5 wt.% here.

Table 5. Composition used in this study.

Elements	Zr	U	Pu	Am	Gd	Ce	La
wt.%	10.81	80.60	0.413	0.5	0.005	0.542	0.284

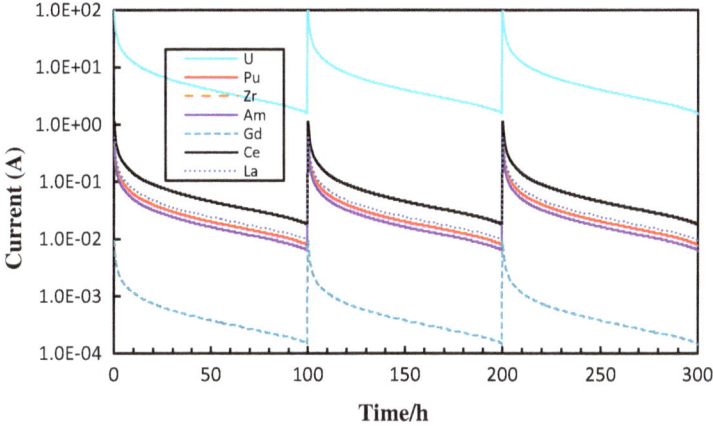

Figure 35. Anode current.

To transport U to solid cathode, the current was set as 100 A and the anode loading contains 250 fuel segments. Each segment was 9.6 g, and the total weight was 2.4 kg. Other conditions were the same as run #4 by Koyama *et al.* [148]. When the concentration of Pu in LiCl–KCl electrolyte reached 3 wt.%, the solid cathode was switched to liquid Cd cathode to co-deposit the Pu with other actinides. The process was run in three cycles with 100 h for each cycle to dissolve all the U in the anode. The initial concentration of U was 6 wt.% for each cycle. The current was decreased by 5% each time the actinide could not support the current to avoid the dissolution and deposition of Zr [145]. Figures 35 and 36 show the anode and cathode currents, respectively, in three cycles. Only U is deposited on the solid cathode by controlling the current as indicated in Figure 37. Figure 39 plots the anode and cathode potentials. Basically, they almost keep constant due to the lack of dissolution and deposition of Zr. The concentrations of species are shown in Figure 38 and are listed in Table 6, which were used as the input when switching to liquid Cd cathode.

For electrorefining using liquid Cd, in order to remove Pu as much as possible, an inert anode electrode was applied to stop the U supplement

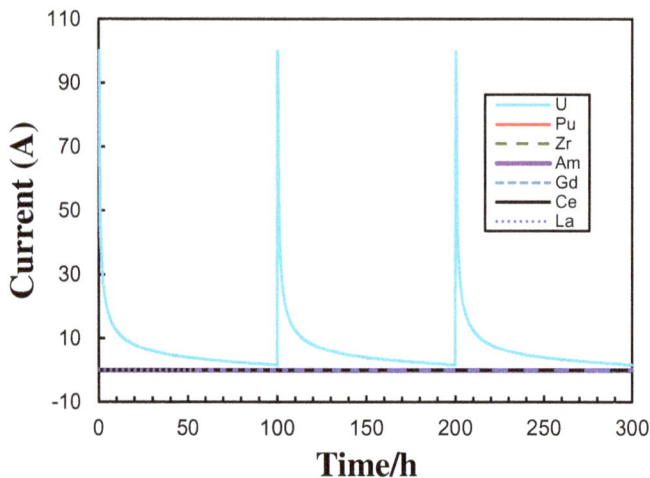

Figure 36. Cathode current (all currents are zero except that of U).

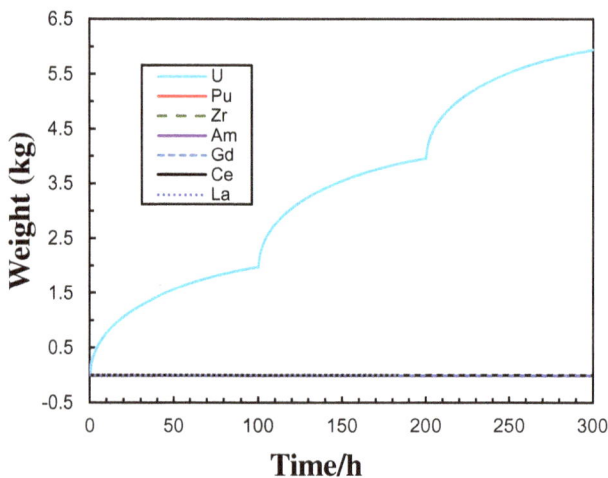

Figure 37. Deposition on the cathode (all weights are zero except that of U).

from the anode so that Pu can be removed efficiently. Here, all the conditions were set to be the same as the run #2 by Koyama [148] and the mass of Cd was 1000 g to avoid any saturation. The simulation duration was 35 h because of the rapid deposition of Ce after that. Figure 40 shows

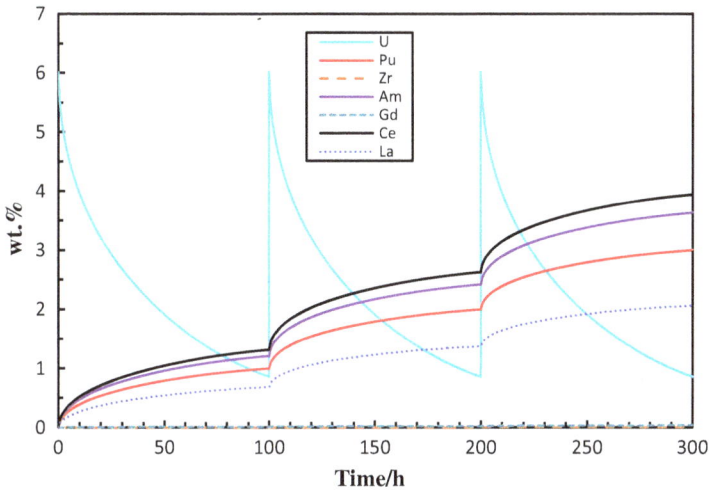

Figure 38. Concentration in electrolyte.

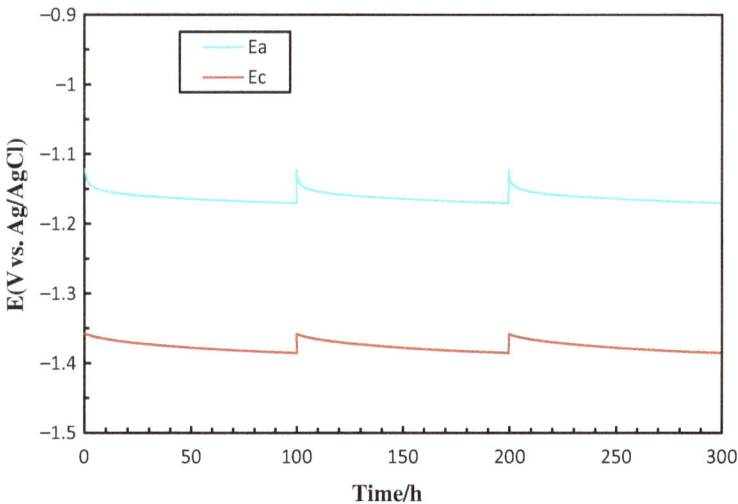

Figure 39. Anode and cathode potential.

the cathode current. It can be observed that the main deposition was the Pu, Am, and U, which can also be seen from Figure 41. Figure 42 plots the concentration of each species in molten salt electrolyte. Figure 43 is the cathode potential. It was maintained at an almost constant level because

W. Zhou and J. Zhang

Table 6. Concentration of each species in electrolyte after three cycles.

Elements	Zr	U	Pu	Am	Gd	Ce	La
wt.%	0	0.855	3.003	3.635	0.036	3.940	2.065

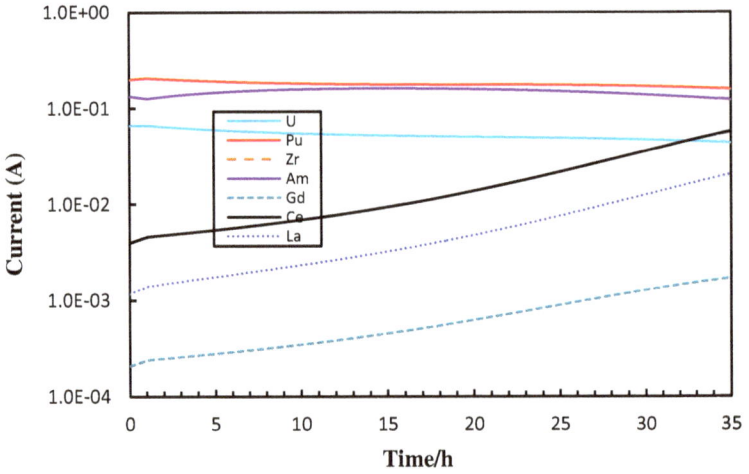

Figure 40. Cathode current.

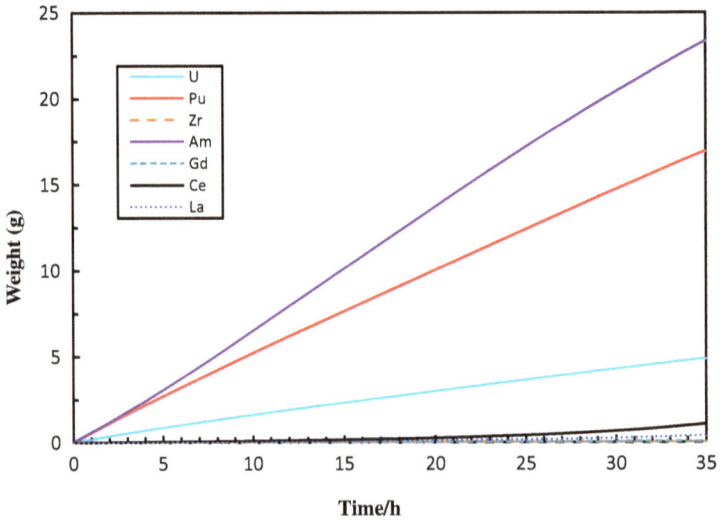

Figure 41. Deposition on the cathode.

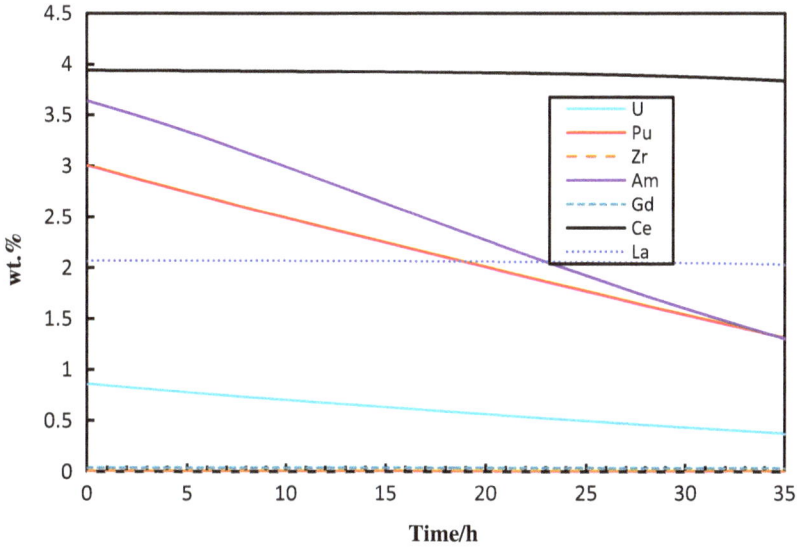

Figure 42. Concentration in electrolyte.

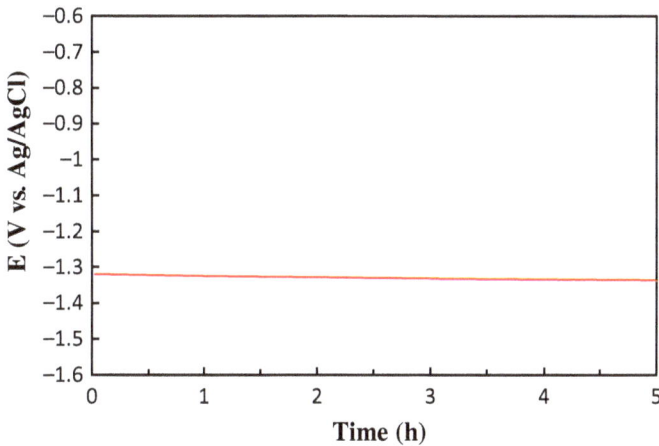

Figure 43. Cathode potential for Pu.

all the species started to deposit on the cathode at the beginning. Table 7 lists the concentrations of species at the end of the simulation.

The separation factor of each species and the composition of the deposition product are shown in Figure 44. More than half the actinides were

Table 7. Concentration of each species in the electrolyte after removal of Pu.

Elements	Zr	U	Pu	Am	Gd	Ce	La
wt.%	0	0.370	1.312	1.303	0.031	3.834	2.029

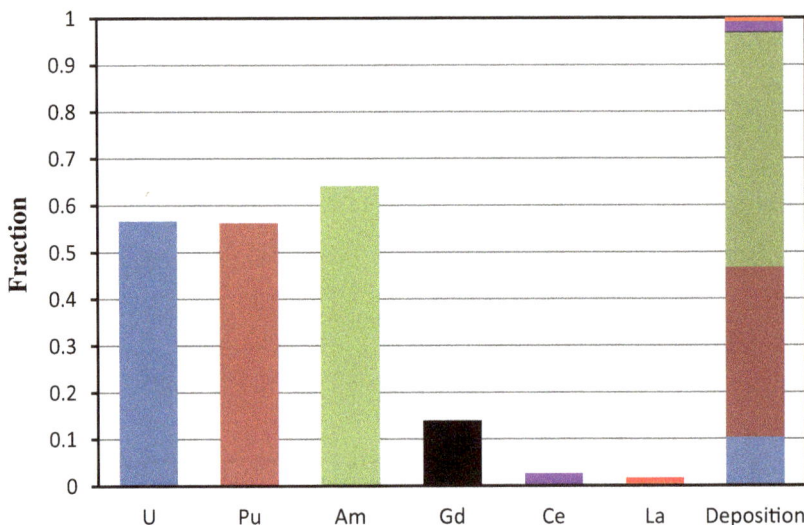

Figure 44. Separation factor of each species and the composition of deposition products.

deposited into liquid Cd cathode. The deposition product was a mixture of actinides and a small amount of lanthanide, which was around 3 wt.% totally.

Then the molten salt with lanthanides and residual actinides was electrolyzed to drawdown the actinides first. The applied potential was $-2.015\,V$ to maximize the separation of Am and minimize the separation of lanthanides. Other conditions were the same with the run #4 by Koyama [148]. Figure 45 shows the cathode current. Ce but not Gd would deposit with the actinides. That is due to the high concentration of Ce, which increases its redox potential. Figures 46 and 47 give the deposition amount on the cathode and concentration in molten salt electrolyte, respectively. Figure 48 plots the separation factor of each species and the composition of the deposition products. All the U and Pu but only 73 wt.% of Am were separated. The deposition product comprised 95% actinides and 5%

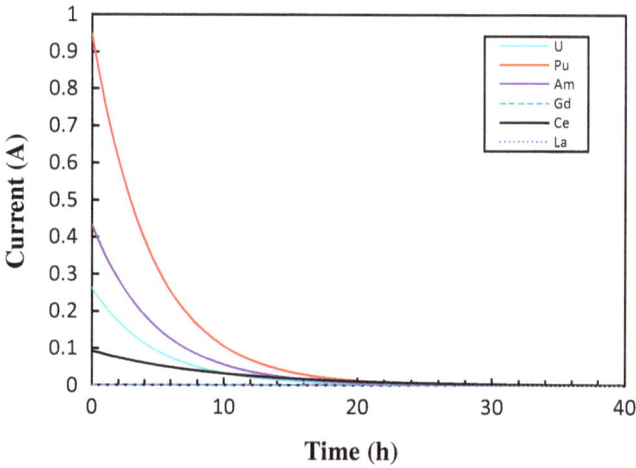

Figure 45. Cathode current.

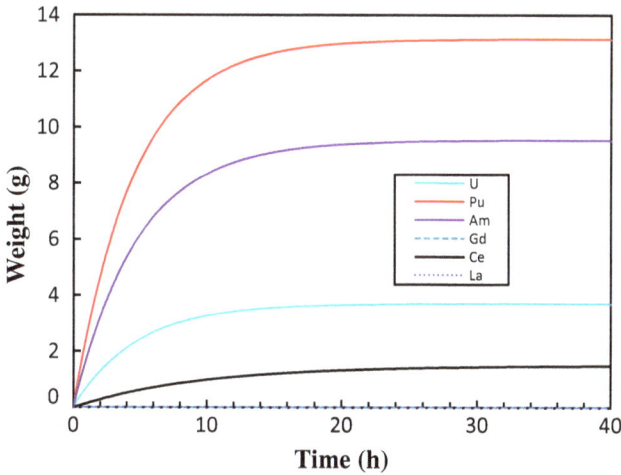

Figure 46. Deposition on the cathode.

lanthanides. Am cannot be separated completely without the deposition of lanthanides. If a more positive potential is applied, actinides and lanthanide can be separated better from each other while more Am will be left in the electrolyte. Table 8 lists the concentration of each species after the actinide drawdown which was used as the input for the lanthanide drawdown.

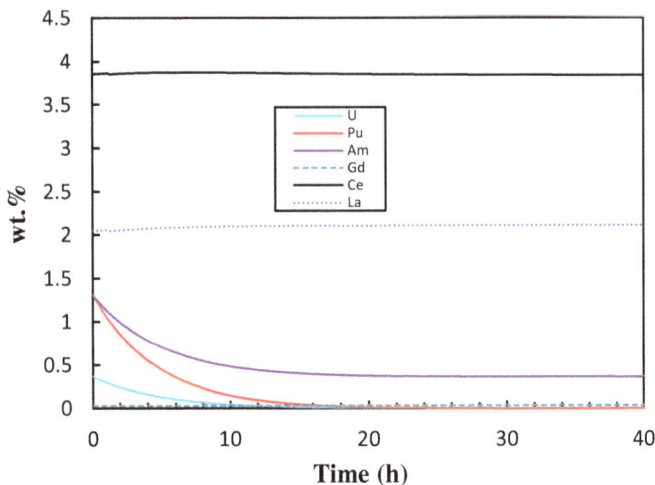

Figure 47. Concentrations in molten salt electrolyte.

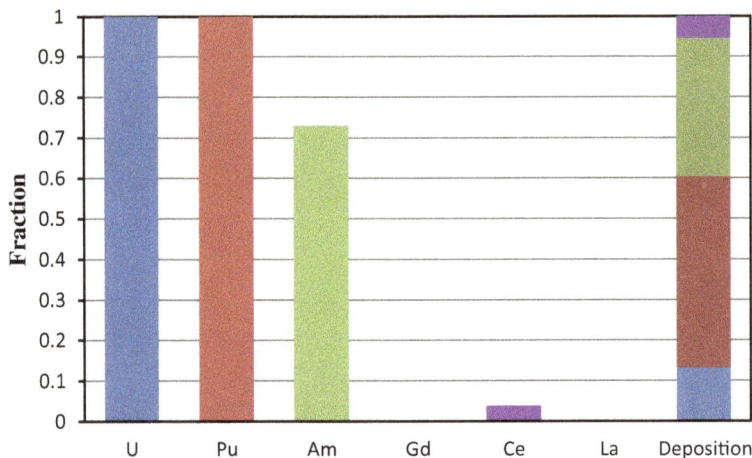

Figure 48. Separation factor of each species and the composition of deposition products.

For the lanthanide drawdown, a negative enough potential of $-2.38\,\text{V}$, which is just above the Li^+/Li reduction potential, was applied [149]. All other conditions were the same as the work by Koyama [148]. Figure 49 shows the cathode current. Figures 50 and 51 give the deposition amount on the cathode and concentration in the electrolyte. Basically, all the remaining Am, Gd, Ce, and La were separated completely from the molten salt.

Table 8. Concentration of each species after the actinide drawdown.

Elements	Zr	U	Pu	Am	Gd	Ce	La
wt.%	0	0	0	0.350	0.031	3.685	2.029

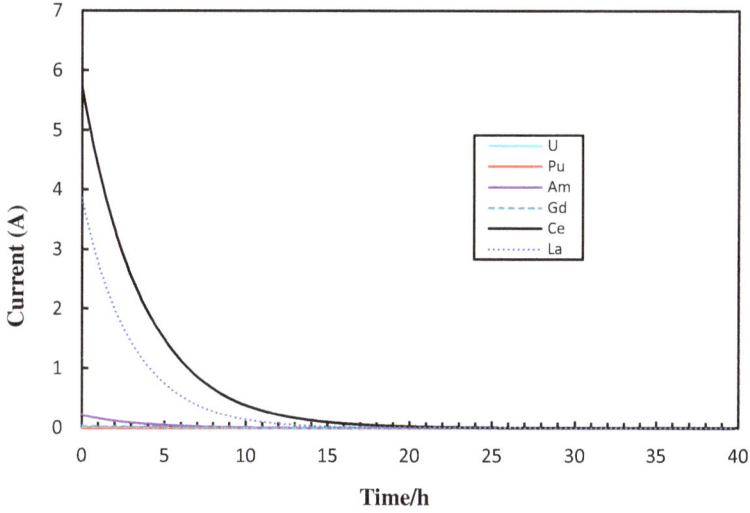

Figure 49. Cathode current.

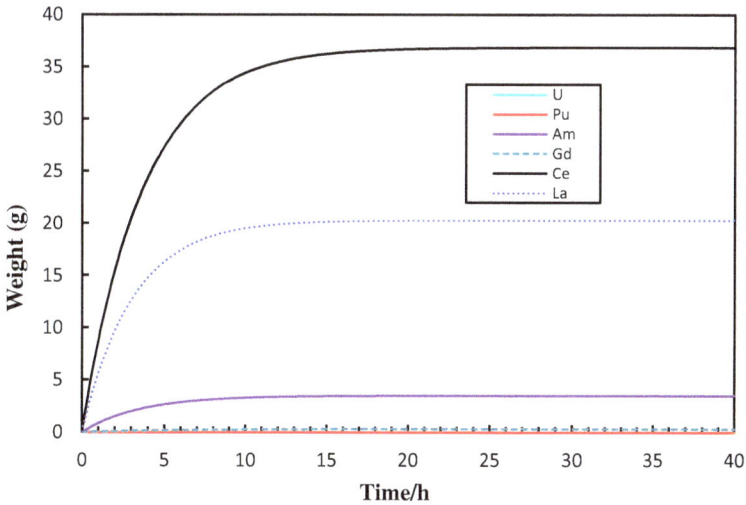

Figure 50. Deposition on the cathode.

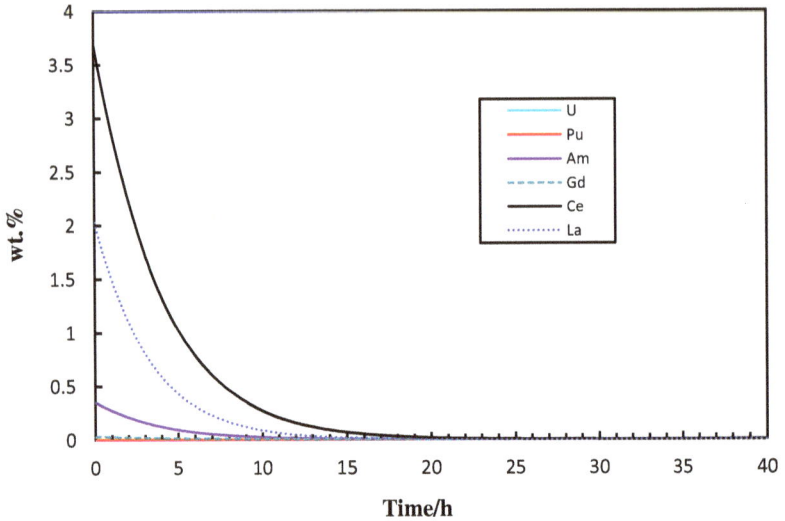

Figure 51. Concentration in the molten salt electrolyte.

9. Summary

How to safeguard a pyroprocessing facility is one of the main barriers preventing it from being maturely developed and commercialized. Due to the special features and designs of pyroprocessing, previous technologies developed for the aqueous separation process may not be readily applicable. Basically, for the safeguards of the pyroprocessing facility, one needs to develop accurate methods to measure the Pu in SNF and online assay techniques to measure inventory of Pu in the electrorefiner. Knowledge of the initial amount of Pu in SNF is crucial to the material accountancy in downstream processes. With it, the Pu can probably be tracked by the gross neutron detection integrated with the DA or NDA on U/TRU product, but the premise of similar properties between Cm and Pu has to be verified first. For the material balance in electrorefiner, the electrochemical methods have the advantage of being readily applicable due to the intrinsic properties of electrorefining. In addition, kinetic model integrating transport and electrochemical processes can also be a promising method. It not only has the ability to safeguard the pyroprocessing but can also monitor the separation process and instruct the operation of pyroprocessing facilities via its output signals, which mainly include the potential, current, and species concentration. Potential determines the quality of the deposition on the cathode. Sudden decrease or increase of the potential at the cathode indicates new element deposition and may serve as a signal of SNM diversion. By adjusting the potential, purity of the deposition can be controlled. Current can be used to measure the quantity of deposition by integrating it with respect to the time. So the deposition amount and throughput can be predicted. Current is also an indicator of material transport and can be used to verify the material distribution obtained by other methods. Species concentration in molten salt is the direct variable to close the material balance. Monitoring of it is the most straightforward method to detect any scenario of SNM diversion to safeguard the pyroprocessing facilities. However, for a more accurate prediction and monitoring, fundamental data of elements in LiCl–KCl molten salts needs further understanding, especially by considering the effects of solute concentration and presence of other elements. Briefly, various new approaches have been proposed and investigated especially for the material accountancy in electrorefiner, but most of them are still under preliminary study. More efforts are still merited to demonstrate their accuracy and feasibility, especially in harsh environments and multicomponent systems.

References

1. U.S. Department of Energy, Office of Nuclear Energy, Science and Technology, The history of nuclear energy, No. DOE/NE-0088 (1995).
2. R. Rhodes, *Making of the Atomic Bomb*, Simon and Schuster, pp. 251–254 (2012).
3. L. Meitner and O. R. Frisch, Disintegration of uranium by neutrons: a new type of nuclear reaction, *A Century of Nature: Twenty-One Discoveries that Changed Science and the World*, University of Chicago Press, Chicago IL, p. 70 (2003).
4. E. Fermi, Fermi's own story, *Chicago Sun-Times*, 21–26 (November 23, 1952).
5. E. Fermi, The development of the first chain reacting pile, *Proc. Am. Philosoph. Soc.* **90**(1), 20–24 (1946).
6. U.S. Department of Energy, The first reactor, December (1982).
7. Nuclear technology review 2016, Report by the Director General, GC(60)//INF/2, June (2016).
8. Nuclear energy data 2016, OECD & NEA, NEA No. 7300 (2016).
9. Power reactors, NRC, https://www.nrc.gov/reactors/power.html, retrieved at February (2017).
10. Fuel, spent nuclear. Accumulating quantities at commercial reactors present storage and other challenges, GAO-12-797, US Government Accountability Office, 2012.
11. Spent fuel reprocessing options. IAEA-TECDOC-1587, August (2008).

12. K. Tucck, Neutronic and burnup studies of accelerator-driven systems dedicated to nuclear waste transmutation, Ph.D. Thesis, Royal Institute of Technology, Diss. Fysik, 2004.

13. P. F. Peterson, The pros and cons of nuclear fuel recycling, *Science* **294**(5549), 2093–2094 (2001).

14. R. S. Hartenberg (ed.), *National Historic Mechanical Engineering Landmarks*, American Society of Mechanical Engineers, New York, NY (1979).

15. DOE, US. Plutonium the first 50 years, Washington, DC, February (1966).

16. The Hanford Site: An anthology of early histories, Energyunder Contract DE-ACO6-STRL10930, US Department (1993).

17. Bismuth phosphate process for the separation of plutonium from aqueous solutions, U.S. Patent 2,785,951, issued March 19 (1957).

18. C. C. Kelly (ed.), *Remembering the Manhattan Project: Perspectives on the Making of the Atomic Bomb and its Legacy*, World Scientific, Singapore, p. 146 (2005).

19. T. F. Evans and R. E. Tomlinson, Hot semiworks REDOX Studies. No. HW-31767, General Electric Co., Richland, WA, USA, Hanford Atomic Products Operation (1954).

20. R. E. Gephart, A short history of hanford waste generation, storage and release, Technical Report PNNL-13605, Rev. 4, Pacific Northwest National 360 Laboratory (2003).

21. R. Jubin, Spent fuel reprocessing, Introduction to Nuclear Chemistry and Fuel Cycle Separations Course, Consortium for Risk Evaluation With Stakeholder Participation, http://www.cresp.org/education/courses/shortcourse (2008).

22. L. W. Gray, From separations to reconstitution — A short history of Plutonium in the US and Russia, Document ITT99-017, UCRL-JC-133802, Lawrence Livermore National Laboratory, Livermore, CA 94551, pp. 21–25 (1999).

23. M. F. Simpson and J. D. Law, Nuclear fuel, reprocessing of, In *Nuclear Energy: Selected Entries from the Encyclopedia of Sustainability Science and Technology*, Tsoulfanidis, N., ed. Springer Science & Business Media, pp. 153–174 (2012).

24. A. Andrews, Nuclear fuel reprocessing: US policy development, Library of Congress Washington DC Congressional Research Service (2006).

25. C. Jimmy, Nuclear power policy statement on decisions reached following a review, April 7 (1977).

26. N. Xoubi, The politics, science, environment, and common sense of spent nuclear fuel reprocessing, 3 Decades Later, In *Symp. Technology of Peaceful Nuclear Energy*, Irbid, Jordan, October 14–16 (2008).

27. J. D. Werner, US spent nuclear fuel storage, Congressional Research Service, Library of Congress (2012).

28. President signs Yucca Mountain bill, Statement by the Press Secretary, The White House, July 23 (2002).

29. Restarting the Yucca Mountain Project: The Case For and Against. Bipartisan Policy Center, August 15, 2015.

30. Waste, Commercial Nuclear. Effects of a Termination of the Yucca Mountain Repository Program and Lessons Learned. GAO-11-229, US Government Accountability Office, 2011.

31. L. J. Koch, *Experimental Breeder Reactor II: An Integrated Experimental Fast Reactor Nuclear Power Station*, La Grange Park, IL: American Nuclear Society, pp. 1–8 (2008).

32. C. E. Stevenson, The EBR-II fuel cycle story (1987).

33. T. Todd, Dry (non-aqueous) processing of used nuclear fuel, *Waste Management 2012* (2012).

34. G. F. Kessinger, Evaluation of the electrorefining technique for the processing of radioactive scrap metals. No. WINCO-1165. Westinghouse Idaho Nuclear Co., Inc., Idaho Falls, ID (United States), 1993.

35. Electrometallurgical techniques for DOE spent fuel treatment. *Committee on Electrometallurgical Techniques*, National Academies Press, Washington, DC (2000).

36. S. X. Li *et al.*, Electrorefining Experience for Pyrochemical Reprocessing of Spent EBR-II Driver Fuel. No. INL/CON-05-00305. Idaho National Laboratory (INL), (2005).

37. M. F. Simpson, Developments of spent nuclear fuel pyroprocessing technology at Idaho national laboratory. Idaho National Laboratory, INL/EXT-12-25124, Idaho Falls, ID, USA, 2012.

38. K.-C. Song *et al.*, Status of pyroprocessing technology development in Korea, *Nuclear Eng. Technol.* **42**(2), 131–144 (2010).

39. T. Inoue, T. Koyama, and Y. Arai, State of the art of pyroprocessing technology in Japan, *Energy Procedia* **7**, 405–413 (2011).

40. H. Boussier, R. Malmbeck, and G. Marucci, Pyrometallurgical Processing Research Programme "PYROREP". Final technical report. European commission: Nuclear science and technology, Louxembourg (2003).

41. Y. I. Chang, The integral fast reactor, *Nuclear Technol.* **88**(2), 129–138 (1989).

42. W. E. Miller and Z. Tomczuk, Pyroprocess for processing spent nuclear fuel. U.S. Patent No. 6,461,576. 8 October (2002).

43. Y. Chang, Il, Integral fast reactor and associated fuel cycle system: Part 3. Status of pyroprocess development, IAEA/ICTP school on physics and technology of fast reactor systems, Trieste, Italy, November 9–20, 2009.

44. E.-Y. Choi and S. M. Jeong, Electrochemical processing of spent nuclear fuels: An overview of oxide reduction in pyroprocessing technology, *Prog. Nat. Sci.: Mater. Int.* **25**(6), 572–582 (2015).

45. T. Koyama *et al.*, An experimental study of molten salt electrorefining of uranium using solid iron cathode and liquid cadmium cathode for development of pyrometallurgical reprocessing, *J. Nucl. Sci. Technol.* **34**(4) 384–393 (1997).

46. S. X. Li and M. F. Simpson, Anodic process of electrorefining spent nuclear fuel in molten LiCl-KCl-UCl3/Cd system, *Meet. Electrochem. Soc.* **201**, 1–13 (2002).

47. D. Yoon, Electrochemical studies of cerium and uranium in LiCl-KCl eutectic for fundamentals of pyroprocessing technology. Ph.D. Thesis, Virginia Commonwealth University, Richmond, VA, 2016.

48. R. O. Hoover, Uranium and zirconium electrochemical studies in LiCl-KCl eutectic for fundamental applications in used nuclear fuel reprocessing. Ph.D. Thesis, University of Idaho, Idaho Falls, ID, 2014.

49. Y. Sakamura *et al.*, Measurement of standard potentials of actinides (U, Np, Pu, Am) in LiCl–KCl eutectic salt and separation of actinides from rare earths by electrorefining. *J. Alloys Compounds* **271**, 592–596 (1998).

50. D. Morgan and J. Eapen, Modeling Solute Thermokinetics in LiCl-KCl Molten Salt for Nuclear Waste Separation. Final report for project No. 10-938, Nuclear Energy University Programs, U.S. Department of Energy (2013).

51. T. Inoue and T. Koyama, An overview of CRIEPI pyroprocessing activities, In *Actinide and Fission Product Partitioning and Transmutation*, Ninth Information Exchange Meeting, Nimes, France, September 25–29, p. 187 (2006).

52. J.-H. Jang *et al.*, Development of continuous ingot casting process for uranium dendrites in pyroprocess, *J. Radioanal. Nuclear Chem.* **295**(3), 1743–1751 (2013).

53. M. A. Williamson and J. L. Willit, Pyroprocessing flowsheets for recycling used nuclear fuel, *Nuclear Eng. Technol.* **43**(4), 329–334 (2011).

54. G. Kessler, *Sustainable and Safe Nuclear Fission Energy: Technology and Safety of Fast and Thermal Nuclear Reactors*. Springer-Verlag, Berlin, Heidelberg, p. 255 (2012).

55. F. Gao *et al.*, Criticality safety evaluation of materials concerning pyroprocessing, *J. Nuclear Sci. Technol.* **48**(6), 919–928 (2011).

56. M. Iizuka *et al.*, Development of plutonium recovery process by molten salt electrorefining with liquid cadmium cathode. 6th information exchange meeting on actinide and fission product P&T, Madrid, Spain, 2000.

57. INFCIRC, IAEA. "153." The structure and content of agreements between the agency and states required in connection with the treaty on the non-proliferation of nuclear weapons, 1971.

58. IAEA, *IAEA Safeguards Glossary*, 2001 Edition, Vienna, Austria, June 2002 (2001).

59. W. Bekiert *et al.*, Safeguarding Pyroprocessing Related Facilities in the ROK (IAEA-CN–220). International Atomic Energy Agency (IAEA) (2015).

60. B. B. Cipiti *et al.*, Modeling and design of integrated safeguards and security for an electrochemical reprocessing facility. Sandia National Laboratories, SAND2012-9303, Albuquerque, NM, USA, 2012.

61. D. Rappleye, Developing safeguards for pyroprocessing: Detection of a plutonium co-deposition on solid cathode in an electrorefiner by applying the signature-based safeguards approach. Master Thesis, North Carolina State University, Raleigh, North Carolina, 2012.

62. R. Wigeland, T. Bjornard, and B. Castle, The concept of goals-driven safeguards. No. INL/EXT-09-15511. Idaho National Laboratory (INL), 2009.

63. J. Zhang, Kinetic model for electrorefining, part II: Model applications and case studies, *Prog. Nuclear Energy* **70**, 287–297 (2014).

64. P. Lafreniere, Identification of electrorefiner and cathode processing failure modes and determination of signature-significance for integration into a signature based safeguards framework for pyroprocessing, The University of New Mexico, Albuquerque, New Mexico, 2015.

65. P. C. Durst *et al.*, Advanced safeguards approaches for new reprocessing facilities. No. PNNL-16674. Pacific Northwest National Laboratory (PNNL), Richland, WA, USA, 2007.

66. K. Budlong-Sylvester *et al.*, International safeguards for pyroprocessing: options for evaluation. LA-UR-03-0986, January 24, 2003 (Los Alamos, NM: Los Alamos National Laboratory, 2003).

67. P. Masset *et al.*, Electrochemistry of uranium in molten licl-kcl eutecticm, *J. Electrochem. Soc.* **152**(6), A1109–A1115 (2005).

68. S. A. Kuznetsov *et al.*, Electrochemical behavior and some thermodynamic properties of UCl4 and UCl3 dissolved in a LiCl-KCl eutectic melt, *J. Electrochem. Soc.* **152**(4), C203–C212 (2005).

69. P. Masset *et al.*, Thermochemical properties of lanthanides (Ln= La, Nd) and actinides (An= U, Np, Pu, Am) in the molten LiCl–KCl eutectic, *J. Nucl. Mater.* **344**(1), 173–179 (2005).

70. O. Shirai, H. Yamana, and Y. Arai, Electrochemical behavior of actinides and actinide nitrides in LiCl–KCl eutectic melts, *J. Alloys Compounds* **408**, 1267–1273 (2006).

71. J. J. Roy, *et al.* Thermodynamic properties of U, Np, Pu, and Am in molten LiCl–KCl eutectic and liquid cadmium, *J. Electrochem. Soc.* **143**(8), 2487–2492 (1996).

72. R. O. Hoover, *et al.*, Electrochemical studies and analysis of 1–10wt.% UCl 3 concentrations in molten LiCl–KCl eutectic, *J. Nucl. Mater.* **452**(1), 389–396 (2014).

73. O. Shirai *et al.*, Electrode reaction of the U 3+/U couple at liquid Cd and Bi electrodes in LiCl–KCl eutectic melts, *Anal. Sci./Suppl.* **17**, i959–i962 (2002).

74. S. Ghosh *et al.*, Experimental investigations and thermodynamic modelling of KCl–LiCl–UCl 3 system, *Calphad* **45**, 11–26 (2014).

75. D. Inman and J. O'M. Bockris, The reversible electrode potential of the system U/UCl3 in molten chloride solvents, *Canad. J. Chem.* **39**(5), 1161–1163 (1961).

76. D. M. Gruen and R. A. Osteryoung, Measuremekt of the uranium–uranium (III) potential in LiCl–KCl eutectic, *Ann. NY Acad. Sci.* **79**(1), 897–907 (1960).

77. D. Inman *et al.*, Electrode reactions in molten salts: the uranium+ uranium trichloride system, *Trans. Faraday Soc.* **55**, 1904–1914 (1959).

78. S. A. Kuznetsov *et al.*, Determination of uranium and rare-earth metals separation coefficients in LiCl–KCl melt by electrochemical transient techniques, *J. Nucl. Mater.* **344**(1), 169–172 (2005).

79. L. Martinot, Some thermodynamic properties of dilute solutions of actinide chlorides in (Li–K) Cl and in (Na–K) Cl eutectics, *J. Inorganic Nucl. Chem.* **37**(12), 2525–2528 (1975).

80. F. Caligara, L. Martinot, and G. Duyckaerts, Chronopotentiometric determination of U (III), U (IV), UO2 (VI) and Np (IV) in molten LiCl–KCl eutectic, *J. Electroanal. Chem. Interf. Electrochem.* **16**(3), 335–340 (1968).

81. B. P. Reddy *et al.*, Electrochemical studies on the redox mechanism of uranium chloride in molten LiCl–KCl eutectic, *Electrochimica Acta* **49**(15), 2471–2478 (2004).

82. Yamada, D. *et al.*, Diffusion behavior of actinide and lanthanide elements in molten salt for reductive extraction, *J. Alloys Compounds* **444**, 557–560 (2007).

83. T. Kobayashi *et al.*, Polarization effects in the molten salt electrorefining of spent nuclear fuel, *J. Nucl. Sci. Technol.* **32**(7), 653–663 (1995).

84. M. M. Tylka *et al.*, Application of voltammetry for quantitative analysis of actinides in molten salts, *J. Electrochem. Soc.* **162**(12), H852–H859 (2015).

85. J. Serp *et al.*, Electrochemical behaviour of plutonium ion in LiCl–KCl eutectic melts, *J. Electroanal. Chem.* **561**, 143–148 (2004).

86. Y. Sakamura *et al.*, Distribution behavior of plutonium and americium in LiCl–KCl eutectic/liquid cadmium systems, *J. Alloys Compounds* **321**(1), 76–83 (2001).

87. O. Shirai, M. Iizuka, and I. W. A. I. Takashi, Electrode reaction of Pu3+/Pu couple in LiCl–KCl eutectic melts: Comparison of the electrode reaction at the surface of liquid Bi with that at a solid Mo electrode, *Analy. Sci.* **17**(1), 51–57 (2001).

88. D. A. Nissen, Electrochemistry of plutonium (III) in molten alkali chlorides, *J. Inorganic Nucl. Chem.* **28**(8), 1740–1743 (1966).

89. L. Martinot and G. Duyckaerts. Electrochemistry of Pu (III) in molten LiCl–KCl eutectic, *Anal. Lett.* **4**(1), 1–11 (1971).

90. I. Choi *et al.*, Determination of Exchange Current Density of U3+/U Couple in LiCl–KCl Eutectic Mixture. In *Proceedings of the GLOBAL 2009 Congress — The Nuclear Fuel Cycle: Sustainable Options and Industrial Perspectives*, France, p. 567 (2009).

91. M. A. Rose, M. A. Williamson, and J. Willit, Determining the exchange current density and Tafel constant for uranium in LiCl/KCl eutectic, *ECS Electrochem. Lett.* **4**(1), C5–C7 (2015).

92. K. H. Lim, S. Park, and J.-Il Yun, Study on exchange current density and transfer coefficient of uranium in LiCl-KCl molten salt, *J. Electrochem. Soc.* **162**(14), E334–E337 (2015).

93. S. Ghosh *et al.*, Anodic dissolution of U, Zr and U–Zr alloy and convolution voltammetry of Zr 4+|Zr 2+ couple in molten LiCl–KCl eutectic, *Electrochimica Acta* **56**(24), 8204–8218 (2011).

94. P. Bagri and M. F. Simpson, Determination of activity coefficient of lanthanum chloride in molten LiCl–KCl eutectic salt as a function of cesium chloride and lanthanum chloride concentrations using electromotive force measurements, *J. Nucl. Mater.* **482**, 248–256 (2016).

95. W. Zhou and J. Zhang, Direct calculation of concentration-dependent activity coefficient of UCl3 in molten LiCl–KCl, *J. Electrochem. Soc.* **162**(10), E199–E204 (2015).

96. W. Zhou and J. Zhang, Chemical diffusion coefficient calculation of U 3+ in LiCl-KCl molten salt, *Prog. Nucl. Energy* **91**, 170–174 (2016).

97. Y. Wang, W. Zhou, and J. Zhang, Investigation of concentration-dependence of thermodynamic properties of lanthanum, yttrium, scandium and terbium in eutectic LiCl–KCl molten salt, *J. Nucl. Mater.* **478**, 61–73 (2016).

98. P. Bagri and M. F. Simpson, Activity measurements of gadolinium (III) chloride in molten LiCl–KCl eutectic salt using saturated Gd/GdCl3 reference electrode, *J. Electrochem. Soc.* **164**(8), H5299–H5307 (2017).

99. M. F. Simpson, Process monitoring of molten salt-based nuclear fuel processing systems based on electrochemistry. Seminar in Ohio State University, March 9, 2016.

100. D. W. Hahn and N. Omenetto, Laser-induced breakdown spectroscopy (LIBS), Part I: Review of basic diagnostics and plasma–particle interactions: Still-challenging issues within the analytical plasma community, *Appl. spectr.* **64**(12), 335A–366A (2010).

101. D. W. Hahn and N. Omenetto, Laser-induced breakdown spectroscopy (LIBS), Part II: Review of instrumental and methodological approaches to material analysis and applications to different fields, *Appl. Spectr.* **66**(4), 347–419 (2012).

102. A. W. Miziolek, V. Palleschi, and I. Schechter (eds.), *Laser Induced Breakdown Spectroscopy*. Cambridge University Press, New York, NY (2006).

103. T. Hussain and M. A. Gondal, Laser induced breakdown spectroscopy (LIBS) as a rapid tool for material analysis, *J. Phys.: Conf. Ser.* IOP Publishing, **439**(1) 012050 (2013).

104. C. M. Davies *et al.*, Quantitative analysis using remote laser-induced breakdown spectroscopy (LIBS), *Spectrochimica Acta Part B: Atomic Spectroscopy* **50**(9), 1059–1075 (1995).

105. B. C. Castle *et al.*, Variables influencing the precision of laser-induced breakdown spectroscopy measurements, *Appl. Spectr.* **52**(5), 649–657 (1998).

106. A. I. Whitehouse *et al.*, Remote compositional analysis of spent-fuel residues using laser-induced breakdown spectroscopy, *Waste Manage.* **3**, 23–27 (2003).

107. N. A. Smith, J. A. Savina, and M. A. Williamson, Application of laser induced breakdown spectroscopy to electrochemical process monitoring of molten chloride salts, In *Symp. Int. Safeguards: Linking Strategy, Implementation and People*, Vienna, 2014.

108. A. Weisberg *et al.*, Measuring lanthanide concentrations in molten salt using laser-induced breakdown spectroscopy (LIBS), *Appl. Spectr.* **68**(9), 937–948 (2014).

109. P. M. Rinard and H. O. Menlove, Application of curium measurements for safeguarding at reprocessing plants. Study 1: High-level liquid waste and Study 2: Spent fuel assemblies and leached hulls. No. LA-13134-MS. Los Alamos National Lab., NM (United States), 1996.

110. H. R. Trellue *et al.*, Neutron balance for integrated safeguards of a reprocessing facility, *Trans. Am. Nucl. Soc.* **104**, 181 (2011).

111. J. Zhang, Electrochemistry of actinides and fission products in molten salts — Data review, *J. Nucl. Mater.* **447**(1), 271–284 (2014).

112. R. Bean, Project report on development of a safeguards approach for pyroprocessing, Idaho National Laboratory, INL/EXT-10-20057, Idaho Falls, ID, USA, 2010.

113. M. Gonzalez *et al.*, Application of a one-dimensional transient electrorefiner model to predict partitioning of plutonium from curium in a pyrochemical spent fuel treatment process, *Nucl. Technol.* **192**(2), 165–171 (2015).

114. R. M. Cumberland and M.-S. Yim, Development of a 1D transient electrorefiner model for pyroprocess simulation, *Ann. Nucl. Energy* **71**, 52–59 (2014).

115. P. Delahay, High-frequency methods, *New Instrumental Methods in Electrochemistry: Theory, Instrumentation, and Applications to Analytical and Physical Chemistry*. Interscience Publishers, New York, NY (1954).

116. T. Berzins and P. Delahay, Oscillographic polarographic waves for the reversible deposition of metals on solid electrodes, *J. Am. Chem. Soc.* **75**(3), 555–559 (1953).

117. F. G. Cottrell, Residual current in galvanic polarization, regarded as a diffusion problem, *Z. Phys. Chem.* **42**, 385–431 (1903).

118. H. J. S. Sand, III. On the concentration at the electrodes in a solution, with special reference to the liberation of hydrogen by electrolysis of a mixture of copper sulphate and sulphuric acid, *The London, Edinburgh, and Dublin Philosoph. Magaz. J. Sci.* **1**(1), 45–79 (1901).

119. P. Masset *et al.*, Electrochemistry of uranium in molten LiCl-KCl eutectic, *J. Electrochem. Soc.* **152**(6), A1109–A1115 (2005).

120. M. Iizuka *et al.*, Application of normal pulse voltammetry to on-line monitoring of actinide concentrations in molten salt electrolyte, *J. Nucl. Mater.* **297**(1), 43–51 (2001).

121. A. J. Bard *et al.*, *Electrochemical methods: fundamentals and Applications*. Vol. 2, Wiley, New York (1980).

122. D.-H. Kim *et al.*, Real-time monitoring of metal ion concentration in LiCl–KCl melt using electrochemical techniques, *Microchem. J.* **114**, 261–265 (2014).

123. S. X. Li *et al.*, Integrated efficiency test for pyrochemical fuel cycles, *Nucl. Technol.* **166**(2), 180–186 (2009).

124. M. Pałys *et al.*, The separation of overlapping peaks in cyclic voltammetry by means of semi-differential transformation, *Talanta* **38**(7), 723–733 (1991).

125. D. Rappleye, S.-M. Jeong, and M. Simpson, Application of multivariate analysis techniques to safeguards of the electrochemical treatment of used nuclear fuel, *Ann. Nucl. Energy* **77**, 265–272 (2015).

126. I. Johnson, The thermodynamics of pyrochemical processes for liquid metal reactor fuel cycles, *J. Nucl. Mater.* **154**(1), 169–180 (1988).

127. J. P. Ackerman and T. R. Johnson, Partition of actinides and fission products between metal and molten salt phases — Theory, measurement and application to IFR pyroprocess development, *Int. Conf. Actinides.* **93** (1993).

128. J. P. Ackerman and J. L. Settle, Distribution of plutonium, americium, and several rare earth fission product elements between liquid cadmium and LiCl–KCl eutectic, *J. alloys Compounds* **199**(1–2), 77–84 (1993).

129. H. P. Nawada and N. P. Bhat, Thermochemical modelling of electrotransport of uranium and plutonium in an electrorefiner, *Nucl. Eng. Design* **179**(1), 75–99 (1998).

130. S. Ghosh *et al.*, Pragaman: A computer code for simulation of electrotransport during molten salt electrorefining, *Nucl. Technol.* **170**(3), 430–443 (2010).

131. T. Kobayashi and M. Tokiwai, Development of TRAIL, a simulation code for the molten salt electrorefining of spent nuclear fuel, *J. Alloys Compounds* **197**(1), 7–16 (1993).

132. M. Iizuka, K. Kinoshita, and T. Koyama, Modeling of anodic dissolution of U–Pu–Zr ternary alloy in the molten LiCl–KCl electrolyte, *J. Phys. Chem. Solids* **66**(2), 427–432 (2005).

133. M. Iizuka and H. Moriyama, Analyses of anodic behavior of metallic fast reactor fuel using a multidiffusion layer model, *J. Nucl. Sci. Technol.* **47**(12), 1140–1154 (2010).

134. J. Zhang, Kinetic model for electrorefining, Part I: Model development and validation, *Prog. Nucl. Energy* **70**, 279–286 (2014).

135. J. Zhang, Kinetic model for electrorefining, Part II: Model applications and case studies, *Prog. Nucl. Energy* **70**, 287–297 (2014).

136. R. Hoover *et al.*, A computational model of the Mark-IV electrorefiner: Phase I — Fuel basket/salt interface, *J. Eng. Gas Turb. Power* **131**(5), 054503 (2009).

137. R. O. Hoover *et al.*, Development of computational models for the Mark-IV electrorefiner — Effect of uranium, plutonium, and zirconium dissolution at the fuel basket–salt interface, *Nucl. Technol.* **171**(3), 276–284 (2010).

138. R. M. Cumberland and M.-S. Yim, A computational meta-analysis of UCl3 cyclic voltammograms in LiCl-KCl electrolyte, *J. Electrochem. Soc.* **161**(4), D147–D149 (2014).

139. S. Ghosh *et al.*, Exchange current density and diffusion layer thickness in molten LiCl–KCl eutectic: A modeling perspective for pyroprocessing of metal fuels, *Nucl. Technol.* **195**(3), 253–272 (2016).

140. J. Bae *et al.*, A time-dependent electrochemical model of pyrochemical partitioning for waste transmutation, In *Proc. Global 2003*, New Orleans, Louisiana, 2003, p. 784.

141. J. Bae *et al.*, Numerical assessment of pyrochemical process performance for peacer system, *Nucl. Eng. Design* **240**(6), 1679–1687 (2010).

142. S. Choi *et al.*, Three-dimensional multispecies current density simulation of molten-salt electrorefining, *J. Alloys Compounds* **503**(1), 177–185 (2010).

143. B. Krishna Srihari *et al.*, Modeling the molten salt electrorefining process for spent metal fuel using COMSOL, *Separ. Sci. Technol.* **50**(15), 2276–2283 (2015).

144. W. Zhou, Y. Wang, and J. Zhang, Integrated model development for safeguarding pyroprocessing facility, Part I — Model development and validation, *Ann. Nucl. Energy* **112**, 603–614 (2018).

145. W. Zhou, Y. Wang, and J. Zhang, Integrated model development for safeguarding pyroprocessing facility, Part II — Case studies and model integration, *Ann. Nucl. Energy* **112**, 48–61 (2018).

146. D. R. Gabe, The rotating cylinder electrode, *J. Appl. Electrochem.* **4**(2), 91–108 (1974).

147. Z. Tomczuk *et al.*, Uranium transport to solid electrodes in pyrochemical reprocessing of nuclear fuel, *J. Electrochem. Soc.* **139**(12), 3523–3528 (1992).

148. T. Koyama *et al.*, Study of molten salt electrorefining of U–Pu–Zr alloy fuel, *J. Nucl. Sci. Technol.* **39**(Sup 3), 765–768 (2002).

149. M. Gaune-Escard and K. R. Seddon (eds.), *Molten Salts and Ionic Liquids: Never the Twain?* John Wiley & Sons, Hoboken, NJ, pp. 272–273 (2012).

Chapter 5

Waste Disposal

Jean-Francois Lucchini

Los Alamos National Laboratory
Repository Science & Operations, 115N. Main
Carlsbad, NM 88220, USA
lucchini@lanl.gov

1. Introduction

Reprocessing of SNF has three major advantages in terms of waste management [1]:

- A reduction of volume of ultimate high activity waste to be conditioned.
- A reduction of the overall radiotoxicity of the waste, with the removal of plutonium.
- The production of suitable waste packages, which are easier to manage (handling, storage, disposal) than SNF.

Reprocessing activities of 1 t of SNF typically generates [2]:

- ~0.9 t of uranium, and ~10 kg of plutonium. They can be recycled. They account for the intermediate-level long-life waste (IL-LLW).
- ~60 kg of fission products, and a few kg of minor transuranic elements (americium, neptunium, and curium). They constitute the high-level waste (HLW).

When the intermediate-level waste disposal is generally not a problem, final HLW management appears the most controversial issue in the nuclear industry. HLW is highly radioactive and hot: it concentrates over 95% of the total radioactivity and has a thermal power above $2 \, \mathrm{kW/m^3}$, so it requires cooling and shielding.

HLW is therefore the focus of this chapter. A detailed description of HLW resulting from SNF reprocessing is given in Section 2. The preferred option for HLW disposal, which is deep geological disposal in mined repositories, is extensively discussed in Section 4, while other disposal alternatives are presented in Section 3. The alternative options for the management of HLW are complementary, not necessarily opposed.

For simplicity, the discussion is based primarily on HLW generated from reprocessing of uranium oxide (UOX) fuel used in light water reactors (LWRs) using the plutonium and uranium recovery by extraction (PUREX) method. However, most of the concepts and conclusions presented herein can also apply to other types of fuel (e.g., mixed uranium plutonium oxide or MOX, thorium fuels, etc.).

2. Characteristics and Long-Term Behavior of Waste Streams from Reprocessing

2.1. *Types of waste and conditioning*

The PUREX method separates uranium and plutonium, as nuclear materials, from fission products by dissolving the SNF in nitric acid solution [3]. This aqueous method generates some solid waste, mostly resulting from undissolved materials, and large amounts of liquid wastes, which contain most of the fission products, unrecovered uranium, and minor actinides (Am, Np, and Cu).

Radioactive solid wastes can be divided into several types: end caps of fuel structural components generated during the shearing and dissolution processes, and hulls used as fuel cladding, which are both relatively high-level wastes; and combustible or incombustible low-level radioactive miscellaneous solids, which consist of waste components and materials from the plant. They are usually rinsed, go through a volume reduction (e.g., by incineration, compaction) and/or are cemented. The sludges generated by these processes are bituminized or cemented.

Radioactive liquid wastes mainly consist of high-level radioactive effluents (such as tributyl phosphate (TBP), dodecane, sodium carbonate, etc.) and acid effluents (e.g., nitric acid) from the solvent extraction process. They are first solidified directly at the reprocessing plant. The most frequently used solidification process is vitrification. The vitrification process consists of melting the waste products together with a glass material at high temperatures, so that the waste products are incorporated into the glass structure. The characteristics of the vitrified waste differ according to the liquid waste composition (e.g., presence of large amounts of sodium or molybdenum) [3]; it is typically 10–20% waste and 80–90% glass. The melted glass mixture is poured into stainless steel containers, which are sealed by welding and then decontaminated to remove any possible surface contamination.

At the time of fabrication, the radioactivity of vitrified waste is dominated by nuclides with a relatively short half-life, such as ^{90}Sr (29.1 years), ^{137}Cs (30.2 years), and ^{244}Cm (18.1 years), with a total radioactivity level of about 1×10^{16} Bq/t [3]. After the decay of these radionuclides, radioactivity remains relatively high, with the presence of long half-lived radionuclides, such as ^{99}Tc (213,000 years), ^{93}Zr (1.53 million years), and ^{237}Np (2.14 million years). The amount of ^{237}Np, produced by the alpha-decay of ^{241}Am, can be reduced in vitrified HLW when SNF is reprocessed within a few years after being unload from the reactor core, so the production of ^{241}Am is cut from the beta-decay of ^{241}Pu (14.4 years) [3].

Dose equivalent rates at the time of production of vitrified HLW are about 14,000 Sv/h on the waste surface and 420 Sv/h at 1 m apart from the surface [3]. Six years later, a 400-kg HLW glass package has an activity of 16,000 TBq, and estimated dose rates are 500 Gy/h at contact and 80 Gy/h at 1 m distance [1]. These dose rates are several times higher than the lethal dose in 1 h. These values are divided by 10 after 100 years. Because HLW generates such intense levels of both radioactivity and heat, heavy shielding and cooling are required during handling and temporary storage. HLW containers are therefore best stored in specially engineered ventilated vaults, close to the processing unit, for several decades prior to disposal. While stored, both the temperature and radioactivity of the wastes gradually decrease, simplifying their handling and disposal considerably.

The R7T7 glass from the French reprocessing plant of La Hague: An example of vitrified waste package [1]

HLW is calcined, mixed with borosilicate glass powder, and melted in an induction furnace at 1100°C. The steel canister used is 5-mm thick, 1.34-m long, and 0.43-m in diameter. It can hold the entire vitrified HLW (400 kg) generated from the reprocessing of a 1.3 tHM SNF. It then contains 95–99% of the radioactivity of the spent fuel. The levels of radioactivity of the glass package are: 5800 Ci (2.1×10^{14} Bq) alpha; 400,000 Ci (1.5×10^{16} Bq)

(Continued)

(*Continued*)

beta gamma; the dose is 500 Gy/h at contact. Power is 2.5 kW (after conditioning), 1 kW (10 years later), and 0.4 kW (50 years later). The size of the package is chosen so that the temperature at the center of the package (i.e., below the glass transition temperature) is <500°C, with a convective air cooling.

2.2. Long-term behavior of HLW

In terms of nuclear waste management, major requirements for reliable HLW forms are durability and resistance to degradation over a significantly long period of time, particularly regarding high temperature and irradiation, and the presence of water. It has been shown that glass and cementitious materials are satisfactory conditioning matrices for radioactive waste generated from reprocessing activities.

Despite the presence of robust containers, it has been extremely important to investigate and predict the long-term behavior of waste forms for the performance assessment (PA) of a disposal site.

2.2.1. Cements

Cements have been scientifically proven to be a suitable material for the immobilization of a variety of HLW constituents due to their favorable chemical and physical properties [4]. Cements have been extensively modified by reactive admixtures to improve their physical properties to immobilize specific radioactive waste components.

The long-term behavior of cementitious materials in water-saturated media is well understood. The main mechanism is first the leaching of alkalis (Na, K) and then the dissolution of portlandite (which releases calcium ions), and the diffusion of these calcium ions in the cement pore water. This mechanism predicts an altered zone at the interface, and the thickness of the altered zone grows roughly proportionally to the square root of time [1].

An extensive R&D has been going on cement-based materials for waste conditioning, particularly on the effects of radiolysis, and on the chemical and physical perturbations resulting from the incorporation of certain elements in the waste (i.e., phosphates, borates). The goal is to find the most appropriate cement formulations compatible with the different components of the waste, and to ensure a long-term resistance to degradation and leaching.

2.2.2. *Glass*

Vitrification has been a successful method to solidify liquid HLW, because of the excellent leaching resistance, thermal stability, and radiation resistance of glass. Additionally, reprocessing facilities benefit from existing and well-established technologies from the glass industry. However, it is important to collect as many data as possible on glass alteration by water, self-irradiation, and fractures, because these phenomena may cause the release and migration of radionuclides from the glass matrix.

When the underground water reaches the glass package after the breaching of the container, a slow dissolution of the components of the glass will occur through two reactions: (1) the dissolution of silica will slow down the matrix alteration, and (2) a protective gel will form on the glass, which will increase the longevity of the glass. The dissolution rate depends on the glass composition and degree of fragmentation, which determine the reactive surface area. Other parameters, e.g., temperature and the ion content of the groundwater, could also determine the dissolution–precipitation phenomena in the vicinity of the glass. For example, the activation energy of the hydration of glass is relatively high ($\sim$73 J/mol) [1]. However, by the time groundwater reaches the glass matrix in a deep geological repository scenario, the effect of temperature on the glass alteration rate should not be a concern.

Many studies investigated whether self-irradiation would modify the glass structure and alter the long-term durability of the glass matrix [1, 5]. Recoil nuclei from alpha disintegrations indeed cause atomic displacements. Borosilicate glasses doped with a few percent ^{244}Cm were used for research purposes, because they could experience the same number of alpha disintegrations within a few years as an industrial glass during its whole life. No significant influence of the self-irradiation on the glass initial dissolution rate was observed for the fluences encountered in high-activity, long-life waste [5]. A slight modification of the glass density with a small swelling ($<$1%), and an increased (then stabilized) fracture toughness were observed after a cumulative dose of 2×10^{18} alpha disintegrations per gram. Overall, glass dissolution is not significantly affected by the self-irradiation of the matrix or by the irradiation of the surrounding medium.

2.2.3. *Bitumen*

Two phenomena have been considered for the long-term behavior of bituminized packages: radiolysis caused by self-irradiation, and leaching induced by a possible contact to water [1].

Because of their organic composition, bitumen has the property of emitting radiolytic gases under self-irradiation. Depending on the activity incorporated, the accumulated volume of radiolytic gas over 1000 years could be about $1\,m^3$ per drum. First dissolved in the matrix up to saturation (about 1% in volume), gas bubbles then form, and may lead to the swelling of the waste. A suitable package management during the storage period is thus required, e.g., limiting the activity in the package, imposing a void space in the upper part of the drum, and adding cobalt salts [1].

Although pure bitumen is not very permeable to water and dissolved species, water uptake is observed through diffusion and osmosis, because of the presence of salts. On contact with water, the most soluble salts are dissolved locally, and porosity of the glass matrix develops. However, that reaction is slower compared to salt release, due to the low values of effective diffusion coefficients of the solubilized salts ($\sim 10^{-15}\,m^2/s$). This difference results in the swelling of the glass waste, and the leaching front progresses at a rate of the order of $mm/year^{0.5}$.

Globally, bitumen is a good confinement matrix: the rate of release of radionuclides, controlled by the diffusion, is slow and altogether compatible with the requirement of a disposal facility. However, because of concerns regarding radiolysis, gas generation, and swelling, bituminization has been increasingly replaced by cementation as a conditioning matrix [1].

3. Alternatives to Mined Repositories

In any industrial process, storage plays the role of a buffer between two steps. In the case of nuclear waste, the interim storage provides some flexibility in the management of the waste, giving some time for HLW to cool down by radioactive decay, and giving the possibility to wait for an available final destination for the final waste. It acts as a buffer for the waste stream, between production and disposal [1].

Today, geologic disposal is considered the preferred option for the disposal of long-lived waste that must be isolated from the biosphere for protection of human health and the environment [6]. However, no repositories for HLW have been in service at this time. Therefore, the interim storage, although temporary by definition, has been extended in time. Long-term interim storage (>100 years) in surface or subsurface has been proven to be technically feasible, so it appears as a reliable option (see Section 3.1). The main idea behind the long-term interim storage is to give time for the waste to cool down and time to decide on an appropriate waste post-treatment, such as partitioning–transmutation (P&T) and partitioning–conditioning (P&C). The objective of partitioning is to separate long-lived radionuclides (actinides, ^{129}I, ^{99}Tc, ^{135}Cs, ^{90}Sr) from short-life ones in order to manage these categories of waste separately. The general idea of transmutation is to transform the long-lived radionuclides into inactive or short-lived isotopes, using fast reactors or other systems. P&T and P&C are considered a more sophisticated view of reprocessing; they are still at the R&D stage. Ultimately, some materials have been developed for the immobilization of actinides, which could be used as actinide waste forms [7].

Many other different options have been advanced for the disposal of HLW [8]. Numerous options have been investigated in the past; only a few have been implemented (see Section 3.2).

3.1. *"Interim" storage at the surface*

A consequence of huge delays in implementing disposal facilities has been the drastic extension of storage periods [2]. Interim storage thus appears now as a good option for a long-term management of the high-level nuclear wastes.

3.1.1. *Dry storage*

The storage of vitrified waste is already a mature industrial process. Dry storage facilities (vaults, casks, or silos) are often present at the reprocessing plants. Vault storages are concrete structures containing storage wells or pits inside a protective and secure building. Cask storage consists of a concrete or metal structure, which provides containment and shielding. Casks are generally placed on concrete pads, in open air. They can be designed and licensed for both storage and transportation. In silos, canisters are loaded into reinforced concrete modular storage units, and the containment and shielding are provided by a monolithic or modular immovable structure.

In dry storage structures, the containers must be continually cooled in order to reduce the thermal power of the vitrified waste, avoid the thermal stress of the glass, and prevent possible changes in the glass structure. Depending on the kind of storage facility and how much heat needs to be evacuated, stored containers can be cooled by natural air conduction or forced air convection. The radiological impact of vitrified waste is minimized by the shielding from the concrete structure, and criticality control is ensured by the geometry of the structure.

Examples of existing interim storages for vitrified HLW include the French EEV-SE facility, British BNFL's Vitrified Product Store, Japanese

Nuclear Fuel's vitrified waste storage building, and Dutch COVRA's HABOG store [1].

Based on this experience, dry storage facilities are believed to be able to ensure the integrity and safety of vitrified HLW for extended storage periods (i.e., more than 50 years), while geological repositories are being developed.

3.1.2. *Storage versus disposal*

Interim storage and disposal facilities are different in terms of duration, safety, and radiological impact.

While interim storage facilities are already an industrial reality, there is no "return-on-experience" for nuclear waste repositories, since no deep geological repositories are in operation at this time.

The lifetimes of an interim storage facility and of a disposal facility are not the same. The lifetime of storage facilities is the durability of structures made of concrete, and the corrosion of metals. It is on the order of a century, which is conceivable for humans, whereas the lifetime of a repository is so long that it challenges the human imagination.

In normal exploitation, a nuclear waste storage facility has no radiological impact on the environment, since it is designed to totally confine the radioactivity it contains [1]. However, the main weakness of the concept of long-term waste storage is the risk that the society may fail to maintain or may even forget the existence of these installations [2]. Disposal facilities are designed to be safe without any surveillance or maintenance, whereas the interim storage facilities require both. Although no new technical developments are necessary to ensure the safety of stored nuclear wastes, concerns over possible terrorist acts may lead to a reevaluation of the security of the interim storage facilities. Because long-term waste storage facilities offer the possibility to access the waste at any time, they are vulnerable to malicious or accidental human intrusions. Despite the extremely low probability of occurrence of these scenarios, the consequences would be severe, since there is no geological barrier to slow down and dilute released radionuclides. This is the main reason why the duration of interim storage should be limited as much as reasonably possible.

Nevertheless, a big advantage of long-term interim storage is to reduce the cost of disposal [1]. The dimension (and therefore the cost) of a geological disposal facility is roughly proportional to the thermal power produced by the waste. Because the thermal conductivity of the host rock is generally poor, the heat generated by the waste is not easily evacuated in an underground repository. In order to avoid the degradation of the repository performance, the waste packages must be located in such a way that the average power is less than a few tens of watts/m^2. The size and cost of a geological disposal considerably decrease with the cooling time of the waste, typically a factor 2 for a 40-year cooling time [1]. This can be a big incentive for the decision makers to postpone the disposal of the waste.

3.2. *Geological alternatives*

Ultimately, HLW must be permanently isolated from the environment, in order to avoid any chance of radiation exposure to people, or any pollution. Many different disposal options have been investigated worldwide, which seek to provide publicly acceptable, safe and environmentally sound solutions to the long-term management of HLW.

The long timescales over which HLW remains radioactive led to the idea that a deep geological disposal in underground repositories in stable geological formations is the best option for HLW isolation. However, many alternatives to mined repositories have been considered.

3.2.1. *Deep borehole*

From original evaluations [9] by the US National Academy of Sciences (NAS) in 1957 to more recent conceptual evaluations [10, 11], deep borehole disposal of HLW has been considered a credible alternative to mined geologic repositories. The concept involves drilling a borehole about 5 km down into the crystalline basement rock of the Earth. HLW in steel canisters could be emplaced in the lower 2 km, and the upper 3 km could be sealed with materials such as crusher rock backfill, cement, or asphalt, to ensure isolation of the waste from the biosphere. Even though current technology limits the diameter of the borehole to less than 50 cm, the deep borehole concept could be applied to any amount of waste. The environmental impact of deep boreholes would be small. Crystalline basement rocks are located far below sedimentary rock, far below drinking water aquifers, and far below oil and gas deposits. Deep borehole concepts have been developed (but not implemented) in several countries, including the USA.

3.2.2. *Direct injection*

The direct injection approach consists of injecting liquid radioactive waste directly into a suitable layer of rock deep underground that would confine the waste. Direct injection could, in principle, be used on any type of radioactive waste that is in a liquid form (solution or slurry). This condition would make this approach suitable to HLW generated from SNF reprocessing. Ideally, the chosen layer of rock should be sufficiently porous to accommodate the waste, permeable enough to allow easy injection, but surrounded by impermeable layers that would limit the migration of radionuclides. All these requirements entail a profound knowledge of the subterranean geologic features for thousands of years into the future, since no man-made control exists once the liquid material has been injected. During the 1970s, a project of HLW injection into crystalline bedrock was abandoned in the USA before it was implemented due to public concerns. In Russia, however, millions of cubic meters of low-, intermediate-level radioactive wastes (LLW, ILW), and HLW have been injected [12].

3.2.3. *Exotic alternatives*

Several exotic disposal routes have been discussed and studied in the past, but they are no longer considered, mostly because they would not be internationally acceptable at this time. None of the following disposal options has been implemented anywhere.

3.2.3.1. Extraterrestrial disposal

The space disposal concept would involve launching the HLW packages by rocket or space shuttle to place them in orbits around the Earth or the sun, as new "asteroids" [8]. Even though this option would provide the greatest degree of isolation from man's environment, the drawbacks are prohibitive. There is always the possibility of an aborted mission, with a broad contamination when the waste re-enters the atmosphere. Once investigated by the US National Aeronautics and Space Administration (NASA) [13], this option was abandoned due to high cost, technological complexity, and potential risks of launch failure.

3.2.3.2. Rock melting

The concept of deep disposal by rock melting entails the injection of heat generating waste (such as vitrified HLW) in an excavated cavity or deep borehole in host rocks that would have suitable characteristics to reduce heat loss (such as granitic rocks). The high thermal loading of the waste would initially melt the adjacent rock, and dissolve the radionuclides in the molten material. As the rock cools, the radionuclides would be incorporated thus immobilized in the rock matrix. Alternatively, the use of conventional or nuclear explosion was proposed to immobilize the waste in the rock matrix if the heat from the waste was insufficient to melt the rock. However, because of the major disturbance to the rock mass and groundwater it could cause, and because of sensitive international considerations on nuclear weapon explosions, the use of nuclear explosions was rejected. Apart from laboratory and modeling studies of rock melting [14, 15], there have been no practical demonstrations of the feasibility of the rock melting option.

3.2.3.3. Ice-sheet disposal

In the ice-sheet disposal concept, canisters of waste would be laid on stable ice sheets in Antarctica or Greenland [8]. The idea is that the residual heat of HLW would cause the containers to melt their way down to the solid rock. Two major problems are associated to this concept: the waste would be left "accessible" for a short period, and the waste containers could be transported to the sea in a water layer located between the ice and the rock. Additionally, it would require substantial changes to international legal and political agreements.

3.2.3.4. Sub-seabed disposal

In the sub-seabed disposal scheme, HLW canisters or glass would be buried in a suitable geological formation beneath the deep ocean floor [8]. Burial of radioactive waste in deep ocean sediments would be technically possible using either penetrators or drilling equipment. A major advantage would be the enormous dilution capacity provided by the ocean, should the containment system eventually fail. A variation of this burial option in deep ocean sediments would be to dispose of the waste in a repository constructed below the seabed. This method would therefore provide an additional containment of radionuclides.

Even though promising from the point of view of safety, the sub-seabed disposal of HLW has not been implemented anywhere, and is precluded by the London Dumping Conventions international agreements.

3.2.3.5. Disposal at a subduction zone

Subduction zones are areas where one denser section of the Earth's crust dives beneath another lighter section to be swallowed up by the Earth's hot mantle. As the oceanic plate descends into the hot mantle, parts of it begin to melt. The idea here would be to dispose of wastes in trench regions such that they would be drawn deep into the Earth, where they could naturally decay. This burial process would, however, take a very long time, since subduction occurs at geological rates of 5–10 cm/year [16]. Furthermore, subduction zones are geographically very restricted, and not easily accessible to every waste-producing country. Since this approach is a form of sea disposal, it is not permitted by international environmental laws.

4. Geological Disposal

Whatever the fuel cycles, long-life radioactive wastes cannot be physically destroyed; thus, all fuel cycles require a geological repository to support the definitive disposal of radioactive wastes [6]. The main argument is "do not leave the burden of our waste to future generations" [8]. Though geologic disposal has been implemented for LLW, IL-LLW, and transuranic waste, no burial places exist for HLW generated from SNF reprocessing.

Storage cannot be relied upon in the long-term to provide the necessary permanent isolation of the wastes from man's environment, and future generations should not have to bear the burden of managing radioactive wastes produced today. Seen from this perspective, while disposal of HLW is not an urgent technical priority, it is, however, an urgent public policy issue. These political aspects have led to the need for the nuclear industry in recent years to demonstrate the feasibility and safety of HLW disposal and, in some countries, laws have been implemented to require operational HLW disposal capability in the next 15–50 years.

As the ultimate solution for the final waste that has been adequately separated, conditioned, and cooled down, deep geologic repositories are at the end of the sequence of the waste management process [1]. The geological medium ensures a stable long-lasting barrier against natural and man-made events (land erosion, glaciation, war) [6].

4.1. General principles

The first concept of geological disposal was the direct disposal of high-level radioactive liquid wastes in salt formations proposed in 1957 by the US NAS [9]. Today, R&D on geological disposal systems is underway in more than 30 countries [17].

The concept of HLW geological disposal considerably differs from the usual management-type disposal approaches in the sense that it is a passive approach [9]. The goal of HLW geological disposals is to ensure long-term safety for the required period of time (typically several thousands of years) without relying on human involvement.

The Nuclear Energy Agency of the Organization for Economic Co-operation and Development (OECD/NEA) emphasizes the importance of not only protecting human health and the environment but also minimizing the burden on future generations by using geological disposal systems [18]. This means that the management and disposal of HLW, which need to be isolated from the biosphere for a long period of time, should not be left to future generations, but should be the responsibility of the current beneficiaries. It is thus necessary to build disposal facilities so that the level of risk to be transferred to future generations is deemed acceptable [3].

An additional ethical consideration has emerged in a number of countries: the need to include provisions for waste retrieval [19]. If the ability to retrieve the waste from a geologic repository has been shown to be a major factor in gaining public acceptance [19], the feasibility of waste retrievability at various stages after emplacement (including after final sealing and closure) has some important technical and institutional implications, as well as possible R&D requirements [20]. Even if most repository concepts already have some degree of intrinsic retrievability, the technical complexity and the cost of waste retrieval tend to increase as closure approaches. To some extent, the concept of reversible repository provides a bridge between the interim storage and the geological disposal concepts [1]. A deep geological storage that could be gradually converted into a repository leaves time for debate, and for complementary solutions to emerge. An extended period of opening due to retrievability claims may also provide opportunities for extended monitoring related to long-term performance and confidence in long-term safety [20].

4.2. *Technical principles*

Two major technical principles emerge for deep geological disposal systems:

- Isolate the radioactive waste from the biosphere, by means of multiple barriers protecting the waste package, and delaying radionuclide migration long enough for radionuclides to decay.

- Ensure a significant dilution of the radionuclides that would manage to escape, so that the radiological impact on man's environment is negligible.

The long-term safety of HLW disposal is systematically assessed through predictive modeling of the gradual failure of the engineered barriers (i.e., the waste form, waste package, and backfill — see Section 4.2.1) and the subsequent transport to man's environment of radionuclides through circulating groundwater [6]. Such safety assessments must be based on a solid understanding of the processes involved in the release and transport of radionuclides, as well as those affecting the repository and the geological formation (see Section 4.3). Finally, substantial site investigation efforts, involving the collection of data in situ, are also needed at the proposed repository location (see Section 4.4).

4.2.1. *Multi-barrier concept*

In order to isolate the waste from man, the repository concept consists of three major barriers [1, 3]:

(1) The waste matrix (glass, concrete, bitumen) is in solid form and highly insoluble.
(2) The engineered barrier (container and/or overpack, backfill) protects the waste matrix by restricting the flow of groundwater to the waste forms, and delays the release of radionuclides.
(3) The geologic barrier (host rock) delays the release, and dilutes the radionuclides.

The waste form is typically placed in a metal canister, which may consist of steel, copper, or more advanced corrosion-resistant alloys [21]. In order to control underground water movement through a repository after closure and buffer the water composition, the void space between the waste container and the host rock must be filled with a material, called backfill, compatible with the engineered and geological barriers. For instance, bentonite has been extensively investigated as a backfill material for heat-producing waste canisters. For larger waste volumes, host rock materials and cement can be the backfill material of choice.

Most of the engineered barriers are physical in that they delay the access of water to the waste package, or the release of radionuclides from a breached canister. In some cases, the engineered barriers also have a chemical function, affecting the geochemical environment around the waste. For example, magnesium oxide is emplaced with the transuranic waste in the bedded salt repository at the Waste Isolation Pilot Plant (WIPP) in New Mexico, so it removes carbon dioxide (CO_2) and consequently keeps the solubility of the actinides low [21]. The choice of backfill is of major importance and needs to be adjusted to the natural (geologic) barrier.

The region at the engineered/natural barrier boundary, called the excavation-damaged zone (EDZ) [3], is a region of many disturbances, due to excavation-induced cracking in rock of different sources: thermal, mechanical (decompression, cracks), hydraulic (desaturation, groundwater flow, etc.), and chemical (oxidation). These disturbances generate high gradients (stress, pressure, temperature, and concentration) [1]. The region consisting of the engineered barrier and the EDZ is sometimes called the "near-field."

By opposition, the "far-field" is the natural barrier that is not affected by the installation of the engineered barrier. There, the properties of the rock (permeability, fractures) and the hydrology around the repository allow to slow the travel time of radionuclides to the biosphere [21]. Other factors can delay the mobility of the radionuclides in solution: dilution, sorption onto mineral surfaces, precipitation of secondary phases, and matrix diffusion.

In the multi-barrier approach, each barrier works to contain the radionuclides in a supplementary way. For instance, in the case of a breached corroded overpack, the canister and its corrosion products will keep chemical reducing conditions, so that the release of radionuclides in the groundwater will be minimum [3].

4.2.2. *Lifetime*

It is generally acknowledged that the acceptable time needed for the confinement of radionuclides (before they reach man's environment) should be the time taken by the waste radiotoxicity to decrease to the level of the uranium radiotoxicity found in mined ore. According to the diagram of radiotoxicity versus time given by Bonin [2010], the order of magnitude of the time during which a repository must be able to confine its radioactive content is 10,000 years for vitrified HLW from reprocessing, compared with more than 100,000 years for a direct disposal of SNF (no reprocessing).

The challenge is then to predict the evolution of the repository over the first 100,000 years.

4.2.3. *Choice of a suitable location*

The important criteria that should guide the siting of a deep geologic repository are [1]:

- Hydrogeology
  - Low permeability
  - Small hydraulic gradients
- Site stability
  - No active seismic faults, no volcanism close to the site
  - Limited consequences of glaciations
- Chemistry
  - Good sorption capacities of the host rock
  - Chemical stability, buffering capabilities
- Thermal and mechanical properties
  - Excavations technically feasible
  - Good thermal diffusivity
- Minimal depth
  - Protection against erosion, earthquakes, and human intrusions
- No natural resources to be found underneath

4.2.4. *Choice of the host rock*

For repositories below the water table, the host rock must have a low permeability to limit the access of groundwater to the waste containers [21]. This is the case for the three types of host rock that have been selected and investigated around the world to host a geological repository: crystalline rock (such as granite), clay, and salt.

Granite has been investigated by North European countries (Aspö in Sweden, Olkiluoto in Finland), Switzerland (Grimsel), Japan (Mizunami), and South Korea [1]. It has high mechanical strength, a well-known water chemistry, a very low porosity ($\sim$1%, sometimes less), and good thermal conduction properties. However, the presence of fracture systems may increase the hydraulic conductivity, thus affecting the confinement properties of the geological barrier.

Salt is the host rock of the sole operating repository in the world: the WIPP, located near Carlsbad, New Mexico, and used for military IL-LLW disposal [22]. Salt is also considered in Germany (Gorleben). The advantages of salt are a high thermal conductivity, a low water content, a good plasticity (cracks are self-healing) and an excellent stability (on the scale of millions of years). However, salt has some disadvantages. It is a natural resource, easily accessible to future generations, so there is a potential risk of human intrusion. Salt is also corrosive. The complexity of the chemistry of brines (salt concentration can be as high as $6\,mol/L$!) is also an important challenge, but experimental data on brine chemistry and actinides solubility can be produced [23, 24]. Current thermal, hydrologic, and geochemical considerations suggest that radionuclides from HLW would not migrate from the salt repository horizon [25].

Clays are considered as potential host rocks by France (Bure), Switzerland (Mt Terri), and Belgium (Mol) [1]. The excellent properties of both the clay rocks (very low permeability, strong sorption capacity, nanofiltration capabilities) and the underground water in clays (alkaline, reducing) make the migration of radioelements very difficult. Clays' plastic mechanical properties provide interesting sealing capabilities, but solid tunnel-supporting structures are necessary. Other drawbacks of clay rocks include bad thermal conductivity and strong variability.

Volcanic formations were studied exclusively by the USA on the site of Yucca Mountain, until this project was halted in 2010. The particularity of this proposed site was that it is located above the water table, thus in an unsaturated environment [21, 26].

4.2.5. *Migration of radionuclides*

Many factors can affect the migration of radionuclides (or any solutes) in groundwater through the host rock to the biosphere: advection, dispersion, molecular diffusion, and chemical reactions (sorption, solubility) [1].

Advection describes the transport of solutes simply due to the flow of water. Dispersion is the second important transport mechanism in hydrogeology. Mechanical dispersion refers to the spreading and mixing caused by the variations in velocity with which water moves at different scales, due to dispersing objects of any size in the path of the water. Diffusion is the movement of solutes due to concentration gradients. It is governed by Fick's first law, and uses diffusion coefficients. As an example, the order of magnitude of the diffusion coefficient of cations in a compacted clay is typically 10^{-11} to $10^{-12}\,m^2/s$ [1].

In addition to these physical and physicochemical effects (advection, dispersion, and diffusion), chemical interactions between the solutes (also called tracers) and the porous medium have to be taken into account.

The sorption capacity of a tracer onto a mineral surface is defined by its partition (or distribution) coefficient K_d, which is the ratio between the concentration of the tracer species sorbed (i.e., immobilized) onto the medium free surface and the concentration of the tracer species left in solution. It is greatly influenced by the electrostatic charge of the mineral and the tracer. Sorption plays an important role in the delay of radionuclides, particularly actinides, in the host rock. Some laboratory experiments suggested that actinides are very efficiently sorbed (almost immobile) onto many clay minerals [27]. The understanding and the description of ion sorption phenomena onto rock minerals at the microscopic scale remains an important topic of research for geochemists [1].

Speciation is also a factor that limits the solubility of a chemical element in a geological environment. For instance, in reducing conditions, which are typical of underground environments, the actinides are in a rather immobile form (oxide or hydroxide), whereas in oxidizing conditions, they are in the +V or +VI oxidation states, thus more soluble and mobile. However, the presence of colloids in groundwater can enhance the transport of non-soluble contaminants through sorption [28–30]. Actinide migration via colloids is still an active field of investigation.

4.3. *Performance assessment*

Safety is the paramount concern to be addressed when a HLW geological disposal is planned. Estimation of the isolation capability of the repository, particularly during the post-closure period, is given in the PA of the site. The objective of PA is to evaluate the order of magnitude of the radiological impact of the site to the public and the environment.

The basic structure of PA has been formulated in details by the International Atomic Energy Agency (IAEA) jointly with the OECD/NEA [31]. The method to estimate the radiological impact of a geological disposal is based on these two following points [1]:

(1) Define and develop the scenarios for the evolution of the repository.
(2) Model the migration of the radionuclides from waste to man.

4.3.1. *Structure*

A scenario is a chronological sequence of elementary phenomena, called features, events, and processes (FEPs), which can affect the performance of the repository. PA is highly reliable when all the possible and conceivable FEPs are systematically and exhaustively listed and developed [32, 33]. Scenarios can come from natural (somewhat predictable) processes (groundwater transport, earthquakes, volcanic activities, erosion, etc.), or involve "abnormal" events, such as human intrusions, failure of one or several barriers, etc. Concerning human intrusions, wastes could be unintentionally reached because records about the existence of the repository could be lost or no longer available in the future. Also, for example, excavation activities could be carried out in the future for the purpose of mineral, water, or energy resources development or surveys [3].

The scenarios are then classified into highly probable, realistic scenarios, and low-frequency or rare scenarios, using a risk-based approach. The risk-based approach evaluates the impact of a scenario from the standpoint of radiation hazard prevention, by taking into account the possibility of occurrence. For instance, the transport of the radionuclides out of the repository to the biosphere through groundwater flow is one the most important and realistic scenarios in safety assessment. Therefore, the probability of occurrence of radionuclides migration through groundwater is set to 1.

When scenarios are ready, models are constructed on the basis of the scenarios, and sophisticated computer codes are created [3]. Numerical simulations include engineered barrier deterioration, nuclide release from waste forms, nuclide migration in engineered and natural barriers, nuclide migration to the biosphere, and human intake. Even though FEPs, particularly in an engineered barrier, are in a coupled state, meaning that they affect each other, they can be first solved separately, and then connected again. For example, since the conditions in an engineered barrier region, a natural barrier region and the biosphere differ considerably, radionuclide migration analysis is conducted separately in each of these regions, and the results are subsequently connected to evaluate the entire system [3]. All the detailed models are assessed using a conservative assumption approach. Conservative assumptions are made so that the safety of a repository is underestimated. This approach allows a sufficient margin of safety in the overall safety evaluation. As a result, an operational model based on clever simplifications of a very complex reality is developed, ensuring that predictions will always remain conservative despite conceptual and numerical uncertainties [34].

Using the PA model thus obtained, the performance indices are calculated, and eventually, the safety of the repository is compared with the safety standards set by the regulatory authorities [3]. Safety standards are established according to social needs and by comparison with the safety and risks of other engineering systems in our society. The radiological impact of an underground repository to the population should not exceed a predefined limit, which is usually a small fraction of natural exposure dose rate (i.e., ∼1 mSv/year). The IAEA recommends a dose rate limit of 0.3 mSv/year [31].

Ultimately, the PA results are used in safety, technology development, public consensus building process, and policy decision making (e.g., site selection) [3].

4.3.2. Methodology

The three main indicators to evaluate the impact of a radioactive waste underground disposal are the radioactivity (in Bq/m^3) in the repository, the radioactivity flux (in Bq/s) in the geosphere, and the collective dose rate (in Sv/year) to the population [1]:

These are calculated using the following three steps:

- The source term
- The migration of radionuclides in the far-field
- The radiation exposure to man

4.3.2.1. Source term

Source term is the mathematical model used in PA to simulate the release of radionuclides from the near-field (engineered barriers) to the far-field (natural barrier). Radioactive wastes buried in the repository are the source of radionuclides that migrate in groundwater [6]. Multiple barriers are present (see Section 4.2.1) to confine the radionuclides, prevent the infiltration of groundwater, and delay the movement of radionuclides. These processes are modeled in the source term model [3].

However, in the near-field, many physicochemical phenomena, all intricately coupled, will occur (see Section 4.2.5): thermal, hydraulic, mechanical, chemical, and possible biological effects are expected, which make the modeling of the evolution of the near-field difficult.

4.3.2.1.1. Thermal behavior of HLW

The thermal power generated by vitrified HLW is ∼1 kW/t [1]. After 50 years of interim storage on the surface, the maximum temperature increase in vitrified waste packages that have been disposed of is expected between 160°C and 200°C. Because rocks typically have a small thermal diffusivity (∼10^{-6} m^2/s), most of the heat generated by the waste will remain at the proximity of the waste for very long times, until depletion of the radioactive content.

Thermal calculations are thus essential to avoid uncontrolled heating of the waste (that could damage the waste form) and of the surrounding rock.

4.3.2.1.2. *Hydraulic effects*

During the operation of the waste disposal, excavation activities can desaturate the rock vicinity by evaporation. After repository closure, the host rock will saturate again. These phenomena of saturation–desaturation are important disturbances of the water circulation underground, because they generate hydraulic gradients much greater than the natural pre-existent gradients. For instance, the capillary pressure in clay could reach 100 MPa [1].

4.3.2.1.3. *Mechanical effects*

Tunnel excavation activities also create perturbation of the stress field in the host rock, such as fractures in the near-field (source of the EDZ), modification of existing fractures, and localized modification of the rock permeability. Vertical compression due to the above lying rocks, and horizontal tectonic stress are common mechanical effects.

Thermal effects can amplify mechanical effects by the way of dilation phenomena. Coupled hydraulic and mechanical effects are materialized by shrinkage (sometimes cracking) or swelling of the rock due to the remaining groundwater tension during desaturation or saturation, respectively [1]. Swelling is often used to ensure sealing in engineered barriers.

4.3.2.1.4. *Chemical effects*

Many chemical reactions occur in the near-field: redox, dissolution and precipitation, corrosion, radiolysis, etc. These can be amplified by the ambient mechanical stress and thermal gradient [1]. However, these effects are usually slow and much localized.

The underground is generally a reducing environment. Oxidizing conditions introduced during excavations and operations can modify the mineralogical composition of the host rock and the water chemistry at the interface. For example, in the clay rocks exposed to wet air, pyrite minerals readily oxidize and produce sulfuric acid, which causes the acidification of the water inclusions. Also, the introduction of cemented materials in the underground can create an alkaline disturbance that propagates into the rock: smectites, if present, dissolve, and calcium–silicate–hydrate (CHS) phases precipitate [1].

Another example of dissolution and precipitation reactions is the dissolution of portlandite and the precipitation of calcite when concrete is exposed to CO_2. When rock minerals dissolve and different mineral forms are generated in the EDZ, modifications in porosity, permeability, and heat diffusivity at the interface of the host rock are expected [1].

Corrosion of the metal containers and overpacks will occur. Corrosion products can locally modify the porosity of the solid interface, the groundwater chemistry, and the radionuclides transport. After closure, anoxic corrosion of steel, and water radiolysis, could generate a significant amount of hydrogen, which could cause pressure buildup and new fractures. Radiolysis of groundwater can also generate very reactive radicals, which can influence the chemistry at the interface waste/groundwater [35].

4.3.2.1.5. *Biological effects*

Microorganisms and microbial metabolic activity can naturally exist in a HLW underground repository, so biological effects cannot be ruled out. Microorganisms could affect the corrosion of the waste containers, the integrity of engineered barriers, and the fate of radionuclides in the near- and far-fields [36].

4.3.2.2. Migration of radionuclides in the far-field

The rate of release of the radionuclides depends on all the interlinked phenomena taking place in the near-field [1, 3].

If they have not decayed yet, radionuclides will slowly be released from the package through the near-field (altered package, engineered barrier) to the far-field (host rock). Their transport through the host rock to the biosphere will be governed by the chemical conditions prevalent in the geological environment [1].

In the modeling exercise of the migration of radionuclides in the far-field, the geological barrier is segmented into a few permeable and homogenized domains, which display constant properties in time. In the far-field, the migration of radionuclides is governed by a linear sorption model (using K_d values) and solubility of the radionuclides.

Migration models are often validated using the results of tracer experiments. In tracer experiments, a cocktail of tracers is injected into a borehole of the host rock. After some time (typically 1 year), different layers of rock around the injection point are analyzed to determine the distance that the different tracers migrated. Data from these experiments cannot provide a complete understanding of the efficiency of the natural barrier, because the experiments are performed in a short time and a small scale. Natural analogues (such as Oklo) and natural tracers can sometimes provide data on the chemical species of some tracers, and on their migration in the geosphere on scales of time and space that are more representative of an underground repository scenario (see Section 4.4.2) [34].

Experimental and modeling data show that the migration rate for non-reactive species through a geological barrier is very small (e.g., a few cm/year in clays) [1, 3]. Sorbing tracers are orders of magnitude slower.

Because of their high sorption capacity on host rocks and their low solubility in groundwater, the most radiotoxic nuclides, i.e., the actinides, have extremely low release rates, and are not expected to reach the biosphere in most repository environments [3, 6, 37].

4.3.2.3. Radiation exposure to man

The evaluation of a radiation dose to man requires the modeling of radionuclides through the biosphere (radioecology) [1].

The biosphere is modeled in several compartments, which can exchange radioactivity by using transfer coefficients. Compartments include rivers, lakes, air, soil, seas, plants, and animals. The top compartment is the human population. Radiation exposure to man can occur by radionuclides intakes and external radiation.

This last part of the entire PA evaluation is the most difficult to calculate, particularly because of the large uncertainty associated to the dose. In general, the released radioactivity (in Bq/m^3) and the radioactivity flux (in Bq/s) are more defined parameters to evaluate [2].

Bonin [1] demonstrates the PA methodology with an evaluation of the order of magnitude of the radioactivity that could be released from a simplified repository of vitrified HLW. As a result of his exercise, for a typical repository containing the amount of vitrified waste corresponding to 50 years of activity of a fleet of 60 GW LWRs (2×10^4 TWhe), the maximum radioactivity of ^{135}Cs that could be released to the biosphere,[a] will be of the order of $10 Bq/m^3$. This result is well below the dose brought by the natural radiation background. As a matter of fact, the natural activity of a mineral water can be as high as $10,000 Bq/m^3$!

---

[a] ^{135}Cs is considered the more likely radionuclides to be released from a HLW repository and to migrate to the biosphere, due to its long half-life (2.3×10^6 years), high solubility, low sorption to buffers, and relatively large amount in the waste [3].

Despite multiple conservative assumptions, the PA results of different HLW disposal repository concepts around the world indicate that a HLW geological disposal can achieve a level of safety higher by several orders of magnitude than the level required by the safety standards, particularly thanks to the redundancy and the efficiency of multiple barriers. This is the main technical basis for the present international consensus that geological disposal is safe [3].

4.4. *Underground research laboratories and natural analogues*

Radioactive waste disposal facilities in deep geological formations must safely be sealed for long time periods in order to isolate the wastes from the biosphere for at least several hundred thousand years. Since it is impossible to directly verify the long-term functionality (for times scales foreseen in long-term safety assessments) of a geological repository, an overall understanding of the main chemical and physical processes involved has been developed using modeling calculations and different prediction procedures to extrapolate the safety concepts with a sufficient degree of reliability. Natural analogues and large-scale in situ tests are necessary and integral parts of the procedures for verifying and enhancing this knowledge [34, 38].

4.4.1. *Underground research laboratories*

Several countries have constructed and developed Underground Research Laboratories (URL) in selected geological environments to demonstrate the safety of a geological disposal, to evaluate engineering feasibility, and to develop and refine techniques for site investigation. Laboratories in Belgium (Mol), Germany (Gorleben), Sweden (Aspö), Finland (Olkiluoto), Japan (Horonobe) [39], France (Bure) [27], and Switzerland (Grimsel, Mont Terri) have been the focus of major international collaboration research programs [1, 3, 21, 40].

4.4.2. *Natural analogues: the example of Oklo*

Natural analogues, such as the Oklo natural reactors, can help to validate the models of long-term behavior of radionuclides in a waste repository [1, 41].

In 1972, measurements of the isotopic composition of uranium ore samples from the Oklo mine in Gabon showed a depletion of ^{235}U. Specialists in

Pierrelatte, France, concluded that nuclear chain reactions took place there, about 2 billion years ago. Those natural nuclear reactors started in parts of the deposit where $^{235}U \sim 10\%$, and reached criticality with the presence of water (as a neutron moderator). The amount of fission products that were produced during the activity of the reactors was calculated. The site was investigated as an interesting natural analogue of a waste repository.

The efficiency of the geological barrier for radionuclide confinement was demonstrated for some radionuclides, like uranium, which has been trapped in the mine veins for 2 billion years. However, in 2 billion years, most of the radionuclides generated by the natural reactors decayed, and the site went through numerous geochemical changes. Consequently, the migration of some radionuclides was difficult to establish.

The "Oklo phenomenon" also allowed the discovery of the confinement properties of minerals like apatites.

4.5. *Issues and challenges*

Although the concept of disposal in deep geological formations has been recognized as the most promising form of confinement for HLW [3, 9, 17, 18], to date, no deep geologic repository for HLW is in operation. Every waste disposal program in the world has experienced delays or obstacles [6].

In the USA, an integrated nuclear waste management strategy is still lacking, contributing to a public perception that the radioactive waste problem cannot be readily solved [6, 42]. Yet, the USA have demonstrated that waste disposal facilities for defense-generated radioactive waste (LLW and transuranic) can be effectively and successfully managed. The WIPP is indeed the only licensed, constructed, and operating deep geological repository in the world that has been receiving and burying radioactive waste.

The Waste Isolation Pilot Plant (WIPP) [6, 22]

The United States has been operating a geological repository for defense transuranic wastes near Carlsbad, New Mexico, since 1999. The ultimate WIPP waste inventory in terms of long-lived radioactive materials should be 1–2% of a SNF repository. The siting, construction, and operation of WIPP had challenges that were successfully surpassed.

WIPP was a high priority of the federal government because of the need to dispose of weapons wastes that were accumulating in the nuclear

---

(*Continued*)

(Continued)

weapons complex. The US government provided compensation for hosting such a facility and power-sharing in the form of an oversight role by the host state. State cooperation was also partly influenced by the presence of Los Alamos National Laboratory (LANL) that had a large inventory of transuranic waste that would be disposed of in WIPP. The city of Carlsbad and the surrounding region saw WIPP as an opportunity to provide a long-term stable economy for the area. Over time, WIPP gained community support. The continuous presence of competent technical teams from LANL and Sandia National Laboratories (SNL) supporting the development of WIPP has been greatly helping in dissipating public concerns and enforcing a world-recognized expertise in radioactive waste management. The development of WIPP resulted in other fuel cycle facilities moving to Southeast New Mexico and West Texas — including a uranium enrichment plant and a LLW storage facility.

The history of efforts to build a geological repository for SNF or HLW worldwide validates many of the concerns reported in 1982 by the US Office of Technology Assessment (OTA) [43]. The following lessons have been learned [6]:

(1) Waste program must be continued and should be managed with specialized government or utility organizations with a strong waste generator commitment.
(2) Appropriate, adequate, and available funds are required.
(3) Public transparency and major outreach programs are critical.
(4) Compensation to, involvement and approval of local communities are imperative.
(5) Repositories should be designed to enable long-term waste retrievability, and the safety assessment should be understandable by the public.

In 1990, the National Research Council (NRC) of the US NAS highlighted that the safety of repositories was not only a scientific, technical issue but also a societal issue [44–46].

Even with a strong scientific basis for the safety analysis and the confidence of the scientific community and of the regulatory agency in a geological disposal project, a significant fraction of the public does not share this confidence [2, 21]. Public acceptance should not come only at the end of the site selection process but is rather earned throughout the process from the initial site selection through license approval, even if this strategy could lead to controversy, delays, or failures. In this regard, a robust strategy based on simple physical and chemical principles that can be readily understood by everyone is certainly more useful when communicating to the public, politicians, and stakeholders than a very sophisticated, probabilistic, highly scientific analysis that is less transparent [6].

Despite the scale and complexity of the engineering and science involved in implementing a geological disposal, the biggest controversial technical issue is the credibility of the predictions of the safety assessment for a repository for tens or hundreds of thousands of years into the future [2]. The society has not been ready to deal with this concept of safety over a period longer than 10,000 years [3]. People are generally reluctant to accept any irrevocable waste management solutions, particularly in the case of radioactive waste [6, 19, 20]. In many programs (France, Sweden), the concept of retrievability (or reversibility) has been considered to facilitate confidence building and possibly stimulate public acceptance (see Section 4.1).

5. Conclusions

Management of HLW generated from SNF reprocessing activities is often considered an unresolved issue by the public. However, technical solutions have been developed and implemented [1, 8].

Currently, HLW has been stored in surface dry storage facilities. Storage is, however, a temporary solution. It provides flexibility for the management of the waste. Because the waste cool down, the thermal load of the waste for a final disposal facility will be reduced, and so will the cost of the disposal facility. However, the safety of storage facilities is not as well assured as the safety of an underground disposal facility, because these facilities demand active maintenance and security, and are more vulnerable to human intrusions [1].

Deep underground geological disposal, which does not require a continuous control by the society, seems to be the only long-term solution to HLW management. A general consensus has been reached on this issue, under the auspices of IAEA, OECD/NEA, and other authorities [6, 8, 18, 31, 43, 47]. No other sound alternative solutions have emerged.

The safety of an underground disposal relies on the capacity of the repository and of the multi-barriers to confine radionuclides and isolate them from human environment, until their radiotoxicity is down to an acceptable level. The evolution of the waste packages in underground conditions, and the migration of radionuclides through the engineered and geological barriers have been extensively investigated [4, 5, 7, 23, 25–30, 34, 35, 37, 38, 48]. Results from experimental data, and from a solid and broad risk-based modeling approach, show that the radiological impact of a deep geological waste disposal is expected to be negligible, localized, and delayed, thanks to the efficiency and the redundancy of the multi-barriers.

HLW burial in deep repository is considered a robust concept by the technical experts, even with the option of waste retrievability. However, the main obstacle to the implementation and operation of a HLW repository is the lack of public acceptance [1, 2, 49]. This societal challenge is on the way to be overcome in some countries (e.g., Sweden and Finland), which are moving toward the operation of a repository in the next decades. These facilities could then be used as reference facilities for HLW disposal.

Meanwhile, R&D efforts need to continue in the areas of reducing the radiotoxicity and the volume of the waste (e.g., recycling, partitioning, transmutation), and of improving and developing suitable conditioning forms for specific radionuclides. While conditioning already reached an industrial scale (e.g., vitrification), a lot remains to be done on partitioning and transmutation of some radionuclides (particularly the minor actinides). Decreasing the thermal load and the radiotoxicity of the waste would diminish the size and the cost of a HLW repository. It could also reduce the scientific uncertainties associated with the long timescale predictions for the repository safety. These arguments, along with waste retrievability, could help to convince the public and the politics to accept the idea of a geological disposal to ultimately get rid of the waste, without leaving the burden of the waste to future generations.

References

1. B. Bonin, The scientific basis of nuclear waste management, Chapter 28, In *Handbook of Nuclear Engineering*, D. G. Cacuci, ed., pp. 3257–3419, Springer Science (2010).
2. C. McCombie, International perspectives on the reprocessing storage, and disposal of spent nuclear fuel, *Bridge, Natl. Acad. Eng.* **33**(3), 5–10 (2003).
3. S. Nagasaki and S. Nakayama (eds.), *Radioactive Waste Engineering and Management*, Springer (2011).
4. International Atomic Energy Agency, *The Behaviours of Cementitious Materials in Long Term Storage and Disposal of Radioactive Waste*, IAEA-TECDOC-1701, Vienna (2013).
5. S. Peuget, J. N. Cachia, C. Jegou, X. Deschanels, D. Roudil, V. Broudic, J. M. Delaye, and J. M. Bart, Irradiation stability of R7T7-type brorosilicate glass, *J. Nucl. Mater.* **354**, 1–13 (2006).

6. MIT Interdisciplinary study, *The Future of the Nuclear Fuel Cycle*, MIT (2011).

7. R. C. Ewing, Actinide and radiation effects: Impact on the back-end of the nuclear fuel, *Mineral. Mag.* **75**(4), 2359–2377 (2011).

8. R. L. Murray, *Understanding Radioactive Waste*, 5th edition, Batelle Press, USA (2003).

9. US National Academy of Sciences, National Research Council, *The Disposal of Radioactive Waste on Land*, The National Academies Press, Washington, DC (1957).

10. P. V. Brady, B. W. Arnold, G. A. Freeze, P. N. Swift, S. J. Bauer, J. L. Kanney, R. P. Rechard, and J. S. Stein, Deep borehole disposal of high-level radioactive waste, Sandia report, SAND2009-4401, Sandia National Laboratories, Albuquerque, New Mexico (August 2009).

11. K. P. Travis and F. G. F. Gibb, Deep borehole disposal research: What have we learned from numerical modeling and what can we learn? *Mater. Res. Soc. Symp. Proc.* **1744**, 193–203 (2015).

12. R. Chenko, *Deep Injection Disposal of Liquid Radioactive Waste in Russia*, M. G. Foley and L. M. G. Ballou, eds., Batelle Press, USA (1998).

13. R. E. Burns, W. E. Causey, W. E. Galloway, and R. W. Nelson, Nuclear waste disposal in space, Technical Report 1225, NASA, George C. Marshall Space Flight Center, Alabama (May 1978).

14. J. L. Cohen and T. L. Steinborn. The rock-melt approach to nuclear waste disposal in geological media, In *Scientific Basis for Nuclear Waste Management, Volume 1, Proc. Symposium on "Science Underlying Radioactive Waste Management"*, Materials Research Society Annual Meeting, Boston, MA, pp. 261–264 (December 1978).

15. F. E. Heuze, On the modeling of nuclear waste disposal by rock melting, In *Proc. 23rd US Symp. on Rock Mechanics*, Berkeley, CA (August 1982).

16. E. A. Silver, Subduction zones: Not relevant to present-day problems of waste disposal, *Nature* **239**, 330–331 (1972).

17. P. Witherspoon and G. Bodvarsson (eds.), Geological challenges in radioactive waste isolation: Fourth worldwide review, Report LBNL-59808, Lawrence Berkeley National Laboratory, Berkeley (2006).

18. Nuclear Energy Agency (NEA), *Disposal of Radioactive Waste: An Overview of Principles Involved*, Organisation for Economic Cooperation and Development, Paris (1982).

19. International Atomic Energy Agency, *Geological Disposal of radioactive Waste: Technological implications for Retrievability*, IAEA Nuclear Energy Series No. NW-T-1.19, Vienna (2009).

20. Nuclear Energy Agency, *Reversibility and Retrievability in Geologic Disposal of Radioactive Waste, Reflections at the International Level*, Organisation for Economic Cooperation and Development, Radioactive Waste Management, Paris (2001).

21. B. W. D. Yardley, R. C. Ewing, and R. A. Whittleston (eds.), Deep-mined geological disposal of radioactive waste, *Elements* **12**(4) (August 2016).

22. R. Nelson (ed.), WIPP@ *10*, Radwaste solutions, *Am. Nucl. Soc.* **16**(3) (May/June, 2009).

23. J. F. Lucchini, M. Borkowski, M. K. Richmann, and D. T. Reed, Uranium(VI) solubility in carbonate-free WIPP brine, *Radiochim. Acta* **101**, 391–398 (2013).

24. J. F. Lucchini, M. Borkowski, H. Khaing, M. K. Richmann, J. S. Swanson, K. A. Simmons, and D. T. Reed, WIPP actinide-relevant brine chemistry, Los Alamos National Laboratory report LA-UR-14-21612, Los Alamos, New Mexico (2014).

25. F. D. Hansen and C. D. Leigh, Salt disposal of heat-generating nuclear waste, Sandia National Laboratories report SAND2011-0161, Albuquerque, New Mexico (2011).

26. I. R. Triay, A. Meijer, J. L. Conca, K. S. Kung, R. S. Rundberg, B. A. Strietelmeier, C. D. Tait, D. L. Clark, M. P. Neu, and D. E. Hobart, Summary and synthesis report on radionuclide retardation for the Yucca Mountain site characterization project, Los Alamos National Laboratory report LA-13262-MS, Los Alamos, New Mexico (1997).

27. ANDRA, *Dossier 2005-Les recherches de l'Andra sur le stockage geologique des dechets radioactifs a haute activite et a vie longue, resultats et perspectives*, In French, Agence nationale pour la gestion des dechets radioactifs, Paris, France (2005).

28. A. B. Kersting, D. W. Efurd, D. L. Finnegan, D. J. Rokop, D. K. Smith, and J. L. Thompson, Migration of plutonium in groundwater at the Nevada test Site, *Nature* **397**(6714), 56–59 (1999).

29. A. P. Novikov, S. N. Kalmykov, S. Utsunomiya, R. C. Ewing, F. Horreard, A. Merkulov, S. B. Clark, V. L. Tkachev, and B. F. Myasoedov, Colloid transport of plutonium in the far-field of the Mayak Production Association, Russia, *Science* **314**(5799), 638–641 (2006).

30. J. I. Kim, Actinide colloid generation in groundwater, *Radiochim. Acta* **52/53**, 71–81 (1991).

31. International Atomic Energy Agency, Geological disposal of radioactive waste, IAEA Safety Requirements No. WS-R-4, jointly sponsored by the IAEA and the OECD/NEA (2006).

32. Nuclear Energy Agency (NEA), *Scenario Development Methods and Practice: An Evaluation Based on the NEA Workshop on Scenario Development*, Madrid, Spain, May 1999 (2001).

33. Nuclear Energy Agency (NEA), *Features, Events, and Processes (FEPs) for geological disposal of Radioactive Waste: An International Database*, Organisation for Economic Cooperation and Development, Paris (2000).

34. C. Poinssot and S. Gin, Long-term Behavior Science: The cornerstone approach for reliably assessing the long-term performance of nuclear waste, *J. Nucl. Mater.* **420**, 182–192 (2012).

35. A. K. Pikaev, G. N. Pirogova, I. M. Kosareva, A. V. Gogolev, and V. P. Shilov, Radiation-chemical aspects of radioactive waste management, *High Energy Chem.* **37**(2), 60–72 (2003).

36. T. Behrends, E. Krawczyk-Barsch, and T. Arnold, Implementation of microbial processes in the performance assessment of spent nuclear fuel repositories, *Appl. Geochem.* **27**, 453–462 (2012).

37. J. I. Kim and B. Grambow, Geochemical assessment of actinide isolation in a German salt repository environment, *Eng. Geol.* **52**, 221–230 (1999).

38. R. Mauke and H. J. Herbert, Large scale in-situ experiments on sealing constructions in underground disposal facilities for radioactive wastes — Examples of recent BfS- and GRS-activities, *Prog. Nucl. Energy* **84**, 6–17 (2015).

39. Nuclear Energy Agency, *Japan's Siting Process for the Geological Disposal of High-level radioactive Waste — An International Peer Review*, Organisation for Economic Cooperation and Development, Radioactive Waste Management, Paris (2016).

40. Nuclear Energy Agency, *Underground Research Laboratories (URL)*, Organisation for Economic Cooperation and Development, Radioactive Waste Management, Paris (2013).

41. G. A. Cowan, A natural fission reactor, *Scientific American*, pp. 36–47 (July 1976).

42. Nuclear Energy Agency, *Local Communities' Expectations and Demands on Monitoring and the Preservation of records, Knowledge and Memory of a Deep Geological Repository*, Organisation for Economic Cooperation and Development, Radioactive Waste Management, Paris (2014).

43. Office of Technology Assessment, Congress of the United States, *Managing Commercial High-Level Radioactive Wastes* (1982).

44. US National Academy of Sciences, National Research Council, *Rethinking High-Level Radioactive Waste Disposal: A Position Statement of the Board on Radioactive Waste Management*, The National Academies Press, Washington, DC (1990).

45. Nuclear Energy Agency (NEA), Post-closure safety case for geological repositories: Nature and purpose, NEA No. 3679, Organisation for Economic Cooperation and Development, Paris (2004).

46. International Atomic Energy Agency, Factors affecting public and political acceptance for the implementation of geological disposal, IAEA-TECDOC-1566, Vienna (2007).

47. Nuclear Energy Agency (NEA), *Geological Disposal of radioactive Waste in Perspective*, Organisation for Economic Cooperation and Development, Paris (2000).

48. B. Kursten, E. Smallos, I. Azkarate, L. Werme, N. R. Smart, G. Marx, M. A. Cunado, and G. Santarini, Corrosion evaluation of metallic HLW/spent fuel disposal containers — Review, In *Proc. of the 2nd International Workshop on Prediction of Long Term Corrosion Behavior in Nuclear Waste Systems (Eurocorr 2004)*, pp. 153–161, Nice, France (2004).

49. US National Academy of Sciences, National Research Council, *Disposition of High-Level Waste and Spent Nuclear Fuel: The Continuing Societal and Technical Challenges*, The National Academies Press, Washington, DC (2001).

Chapter 6

Spent Fuel Interim Dry Storage System and Chloride-Induced Stress Corrosion Cracking

Yi Xie and Jinsuo Zhang*

Nuclear Engineering Program, Department of Mechanical Engineering
Virginia Tech (VT), Blacksburg, VA 24060
*zjinsuo5@vt.edu

1. Background and Introduction

Spent fuel is stored in spent fuel pool when moved out from the reactor core, known as wet storage, which requires a constant power supply to keep the cooling water pumps operating. Avoid filling up with stored spent fuel, after several years, the old fuel must be transferred to independent spent fuel storage installations (ISFSI) which is away from the reactor, mostly being near to the river or lake side or coastal. The system that is used for the transportation and store is called dry storage system. When the fuel is moved to the dry storage system after several years in the spent fuel pool, the decay heat given off by the fuel has significantly decreased. With the absence of the long-term spent fuel repository, ISFSIs are used for the interim purpose. They are licensed to provide safe storage for 60 years and are likely well beyond the period until the fuel is either reprocessed or entombed in a long-term repository [1]. Potential degradation mechanisms of interim dry storage systems caused by the time need to be identified and addressed.

Chloride ions can easily penetrate the protective oxide layer formed on the canister material, and take electrochemical reactions with the metal matrix, the process is called pitting corrosion, which is broadly recognized as the initiation of stress corrosion cracking (SCC). When the growth rate of pitting corrosion becomes less than the growth rate of cracking, SCC takes place of pitting corrosion. The SCC-accelerated tests taken under the SCC test system have been conducted by many researchers. The chloride-induced stress corrosion cracking (CISCC) growth rates are suggested to be in the range of 10^{-12} to 10^{-9} m/s; however, the actual CISCC rate of the in-service canister is unknown. Although salt concentration would affect the susceptibility to CISCC, a multitude of studies confirmed that the corrosion resistance is critically related to the types of salts but not sensitive to the concentration of ions.

The dry storage system addressed in this chapter is the canister-based system. The system is typically composed of a thick concrete overpack and a thin-walled welded stainless steel (SS) inner canister. The in-service interim canister surface temperature is always higher than the ambient temperature because of the spent fuel decay heat. The temperature distributions are nonhomogenous on the surface due to the distributions of spent fuel and the flow direction of cooling air. Located near coastal areas, the surface deposits include sodium–magnesium chloride sulfate salts and dust. The salts will be deliquescent when the relative humidity (RH) on the canister surface reaches the limiting humidity of salt deliquescence, making a highly chloride-concentrated moisture film form on the surface, therefore being corrosive.

Section 2 introduces the design of the dry storage system, including canister-based system (typically addressed in this chapter) and bare-fuel system. Section 3 describes the environment of the dry storage system that's exposed to river or marine coastal and explains the effects on canister corrosion. Section 4 investigates the mechanism of pitting corrosion as well as the previous studies on this issue. Section 5 obtains from Ref. 2, showing the results of the experimental study which focuses on the pitting corrosion behavior of SS exposed to highly chloride concentration solution at the elevated temperature. Section 6 reviews the mechanism of CISCC and the previous experimental studies. Section 7 obtains from Ref. 3, displaying the experimental method of measuring crack growth rate (CGR) and results of accelerated SCC test of canister material exposed to the simulated canister surface environment.

2. Types of Dry Storage System

Various dry storage system designs have been approved by the United States Nuclear Regulatory Commission (U.S. NRC) [4]. Based on the approach for loading spent fuel from the reactor to the storage system, the category of dry storage systems can be divided into canister-based system and bare-fuel system [5].

Canister-based system consists of a thin-walled (1/2 inches or 5/8 inches) SS canister within a reinforced overpack. Spent fuel assemblies are loaded into baskets (criticality control is provided by the basket that holds the spent fuel assemblies within individual compartments and may contain boron-doped metals to absorb neutrons) integrated into the SS cylinder. The canister is sealed with a welded lid. Like the NUHOMS® (NuTech Horizontal Module System) design, the canister shell is typically constructed using two pieces of rolled plates creating two half shell length longitudinal welds and a single circumferential weld in the middle. The shell welds are full penetration butt welds [7]. The canister provides confinement and criticality control for the storage and transfer of irradiated fuel that has already been cooled in the spent fuel pool for years. The canister provides an inert environment (filled by helium gas) for the spent fuel.

Depending on the design, the overpack (or vessel) is either concrete or a hollow carbon steel shell filled with concrete, which can shield gamma radiation and neutrons. The overpack both protects the canister from the weather and provides shielding as that high radiation fluxes pass through the thin canister. Some of the system designs can be used for both storage and transportation if the canister is placed within a suitable overpack. The overpack is closed with a bolted lid. The overpack has air inlets and outlets on the cylinder body to make air move the decay heat of spent fuel. Fuel cooling is passive as the cooling replies a combination of heat conduction through solid materials and natural convection or thermal radiation through air to move decay heat from the spent fuel to the ambient environment. When the fuel is moved from the spent fuel pool to the dry storage system, the decay heat given off by the fuel has significantly decreased. Dry storage systems store the spent fuel in either a horizontal or a vertical position.

Bare-fuel system is the second type of the dry storage system. Spent fuel assemblies are placed directly into a thick-walled basket that's integrated into the cask itself. Casks do not have an overpack and are generally bolted shut, as opposed to being welded. Casks are made of various materials, being thick-walled and self-shielding. Similar to the canister-based system, the bare-fuel system can be used either for storage only or for both transportation and storage [1].

For those addressed in this chapter, the dry storage system refers to the canister-based system with the inner welded metal canister, as which is susceptible to pitting corrosion (Sections 4 and 5) and CISCC (Sections 6 and 7).

3. Canister Surface Environment

Being located at coastal sites and loaded spent nuclear fuels, the canister surface has salt deposit with elevated temperature. This section is to understand the environment that canister exposed to. Three factors are introduced in this section: surface deposit, humidity, and temperature.

3.1. *Surface deposit*

Canister surface is inevitable to accumulate a level of salt, dust, and other contaminations being located at coastal. Dew, ocean spray, rainwater, fog, and other forms of water splashing, which are dominated by sodium–magnesium chloride sulfate salts [6], are the main source of deposit. Salt phases, such as Cl, S, Ca, Mg, and Na, are the priority of concern on the material degradation. Salt species and density are varied by geographical location and accumulation time. Salt deposits on the dry storage system are not pure sea-salt. Chloride is the most aggressive phase to SSs as the ion [Cl^-] penetrates the oxide layer of SSs and reacts with the metal underneath, being corrosive and CISCC susceptive.

The surface deposit has been studied on the dry storage systems at Calvert Cliffs and Hope Creek. Inspections on the Calvert Cliffs SS dry storage canister surface deposits have showed that the major anion species were sulfate (SO_4^{2-}) and nitrate (NO_3^-) after 15.6-year deposition [6]. The major cation species were Ca^{2+} with less amounts of K^+, Na^+, and Mg^{2+}. Considering the sea water is dominantly a Na–Mg–Cl–SO_4 assemblage [7] and the inland rainwater is dominantly a NH_4–Ca–NO_3–SO_4 [8], the species of deposits more closely resembled inland rainwater than standard seawater. In addition, the actual chloride concentration was well below $0.1\,g/m^2$. As the measured surface temperature was in the range of 40–52°C, the salts were deliquescent and form a moisture layer on the surface. According to lab tests, performed on U-bend SS samples using different concentrations of deliquescent salts (0.1, 1.0, $10.0\,g/m^2$ at 35–45°C), this level of deposit concentration probably is out of the susceptibility range of chloride concentration for SCC occurrence. The inspection on the Hope Creek unused canister suggested that the salts were dominantly sulfates, with rare chlorides and nitrates, after 7-year deposition [9]. Ca–SO_4 and Na–Al–SO_4 were the most common sulfates, but the phases of Na–SO_4 and K–SO_4 were also present. The concentrations are not included in the report [9].

3.2. *Surface humidity*

Ambient humidity represents the vapor density in air, being 30–$35\,\mathrm{g/m^3}$ at coastal. Affected by the heat on canister surface, the vapor density, which is very close to the surface, is different from the vapor density in air. This vapor density is defined by the RH, being the ratio of partial pressure of vapor in air to the saturated vapor pressure at the same temperature. At atmosphere, the reduction of temperature increases the RH; in other words, with the decreasing of spent fuel decay heat with time, the RH on canister surface is increasing.

The RH in the vicinity of chloride-containing surface salt deposits determines the chloride concentration of the atmospherically induced electrolyte layer and, in the case of sea-salt, its composition too. Previous studies suggested that the limiting RH (the minimum vapor density level) for sea-salt to deliquescence is 15% and the surface temperature must be reduced to around 70°C. Caused by the deliquescence, a highly concentrated chloride-containing moisture film would be formed on the surface. Even if the salt deposit is not dense, the ion concentration could be heavy due to the bit of vapor. However, too much vapor or water drops on the surface would dilute the ion concentration, which supports corrosion.

3.3. *Temperature*

In-service interim canister surface temperature is always higher than the ambient temperature because of the spent fuel decay heat. The temperature reaches to hundred-degree at the start of service when transferring from the wet storage system to the dry. Fuel temperature in storage is limited to 200–300°C, depending on the burnup. The surface temperature is decreasing with time, but extremely slow due to the amount of decay heat.

The surface temperature distribution is significantly varied as the decay heat located inside is not homogeneous and the cooling system is passive in nature. The distribution of spent fuel inside the canister affects the heat convection from the hot fuels to the canister wall. Within the concrete vessel, the cooling air passively flows from the bottom to the top and brings away the decay heat, resulting in large temperature gradients vertically. The bottoms of canisters are generally the locations of lowest temperature through the canisters, and progressively hotter going up the canister side with a temperature difference more than 50°C [6, 9–11].

The effect of temperature on materials corrosion is overwhelmingly complex. There is a range of temperatures over which localized corrosion and

SCC can occur. When temperature is beyond the range, it reduces the surface RH below the limiting. At RH values below the limiting, it will prevent the formation of deliquescent salts, hence corrosion cannot occur. Under a constant humidity, the increase of temperature tends to increase the electrochemical and diffusion kinetics, resulting in a higher corrosion rate [12]. On the other hand, the increase of temperature decreases RH on the surface, preventing the deliquescence of salts and decreasing the evaporation of surface electrolyte, thus decreases the corrosion rate [13]. When temperature is below the range, it will increase the surface RH to a high value and make the moisture layer too dilute to support corrosion.

The canister material has been suggested to be susceptible to pitting corrosion when exposed to the environment as discussed above. The fundamentals of pitting corrosion are introduced in Section 4. The pitting corrosion resistance of canister material exposed to the typical coastal environment as discussed above has been investigated and presented in Section 5.

4. Pitting Corrosion

Canister material is susceptible to pitting corrosion when exposed to chloride-containing environment. The effect of chloride ions is introduced in Section 4.1. Section 4.2 investigates the effect of metallurgy property in relative to pitting corrosion resistance. Pitting corrosion process includes five states: initial state, growth state, declining state, repassivation state, and critical state (Section 4.3). The electrochemical techniques applied on pitting corrosion analysis are explained in Section 4.4. Section 4.5 introduces the modeling development of pitting corrosion.

4.1. *Chloride ion*

SSs are resistant to electrochemical corrosion due to the chromium oxide passive layer (1–3 nm) [14]. Cr is one of the primary elements that play a major role in corrosion resistance of SSs. The presence of Cr in SSs increases the corrosion resistance, hardness, and toughness due to the formation of chromium-rich inner barrier oxide layers [15–19]. Such passive layers, however, are often susceptible to localized breakdown at defect sites in the presence of aggressive chloride ions, which causes pitting corrosion and severe damage to the structure [20].

The pitting corrosion behavior of SSs in chloride-containing environment has been extensively investigated. Chloride is an anion of a strong acid, and many metal cations exhibit considerable solubility in chloride solutions. The fact that chloride ions are small and firmly attracted to each other leads to reactions of the chlorides with other elements readily due to the strong forces of attraction in ionic bonding, which results in the high solubility of chloride salts [19, 21]. The reason for the aggressiveness of chloride ions has been considered, and a number of investigations and examinations were carried out in marine and offshore installations [22]. Most theories believe that the chloride ions could penetrate through the passive layer and come in contact with the metal layer, resulting in film breakdown [23–26]. The pitting corrosion resistance tends to be varied with the concentrations of chloride ions. The relation of concentration of chloride ions and pitting corrosion resistance is discussed in Section 5.

4.2. *Metallurgy property*

Localized corrosion frequently forms at the locations of inhomogeneous steel matrix (typically at weld joints), discontinuous surface films, differential aeration and different pH, as where usually form the macro-galvanic cells [27]. Pits almost always initiate due to chemical or physical heterogeneity at the surface such as inclusions, second phase particles, solute-segregated GBs, mechanical damage, and dislocations [28]. The pitting corrosion resistance depends on the nature and number of undesirable phases, the chemical compositions, and the production conditions of metal [29, 30]. To maintain high resistance to localized corrosion, it is commonly agreed upon that the alloying elements must be homogeneously distributed.

Most engineering alloys have many or all such defects, and a pit tends to form at the most responsive sites first. For example, pits in SSs often associate with manganese sulfide (MnS) inclusions. The role of MnS inclusions in promoting the breakdown and localized corrosion of SSs has been broadly recognized [22, 28, 31]. In addition, the balance of alloying elements can be changed by the precipitation of various secondary phases, caused by welding processes or other applied conditions (e.g., fabrication, improper heat treatment, prolonged exposure to high temperature during their service lives). The microstructural changes in the welding zone endanger the corrosion resistance of SS in alkaline solutions [32, 33]. The presence of precipitates degrades the metal property, particularly corrosion resistance. In the weld joint, coarsening of the $Cr_{23}C_6$ precipitates by the welding thermal cycle render the creep rupture strength of the weld joint significantly lower than the as-received [34]. The precipitates also make a depletion zone formed around the GB. The study has found that the IG crack occurred in the stabilized SS was caused by Cr depletion due to the segregation of the solute Cr atoms in the GBs [35]. The depletion zone weakens the corrosion resistance.

The alloy compositions and microstructure have strong effects on the pitting corrosion resistance. An existing pit can also be repassivated if the material contains a sufficient amount of alloying elements such as Cr, Mo, Ti, and so forth. Pitting potential (i.e., E_{pit}, see Section 4.4) was correspondingly found to increase dramatically as the Cr content increased above 13%, which is a critical value to create SSs. Increasing concentration of Ni, which stabilizes the austenitic phase, moderately improves the pitting resistance of Fe–Cr balanced alloy. Small increases in certain minor alloy elements include Mo and N can greatly reduce pitting susceptibility [28]. Mo is practically effective but only in the presence of Cr through enhancing the enrichment of Cr in the oxide film.

4.3. *Pitting corrosion state [36]*

Pitting corrosion is a stochastic process, being a consequence of the breakdown of the oxide passive layer caused by random fluctuations in local conditions. This process can be divided into five states which include (1) initial state, (2) growth state, (3) declining state, (4) repassivation state, and (5) critical state. Figure 1 shows the transitions of pitting corrosion states. In the initial state, the surface has no pits and is protected by a passive oxide layer. Probability exists that pits will form and grow through the diffusion of aggressive ions. Metastable pits are pits that initiate in micro size (above the level of detectability of the modern microscopy technology) and grow for a limited period before repassivating, or entering a state of stable pit growth [22]. Repassivation occurs when the oxygen diffuses and oxide passive layer regrows at the pit location, returning the surface to a passive condition. Accordingly, three states can be used to describe the metastable

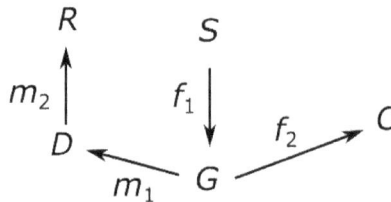

Figure 1. A physics-based pitting corrosion states transition diagram. S: initial state. G: growth state, D: declining state, R: repassivation state, C: critical state [3].

pits: growth state, declining state, and repassivation state. Declining state indicates that the pit depth is decreasing because of the repassivation property of pitting corrosion, which is an oxidation process. Thus, a metastable pit can either enter a state of stable growth or decline to a passive state. In the stable growth stage, the growth rate depends on material composition, pit electrolyte concentration, and pit bottom potential [22]. Note that even what appear to be small stable pits may in fact be metastable [22]. Hence, the actual micro size and limited period of a metastable pit is dependent on the behavior of the pit. The criterion used to determine whether a pit is stable is the achievement of a critical depth beyond which the pit will continue to grow. Below the critical depth, the pits are in the metastable pitting stage in which they may either be growing or declining. If the pit achieves the critical depth; it enters a state of stable growth. The critical depth is empirically determined and depends on the material composition and environment. With micro-characteristic analysis techniques, all states can be observed. Some other techniques, such as the current noise spectrum detected by the electrochemical technique, also can reveal metastable pit formation (i.e., pit generation and repassivation) [37].

4.4. *Electrochemical technique*

Previous studies used various techniques to quantify the pitting corrosion resistance. These methods include analyzing the local chemistries, the potentiostatic characterization, and potentiodynamic characterization. Figure 2 shows the anodic and cathodic electrochemical reactions that comprise corrosion spatially separate during pitting. The metal, M, is being pitted by an aerated NaCl solution. Rapid dissolution of M releases more M^{n+} in the pit, while oxygen reduction takes place on the adjacent metal surfaces. More chloride ions are attracted in the pit to neutralize, since the total charge of a solution must be zero. Oxygen is consumed in the pit and shifts most of the cathodic reaction outside of the pit. The cathodic reaction is

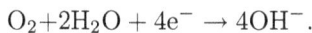

$$O_2 + 2H_2O + 4e^- \rightarrow 4OH^-.$$

The local chemistries were in considerable interest and investigated by a range of techniques. One way to isolate the pit solution is to rapid freeze the electrode in liquid nitrogen, removing the surface excess and subsequent thawing [38]. This approach is mainly used to study the pH in aluminum pits and the chloride ion concentration in pits for SSs, allowing a considerable volume of pit solution to be analyzed. Likewise, it has been

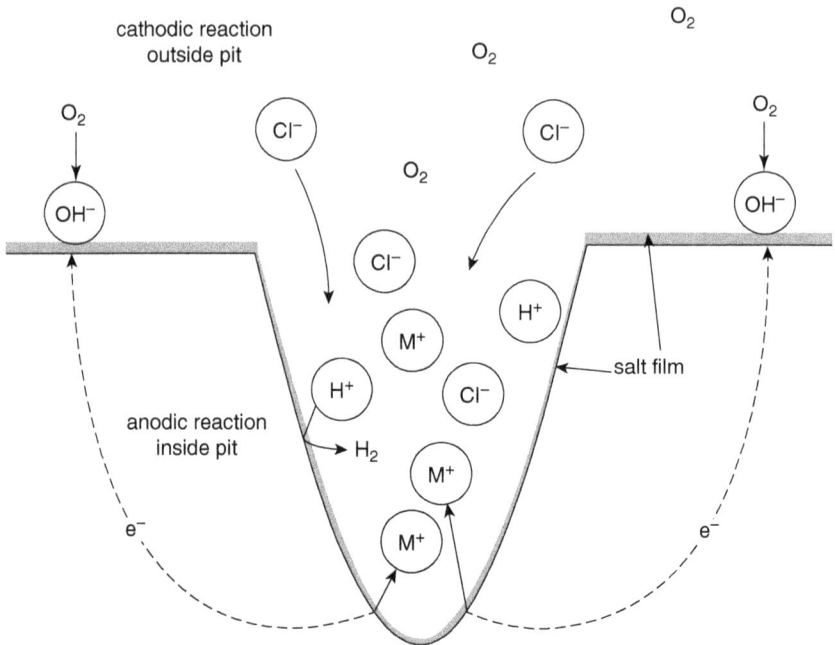

Figure 2. Schematic of anodic and cathodic electrochemical reactions in an initiated pit.

suggested to isolate the pit solution by using artificial pit electrode methods [22]. This is also known as 1-D pit, or lead-in-pencil electrode, which is a wire embedded in an insulator such as epoxy. Another way to isolate the pit solution is by inserting microelectrodes into pits, cracks, and crevices. With this technique, once the local solution composition is fully character-ized, it is possible to reassemble the local environment by reconstituting it in bulk form from reagent-grade chemicals, and then determining the electrochemical behavior of a normal-sized sample extracted from a local environment.

The electrochemical characteristic inside the cavity was also in interest as it would reveal the different aspects of potentials existing within pit-ting corrosion. Figure 3 shows a typical polarization curve displaying the relationship of E_{corr} and E_{pit}, and the region of metastable pits. E_{pit} is principally treated as a standard to compare the pitting resistance among all the metals and as a criterion whether the initiated cavity becomes sta-bly propagated. For instance, metals with low experimentally determined E_{pit} have a higher tendency to form pits naturally at open circuit [39].

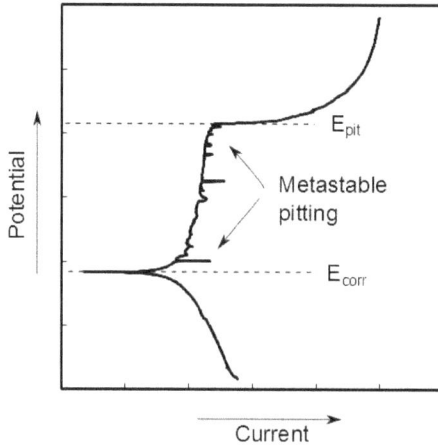

Figure 3. Representative polarization curve with E_{pit} and E_{corr} indicated. Additionally, metastable pits prior to E_{pit} are shown.

The metastable pit, which initiates for a limited period before repassivating, is found that it occurs below E_{pit}, being not an initiation point of pitting.

As indicated in Section 4.1, the aggressive ions take significant effect on pitting corrosion resistance. It is revealed that E_{corr} and E_{pit} decreased, and the current density in the passive region increased with the increase of chloride ion concentrations [40]. A linear relationship between E_{pit} and the chloride ion concentration has been found [40]. The previous anodic potentiodynamic tests of type 304L SS suggested that for the NaCl solution (pH 2) under 4.5% concentration, passivation region had a significant range between the E_{corr} and E_{pit}, and passivation break-down potential E_{pit} shifted toward more positive potentials with the decrease of chloride concentration [40]. A steady increase of current with potential in 4.5% NaCl, which indicated active corrosion, has been observed [40].

It has been agreed that removal of oxidizing agents, e.g., removal of dissolved oxygen, is one powerful approach for reducing susceptibility to localized corrosion; however, it is also supported that sufficient oxygen to the reaction may enhance the formation of an oxide layer and thus repassivate or heal the damage to the passive film, especially the increase in potential associated with oxidizing agents would improve E_{pit} [28].

Repassivation potential (E_{rep}) is the other criteria to interpret the pitting resistance. With the effects of E_{pit} and E_{rep}, cyclic polarization

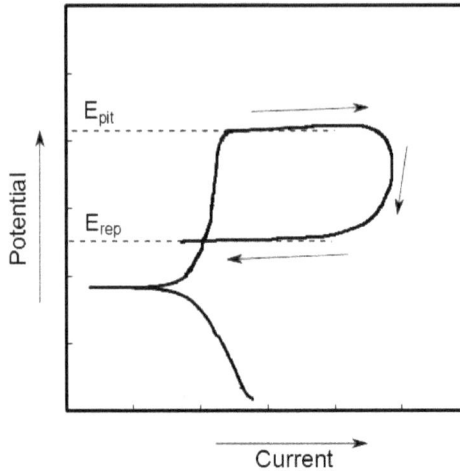

Figure 4. Representative cyclic polarization curve with E_{pit} and E_{rep} indicated.

experiments are used to measure the pitting resistance of metals. The values could be used to determine under what conditions pitting corrosion occurs. Figure 4 illustrates the typical repassivating polarization curve to estimate the susceptibility of the metal alloy to pitting corrosion. The curve is used to find E_{pit} and repassivation potential (E_{rep}). Higher E_{pit} for material in a given environment indicates greater resistance to pitting. Similarly, if the potential reduces below E_{pit}, the surface may repassivate and pit growth can stop. However, if the potential is between E_{pit} and E_{rep}, pitting can be expected. Related studies have been investigated and presented in Refs. [26, 28, 41].

However, using the electrochemical method to compare the pitting corrosion resistance is not absolutely valid. The potentiodynamically determined E_{pit} of most metals exhibits a broad experimental scatter of other of hundreds of millivolts. E_{pit} is insufficient for the development of a fundamental understanding of pitting corrosion mechanism.

The other study interest in pitting corrosion is the influence of pit chemistry on pit growth and stability, which has been provided by Galvele [42]. The concentrations of various ionic species at the bottom of modeled 1-D pit geometry were also studied [22, 42]. The concentration of various ionic species is determined as a function of current density based on a material balance that considered the generation of cations by dissolution, outward diffusion, and thermodynamic equilibrium of various reactions such as cation hydrolysis [22]. Galvele [42] found that a critical value of the factor $x \cdot i$ (where x is pit depth and i is current density) corresponds to a critical pit acidification for sustained pit growth. Current density in a pit is a measure of the corrosion rate within the pit and thus a measure of the pit penetration rate. $x \cdot i$ can be used to determine the current density required to initiate pitting at a defect of a given size. Increasing the pit density increases the ionic concentration in the pit solution, often reaching supersaturation conditions.

4.5. *Modeling of pitting corrosion [36]*

This section is a summary of the important concepts and methods in the analysis of pitting corrosion behavior, which involve deterministic (or mechanistic) and non-deterministic approaches. Section 4.5.1 describes the deterministic models, which are commonly used to simulate the pit growth rate, and compares it with the models that incorporate the stochastic nature of pitting corrosion. Section 4.5.2 discusses the static and dynamic stochastic models, which are two distinct non-deterministic methods of pitting corrosion analysis, and indicates the contributions of the multi-state Markov model.

4.5.1. *Deterministic and non-deterministic models*

The wide range of diverse pitting corrosion models that have been proposed can be categorized as two distinct model types: deterministic (or mechanistic) and non-deterministic. The former is often formulated using partial differential equations based on a set of variables and environmental parameters [43–45]. Different deterministic models use diverse variables to characterize the different aspects of the corrosion mechanism. The partial differential equations for describing pitting corrosion can be based on either reaction kinetics or electrochemistry [46]. Since there are conditions under which further growth of a stable pit is arrested, the deterministic models consider mechanisms that provide limitations on pit size (e.g. [22]). Under freely corroding conditions, a pit acts as the anode and the surrounding material acts as the cathode; a stable pit can continue to grow only as long as its anodic current is matched by cathodic current from the surrounding material [22]. However, stochastically-based pitting models typically do not account for limitations on pit size. Particularly for long-term pitting corrosion, models are needed that address both the stochastic aspects of pit growth but also include a mechanistic treatment of limitations on pit size.

Non-deterministic models utilize the probabilistic characteristics of the data to interpret or evaluate certain aspects of the evolution of pitting corrosion [47–49]. These probabilistic characteristics can be either static or dynamic, which will be further expanded upon in Section 4.5.2. One example of a dynamic probabilistic model divides the metal surface into a 2-D array of hypothetical cells, then assigns the probabilities for the transitions among pitting states to each cell [50]. In the dynamic analysis, nucleation or destruction of a pit embryo is determined probabilistically by taking random draws from assumed probability density functions for pit formation and growth properties. After a pit embryo grows to a certain stage, it becomes a stable pit and follows the stochastic rules for pitting growth. The environmental impacts enter the model based on a combination of theoretical modeling and observed data.

Models that are formulated as non-deterministic in nature are more favorable for the initial modeling effort. Although the deterministic models are better suited for interpretation in terms of actual physical and chemical parameters, they cannot explain the stochastic behavior of a real corrosion process. The non-deterministic models proposed to date only involve simple computations but are likely to generate more realistic results than the deterministic models. Also, existing non-deterministic models are incapable of relating pit growth behavior based on physical or chemical parameters. Besides, the non-deterministic models are mostly locally applicable, not incorporating any global behavior (e.g., [47]).

Mears and Evans stated in 1935 [51] that "from the practical standpoint ... it may ... be more important to know whether ... corrosion is likely to occur at all than to know how quickly it will develop." Thus to date, the deterministic approaches have so far prevailed in corrosion science, even though the stochastic theory has been successfully used to explain the pitting corrosion behavior, such as the probability distribution of both the pitting potential and the induction time. Work by Henshall *et al.* [52] showed that a computational model based on a stochastic approach could describe the pit initiation and growth on SSs, but should involve some deterministic elements. To predict the complex corrosion process, we conclude that we need a computational capability that is probabilistic in formulation but includes mechanistic models.

4.5.2. *Static and dynamic stochastic models*

The static non-deterministic approach involves fitting of a probability distribution function or a combination of several probability distribution functions to a set of observed values of the random variables of pit shapes, i.e., pit depth and radius. The majority of researchers agree on the concept that it is impossible to use a single probability distribution function to simulate the complex pitting corrosion process [53]. Although a combination of several probability distribution functions can provide a good fit for the complicated pitting corrosion process, the limitations are still distinct, including the assumption of nominal "homogeneity" in the system (e.g., random distribution of material microstructure) and the inflexibility in dealing with the long-term changes in operating conditions, environment and the pit shape [54]. Furthermore, there is no static approach that has included all the time-dependent factors (e.g., pH and ion concentration in the pit, pit density, and shape) for naturally induced pitting corrosion. The long-term changes in behavior should also be taken into account, which affect the ultimate severity of pitting damage.

Because of the substantial interest for a flexible approach to deal with the long-term changes of pitting corrosion, the dynamic stochastic methods (DSMs) seem to be better suited for events taking place over time than static probabilistic methods. DSMs can also quantitatively include the time-dependent environmental parameters that influence the "birth and death" of corrosion pits. Other advantages of dynamic stochastic methods are as follows:

- The DSMs were found to be applicable to pit generation events [47, 53].
- The pit initiation and repassivation can be described as a birth and death stochastic process [52].
- The relation between the distribution of pitting potential and the induction time for pit generation can be stochastically derived [55].
- The random current noise generated by pitting corrosion obtained by electrochemical techniques can be modeled as a stochastic process [56, 57].
- DSMs can be applied that include the effects of inclusions and inhibitors [58].

A number of potential corrosion growth models were evaluated from a survey in Ref. [59]: the linear growth model, time-dependent generalized extreme value distribution model, time-independent generalized extreme value distribution model, single-value corrosion rate model (National Association of Corrosion Engineers (NACE) model) and Markov model. It was concluded that the Markov model is the best among them for predicting the corrosion rate distribution with time as it considers the ages and sizes of the corrosion defects as well as the observed dependence of the corrosion defect depth on time [59].

Markov models that have been used to model pitting corrosion have shown to agree well with experimental results. However, improvements in the method are still needed. For instance, Valor *et al.* [53] have treated the pit initiation/generation process as a nonhomogeneous Poisson process, and the pit growth process as a nonhomogeneous Markov process. Although the model obtains the distribution of maximum pit depths resulting from the combination of the initiation and growth processes for multiple pits, it provides neither an approach for estimating the long-term development of pitting corrosion nor the resolution of the pit depth distribution with density. Also most pitting corrosion models are based on a physical description of how a pit is generated and grows on the material surface (e.g., [47, 50]) but do not analyze all the pitting corrosion states or interpret all the pit formation processes. The input parameters for the models relative to the environmental factors, such as temperature, pH, and humidity for different materials, are assumed and suitable for a very limited range of materials and environments. To address these limitations, the multi-state Markov model, which is able to describe the dynamics and determine the input parameters and transition rates accounting for environmental effects (chloride concentration and temperature) for different materials, has been proposed [3, 36] to simulate the pitting corrosion states and pit formation process.

5. Experimental Studies of Pitting Corrosion [2]

Experimental methods have been conducted to explore the corrosion behaviors of SSs exposed to highly concentrated chloride solutions. This section introduces the experimental studies which investigated the pitting corrosion characteristics of as-received 304L and weld 304L exposed to highly concentrated chloride solutions. The experimental setups, including the materials, solution, and test methods, are introduced in Section 5.1. Extensive experimental data have been collected using techniques including electrochemical techniques such as open circuit potential (OCP) measurement (Section 5.2.2), electrochemical impedance spectroscopy (EIS) and potentiodynamic polarization (Sections 5.2.3 and 5.2.4), and optical profilometry (Section 5.2.5).

5.1. *Experimental setup*

5.1.1. *Materials and solution*

The as-received 304L and weld 304L were prepared based on the focus of research. Weld 304L was machined from the FZ of a double-V grooved butt, which was made by GTAW. The hardness of the test material was performed by the Vickers' hardness tester (Maker: Buehler; Model: Micromet II) using a diamond indenter at room temperature. The hardness gauge was HV, the force was 300 gf (2.94 N) and the surface shape was plane.

The outer surface of the weld was exposed to the investigated environments. The specimens were fabricated in two shapes, circular disks and rectangular coupons, for electrochemical testing and immersion testing, respectively. Each circular disk specimen has a 0.196 cm² exposed area (0.5 cm outer diameter) and was assembled with the polytetrafluoroethylene electrode head to be used in the electrochemical testing. The edges of the specimens were protected from crevice corrosion by rubber washers. The exposed surface was mechanically polished up to an 800-grit surface finish by using sandpaper. Each coupon specimen was machined to a 4 cm² (2 cm × 2 cm) and 5-mm-thick specification, and was mechanically abraded and SiC diamond polished to 1 μm.

After polishing, the specimens were ultrasonically cleaned in analytical grade acetone for 3 min, rinsed with ultrapure water (resistivity > 18 MΩ · cm), and dried by analytical grade ethanol. They were kept in a desiccator for 3 days to make air-formed oxide layers grown on the surface. The SSs were rapidly oxidized and formed a well protective Cr-base oxide scales when exposed to dry air [60]. In most early-stage of applications, especially those in nature environment, the steels are protected by the air-formed oxide layers. The purpose of the formation of oxide layers was to make the experiments to approach the reality of applications. Chemical composition of the specimens is displayed in Table 1.

The base solution was made up by dissolving an appropriate amount of analytical grade and chemically pure NaCl and American Society for Testing and Materials (ASTM) standard D1141-98 sea-salt[a] in ultrapure water.

Table 1. Chemical composition of as-received 304L used in the investigations (wt.%).

Fe	Cr	Ni	C	P	Si	Cu	N	Mn	S	Mo	Co
Bal.	18.21	8.06	0.021	0.031	0.44	0.46	0.082	1.65	0.024	0.65	0.11

[a]Chemical compositions (g/L): 24.53 NaCl, 5.20 MgCl₂, 4.09 Na₂SO₄, 1.16 CaCl₂, 0.695 KCl, 0.201 NaHCO₃, 0.101 KBr, 0.027 H₃BO₃, 0.025 SrCl₂, 0.003 NaF, and traces of nitrate compounds.

Table 2. Solution specifications.

Solution series	Solute	Weight concentration (wt.%)	Chloride molarity (M)	Temperature (°C)	pH
A	99.99% pure NaCl	26.7	6.25	40	7.00
B	ASTM D1141-98 sea-salt	10.5	1.5	40	7.32
C	ASTM D1141-98 sea-salt	10.5	1.5	70	7.00

Table 2 shows the solution specifications. Each test solution is marked as A, B, or C in series.

5.1.2. *Electrochemical measurements*

The tests were conducted using a typical three-electrode electrochemical system in a glass cell. Each test included a fresh liter of solution. The circular disk specimen mounted on a PTFE shaft was used as the working electrode (WE). A graphite rod seated inside a fritted glass tube was used as the counter electrode (CE). An Ag/AgCl (4 M KCl) reference electrode (RE) was connected to the cell externally via a Luggin tube. The solution temperature was maintained within ±1°C by using a thermocouple and a controllable hot plate. The solution was exposed to the atmosphere via a condenser, which was also used to avoid the evaporation of water. The solution was brought to a boil and cooled down to the desired temperatures (40°C and 70°C). When the desired temperature was achieved, the solution was maintained at the desired temperature for 3 h for the equilibration with air. All electrochemical measurements were performed using a Gamry Interface 1000^{TM} potentiostat controlled by the Gamry Framework software.

After an hour of immersion, the potentiodynamic polarization measurement was conducted at a scan rate of 0.5 mV/s ranging from −0.2 to 0.8 V vs. OCP. After the measurement, the specimen was released from the WE.

A fresh liter of solution and a new specimen were prepared to take the 43-h electrochemical measurements. The OCP measurement throughout the test was taken at a sample rate of 1 s. The EIS measurements, taken at 1, 4, 9, 17, 25, 34, and 43 h into the test, were carried out at the OCP over a frequency range of 100 kHz to 10 mHz. The sinusoidal voltage applied as the disturbance signal had a 10 mV amplitude. After 43 h of immersion, the potentiodynamic polarization measurement was conducted at a scan

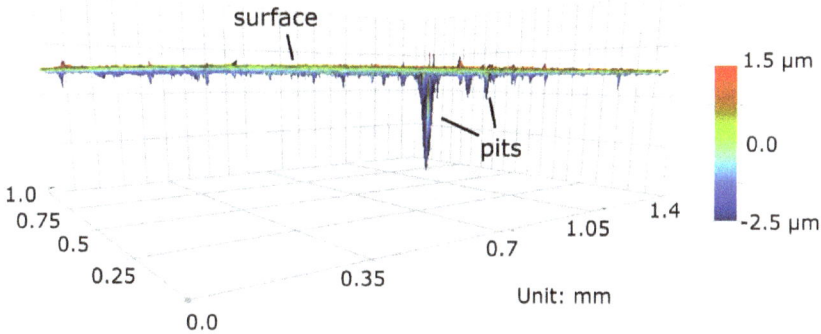

Figure 5. Example 3D morphology profile of pits by an optical profilometry. (For interpretation of the references to color in this figure legend, the reader is referred to Ref. [2].)

rate of 0.5 mV/s ranging from −0.2 to 0.8 V vs. OCP. Replicate tests were conducted, and results were essentially the same.

5.1.3. *Optical profilometry*

The coupon specimens were immersed in the solutions shown in Table 2 at the time intervals of 2, 5, 10, 20, and 30 days. After testing, they were ultrasonically cleaned in analytical grade acetone for 2 min, rinsed with deionized (DI) water, and dried by analytical grade ethanol. The 3D surface morphology of the post-test specimen was observed using a non-contact ContourGT-I 3D optical profilometry; the depth, radius, and density of pits, were analyzed using the Vision64™ software.

Figure 5 is an example 3D morphology profile, showing the pit distribution and geometry on the specimen surface. Since the pits were highly irregular in shape, a pit's maximum depth was used as the depth of the pit, and a pit's diameter at the surface was counted as its diameter.

5.2. **Results and discussion**

5.2.1. *Properties of weld 304*

Before testing, the specimen surface was examined by scanning electron microscope (SEM) in back-scattered electron mode. To characterize the microstructures, the polished specimens were etched in the mixture solution (5 mL hydrochloric acid, 1 g picric acid, and 100 mL ethanol) for several seconds. The differences between as-received 304L and weld 304L are compared in Figure 6. The as-received 304L has the ferrite and austenite

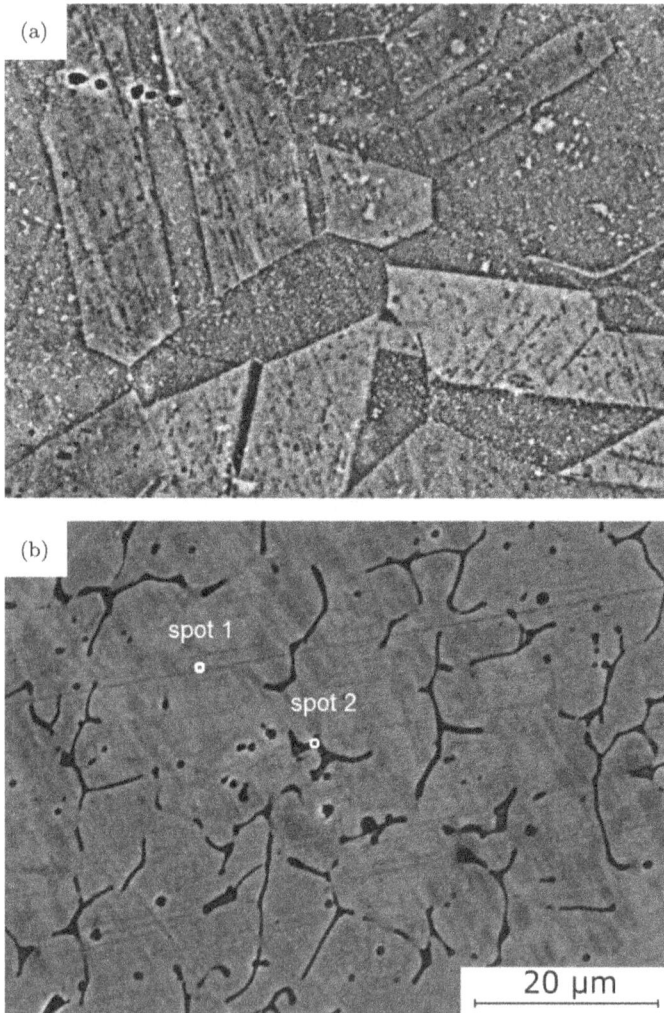

Figure 6. SEM images of the surface of (a) as-received 304 and (b) weld 304L before testing.

phases (Figure 6(a)), and the weld 304L has a dendritic phase (Figure 6(b)). Table 3 displays the compositions of the two phases (marked by spot 1 and spot 2) in Figure 6(b). The phase of spot 2 is susceptible to the formation of Cr–C precipitates due to the higher content of Cr and C and lower content of Ni than the phase of spot 1, which is the same as the as-received.

Table 3. The composition of elements of the spots shown in Figure 6(b).

	Spot 1		Spot 2	
	Atomic %	Uncertainty %	Atomic %	Uncertainty %
Fe	70.43	2.76	60.1	3.01
Cr	19.7	4.02	21.47	4.07
Ni	9.87	7.56	0	0
C	0	0	18.43	18.84

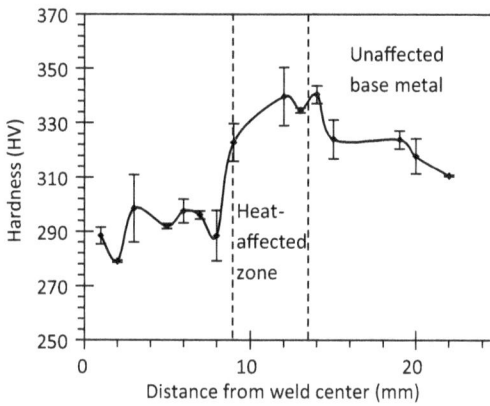

Figure 7. Hardness profile across the weldment.

Figure 7 shows the hardness profile measured by the hardness tester across the weldment. It is found that the area around weld 304L is generally softer than the area around as-received 304, which is believed to be resulted from the WRS, microstructure, and chemical composition. The narrow HAZ obtains more hardness because of the formation of hard, brittle interdendritic phases. As shown in Figure 7, the average hardness of the FZ is 295 HV/0.3; while for the unaffected as-received 304, the average hardness is 323 HV/0.3.

The weld 304L inherently possesses compositional and microstructural heterogeneities. The corrosion resistance of the weld may be inferior to that of properly annealed as-received 304L because of micro-segregation, precipitation of secondary phases, and formation of unmixed zones. The compositional and microstructural heterogeneities of weld may produce an electrochemical potential difference that makes some regions of the weldment more active, and consequently make different corrosion behavior from the as-received. In the following sub-sections, the different corrosion behaviors of the two are discussed in details.

5.2.2. *Evolution of the corrosion potential*

As shown in Figure 8, OCP of the specimens exposed to the investigated environments was measured as a function of time for 43 h. Table 4 summarizes their starting and ending potentials (averaged hourly). An increase of OCP means depolarization of cathode and increase corrosion, a drop in potential is evidence for decreased corrosion [61]. Most of the OCP values fluctuated in the testing, not showing a steady increase or decrease. The rapid changes in OCP indicate the enhancement of the anodic reaction or the formation of a semi-protective film [62]. In other words, they indicate the breakdown and repassivation of oxide layers that alternatively occurred

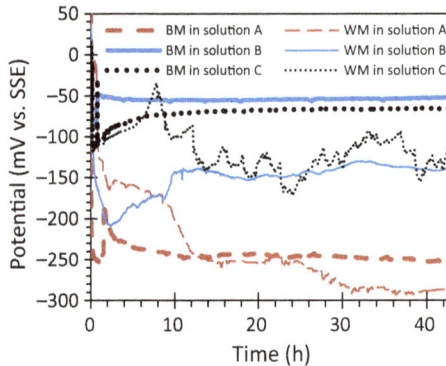

Figure 8. Plots of OCP over 43 h. A, B, and C represent the solutions shown in Table 2. BM stands for as-received 304; WM stands for weld 304. (For interpretation of the references to color in this figure legend, the reader is referred to Ref. [2].)

Table 4. The starting and ending potentials in mV vs. SSE (averaged hourly).

Time	As-received in A	Weld 304L in A	As-received in B	Weld 304L in B	As-received in C	Weld 304L in C
The first hour	−238.3	−40.1	−73.6	−130.5	−101.9	−112.7
The last hour	−251.6	−286.5	−51.8	−139.7	−65.1	−126.9

on the specimen's surface. The duplicate tests showed that the fluctuations of OCP were not always the same, but OCP always reached the same value after 35 h and became relatively stable. The OCP of all specimens became relatively stable after 35 h of immersion, indicating the passivation in the specimens became stable.

OCP of weld 304L was less stable as well as much lower than that of as-received 304L in the same investigated environment. This difference might be caused by the higher anodic dissolution rate at welding zone. The potential differences between the two might cause the severe galvanic corrosion in which case the anodic dissolution of weld 304L will be accelerated while the cathodic reactions take places at the relative large area of as-received. Although the difference between the OCP of the two is small, the corrosion rate could become significantly increased in an exponential manner if assuming potential–current kinetics of the cathodic reaction is similar.

The chloride concentration significantly affected potentials. The specimens exposed to 6.25 M chloride NaCl solution had much lower OCPs than those exposed to 1.5 M chloride sea-salt solution. Temperature has a smaller effect on the potential changes, which was determined by comparing the OCP values of 1.5 M chloride sea-salt solution at 40°C and 70°C.

5.2.3. *Determining the pitting potential*

As indicated in Section 4.3, the pitting corrosion process includes five stages. For the stable pit growth, which is characterized by the pitting potential (E_{pit}), several critical conditions should be met. When the critical conditions are not satisfied, pits will repassivate without further growth, which is called metastable pit growth.

When exposed to the aggressive solution in the early time period, the oxide layer might be unstable. The breakdown and repassivation of the layer alternatively occurred. The results of the potentiodynamic testing at 1 h is shown in Figure 9(a). In the passivation zone, there are significant current density spikes. These spikes can be attributed to the occurrence of metastable pits, explained by the consecutive formation and repassivation of micro-sized pits. The pit nucleation and metastable pit growth appeared as the current spikes on the polarization curves.

Note that the passivation of the specimens became relatively stable after 35 h, as discussed in Section 5.2.2. The potentiodynamic testing at 43 h was conducted to compare with the results of 1 h, shown in Figure 9(b). The current density spikes are no longer seen in the passivation zone. In

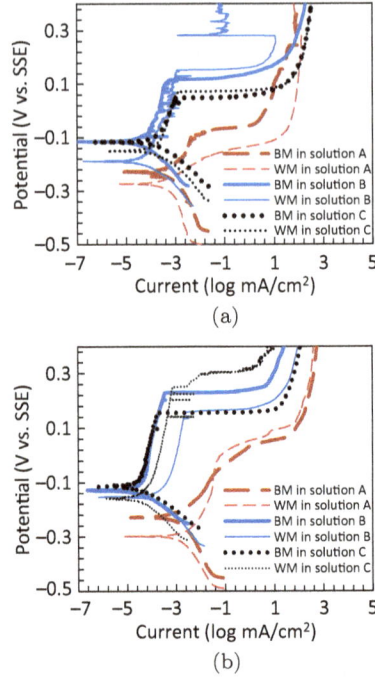

Figure 9. Plots of potentiodynamic polarization curves at (a) 1 h and (b) 43 h. A, B, and C represent the solutions shown in Table 3. (For interpretation of the references to color in this figure legend, the reader is referred to Ref. [2].)

addition, the passivation zones became larger than those obtained after 1 h (see Figure 9(a)).

As shown in Figure 9, weld 304L typically has larger anodic passive current densities and lower cathodic current densities than as-received 304L at the same potential. A larger anodic passive current density is representative of faster metal dissolution rates, therefore the corrosion resistance of weld 304L is depressed by welding. The anodic passive current densities of the specimens exposed to 6.25 M chloride solution were much larger than those of 1.5 M chloride solution. Lower corrosion potential of weld than as-received was also observed, which corresponded to the results of OCP shown in Figure 8.

Table 5 summarizes the electrochemical parameters, including E_{pit}, E_{corr}, β_a (anodic Tafel constant), β_c (cathodic Tafel constant), and i_{corr} (corrosion current density). The consistent pattern of the effect of chloride concentration on E_{corr} and E_{pit} was observed. The potential range of the

Table 5. Electrochemical polarization parameters at (a) 1 h and (b) 43 h of immersion testing.

Test	E_{pit} (mV vs. SSE)	E_{corr} (mV vs. SSE)	β_a (mV/decade)	β_c (mV/decade)	i_{corr} (mA/cm^2)
(a)					
BM-A	−94	−230	105	122	5.46E-04
WM-A	−150	−274	107	346	8.57E-04
BM-B	118	−118	354	126	1.02E-04
WM-B	156	−185	274	74	7.45E-05
BM-C	49	−116	281	79	2.46E-04
WM-C	72	−150	328	78	2.32E-04
(b)					
BM-A	40	−228	114	172	3.50E-07
WM-A	−3	−300	245	373	1.24E-06
BM-B	224	−131	598	68	4.56E-05
WM-B	161	−156	350	132	3.81E-04
BM-C	149	−114	450	58	3.63E-05
WM-C	247	−159	375	113	7.70E-05

passive region is displaced toward more negative potentials and becomes progressively narrower with increased chloride concentration and temperature. The larger i_{corr} and more negative E_{pit} are caused by the adsorption of chloride ions. The pitting potential was significantly lowered due to the increase of chloride concentration for both metals, indicating a higher pitting tendency at higher chloride concentration. It was also found that the pitting potential moved to the negative value when the temperature elevated to 70°C from 40°C, which is coherent with the negative linear temperature dependence of pitting potential reported by Ezuber [63].

Cyclic polarization scans were also applied in the investigated environments. However, the hysteresis loops typically intersect with the cathodic curves, far below the corrosion potentials, indicating little repassivation ability in the investigated solutions.

5.2.4. *Identification of passive layer*

Figures 10(a) and 10(b) show the evolution of the EIS results obtained for the metals exposed to solution A (6.25 M chloride concentration NaCl solution at 40°C). Figure 11(a) is the equivalent circuit used to characterize the EIS results. The equivalent circuit consists of a double-parallel, "R-CPE" arranged in series with the solution resistance. The double-parallel equivalent circuit indicates that the passive films formed as bilayers on the

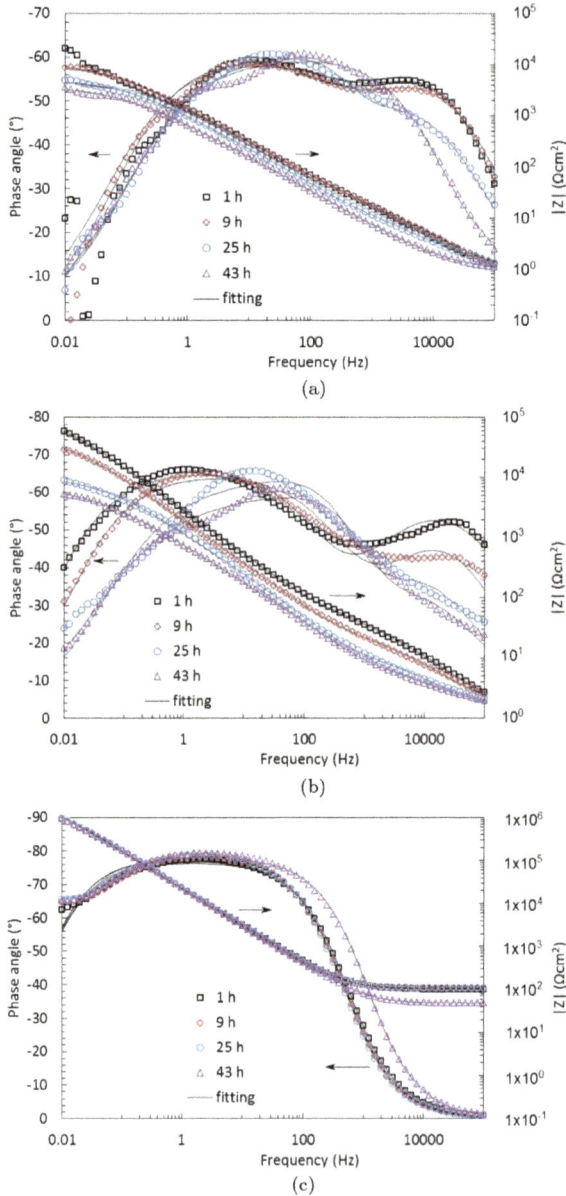

Figure 10. Evolution of the EIS results obtained for (a) as-received 304L exposed to the solution A, (b) weld 304L exposed to the solution A, (c) as-received 304L exposed to the solution B, (d) weld 304L exposed to the solution B, (e) as-received 304L exposed to the solution C, and (f) weld 304L exposed to the solution C at different exposure times: Bode plot representations. (For interpretation of the references to color in this figure legend, the reader is referred to Ref. [2].)

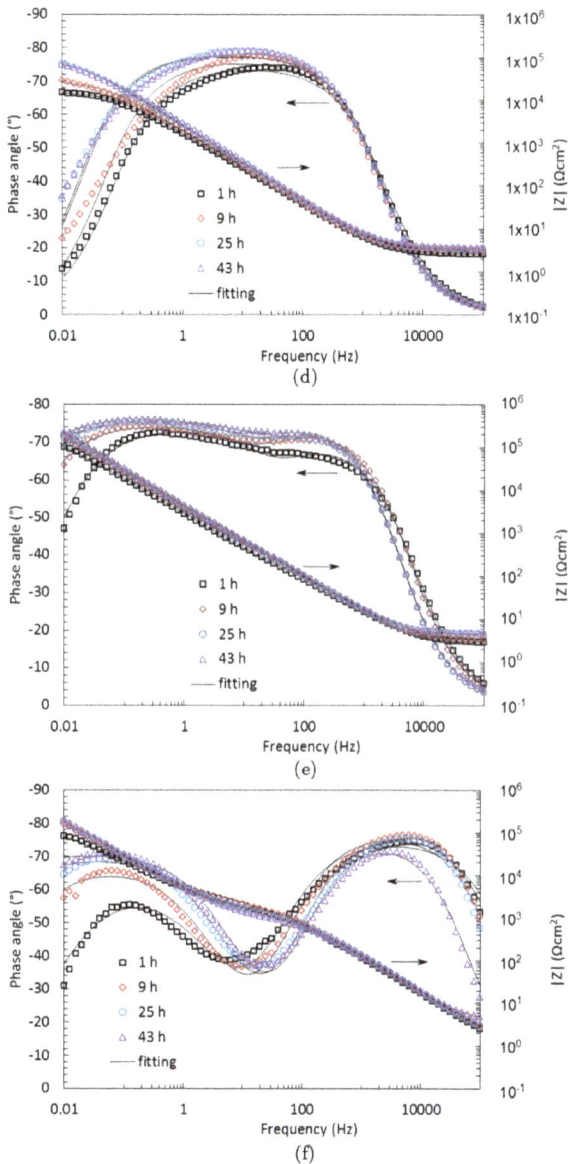

Figure 10. (*Continued*)

SS surface [64]. Each layer formed on the SS surface can be considered to be a parallel circuit of a resistor, due to the conductivity of the film, and a capacitor due to its dielectric properties [65, 66]. "CPE" is used to describe the frequency dependence of the non-ideal capacitive behavior

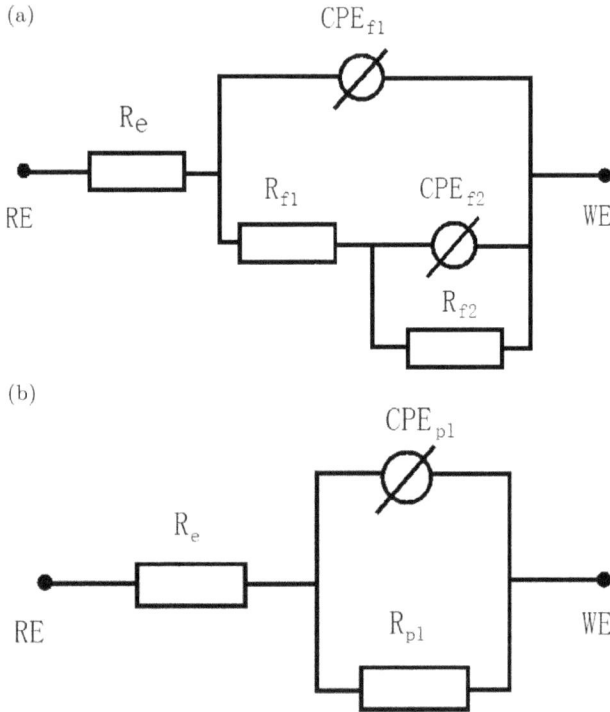

Figure 11. (a) and (b) Equivalent circuits used for modeling the EIS results shown in Figure 10.

in the corroding systems. The impedance representation of a "CPE" is given as

$$Z(\text{CPE}) = \frac{1}{Q_0(j\omega)^n}$$

where Q_0 is a fit parameter. In the ideal case, when the exponential factor $n = 1$, the "CPE" acts as a capacitor with Q_0 equal to the capacitance. In practice, n is less than 1. The CPE behavior arises because microscopic material properties can exhibit a distribution, and the n-value provides the information on the nature of the surface or the passive layer.

On the equivalent circuit (Figure 11(a)), R_e represents the solution resistance identified by the ohmic drop that occurs in the solution, CPE_{f1} represents the capacitance of the outer layer, R_{f1} is the resistance of the outer layer, CPE_{f2} represents the capacitance of the inner layer, and R_{f2}

is the resistance of the inner layer. Tables 6(a) and 6(b) display the evolution of passive elements, which are calculated after fitting the circuit analog proposed in Figure 11(a). The passivation of the anodic dissolution is due to the formation of the inner barrier layer. As shown in Tables 6(a) and 6(b), the resistance of the inner oxide barrier layer (R_{f2}) is much higher than the resistance of the outer porous layer (R_{f1}). The polarization resistance, R_p, which is reciprocal to the corrosion rate, is the sum of R_{f1} and R_{f2} for the bilayer structure. Since the outer layer resistance is negligible, the decrease of the inner layer resistance with time indicates that the barrier layer becoming thinner with time and the corrosion rate becoming higher. This corresponds to the increased anodic current density in solution A (Figure 9) and the declined OCP values observed in Figure 8.

Figures 10(c) and 10(d) show the evolution of the EIS results obtained for the metals exposed to solution B (1.5 M chloride concentration sea-salt solution at 40°C). Figure 11(b) is the equivalent circuit, consisting of a single-parallel "R-CPE" arrangement in series with the solution resistance. R_e is the solution resistance, while R_{pl} and CPE_{pl} represent the resistance of the passive layer and its capacitive behavior, respectively. Tables 6(c) and 6(d) illustrates the evolution of the passive layer's elements, which are calculated after fitting the circuit analog proposed in Figure 10. According to the fitting data in Tables 6(c) and 6(d), the resistance of the passive layer formed on as-received 304L is much higher than that on weld 304, representing better corrosion resistance. In addition, the corrosion resistance and the layer thickness (the inverse of Q_{pl}) of weld 304L both increase with increasing time, coherent with its OCP value displayed in Figure 8.

Figures 10(e) and 10(f) show the evolution of the EIS results obtained for the metals exposed to solution C (1.5 M chloride concentration sea-salt solution at 70°C). The equivalent circuit of the EIS results also can be presented as the same as solution A in Figure 10(a). Tables 6(e) and 6(f) display the evolution of passive elements after fitting. R_{f2} values in Tables 6(e) and 6(f) become larger with the increasing time, representing a slower transportation of the metal ions, and a growth of the passive layer. This is also coherent with the OCP values shown in Figure 8.

5.2.5. *Pit depth and density*

Figure 12 shows the frequency distribution of pit depth in the investigated solutions. Most pits were less than 1 μm deep, and the pit density decreased

Table 6. Values of the elements of the equivalent circuit shown in Figure 11 to fit the impedance spectra in Figures 10(a)–10(f).

t (h)	R_e (Ωcm^2)	R_{f1} (kΩ cm^2)	Q_{f1} (μSs^{n}cm^{-2})	n_{f1}	R_{f2} (kΩ cm^2)	Q_{f2} (μSs^{n}cm^{-2})	n_{f2}
(a)							
1	0.747	0.037	74.949	0.736	11.117	132.551	0.617
4	0.709	0.079	77.347	0.721	11.748	85.969	0.670
9	0.710	0.056	106.327	0.697	9.590	86.429	0.666
17	0.738	0.098	239.592	0.641	6.456	6.117	0.987
25	0.827	0.055	269.235	0.646	5.116	9.541	0.931
34	0.887	0.038	297.296	0.664	3.540	9.791	0.907
43	0.634	0.001	289.694	0.608	3.665	124.745	0.740
(b)							
1	0.855	0.074	41.612	0.696	121.677	49.474	0.762
4	0.977	0.050	47.827	0.692	93.139	67.347	0.736
9	1.229	0.028	55.510	0.697	47.138	79.643	0.743
17	1.369	0.018	174.745	0.649	10.216	37.026	0.858
25	1.363	0.018	211.429	0.639	10.737	25.357	0.899
34	1.274	0.014	335.051	0.606	7.189	28.429	0.873
43	1.292	0.012	369.031	0.604	7.264	27.704	0.879

t (h)	R_e (Ω cm^2)	R_{pl} (kΩ cm^2)	Q_{pl} (μSs^{n}cm^{-2})	n_{f1}
(c)				
1	19.757	529.396	46.128	0.855
4	20.266	511.168	45.561	0.858
9	20.658	534.688	45.969	0.861
17	20.815	553.896	45.903	0.864
25	21.031	566.048	45.490	0.867
34	21.050	568.204	45.184	0.868
43	9.610	562.128	45.061	0.869
(d)				
1	2.505	15.796	130.459	0.829
4	2.775	17.028	133.827	0.835
9	2.956	27.793	106.224	0.848
17	3.283	62.132	84.847	0.862
25	3.303	86.083	78.418	0.868
34	3.314	85.280	74.286	0.869
43	3.526	83.339	75.459	0.865

(*Continued*)

Table 6. (*Continued*)

t (h)	R_e ($\Omega\,cm^2$)	R_{f1} ($k\Omega\,cm^2$)	Q_{f1} ($\mu Ss^n\,cm^{-2}$)	n_{f1}	R_{f2} ($k\Omega\,cm^2$)	Q_{f2} ($\mu Ss^n\,cm^{-2}$)	n_{f2}
(e)							
1	2.836	0.778	60.714	0.810	240.688	15.230	0.866
4	3.367	1.834	51.990	0.824	461.384	12.740	0.852
9	3.256	1.761	43.362	0.840	916.300	13.923	0.813
17	3.454	1.634	40.138	0.848	1460.40	13.112	0.814
25	4.479	1.798	38.918	0.852	1913.35	11.005	0.816
34	4.349	1.962	37.995	0.856	2402.96	9.679	0.820
43	4.483	1.974	36.765	0.857	3087.00	9.327	0.808
(f)							
1	0.784	2.060	5.995	0.835	163.542	57.041	0.704
4	0.958	2.283	4.490	0.853	459.228	50.418	0.721
9	1.104	1.924	3.966	0.861	1103.28	47.173	0.748
17	1.301	1.512	3.811	0.866	2434.32	45.383	0.763
25	1.353	1.307	3.908	0.865	3706.36	44.168	0.772
34	4.036	1.200	4.211	0.858	6134.80	43.122	0.778
43	3.501	1.103	4.348	0.856	10050.9	42.735	0.783

over time, indicating that numerous of the pits formed at the early stage were metastable.

Figure 13 displays the pit density of different sizes as a function of time in the investigated solutions. During the exposure, the pit density of as-received 304L increased in the first 5 days, then declined afterward. The phenomenon represents the repassivation process of metastable pit growth. The pit density of weld 304L declined in the first 5 days, and then increased before decreasing again. The changes in the pit density over time represent the occurrence of metastable pits, supporting the fluctuations of OCP curves shown in Figure 8. The pit densities of the specimens exposed to the investigated solutions after 1 month were determined to be 200–400 pits per cm^2.

The chloride concentration affected the pitting corrosion resistance of the specimens. The pit density of the specimens exposed to the 6.25 M chloride NaCl solution (A) were higher than those exposed to the 1.5 M chloride sea-salt solutions (B and C). Over the range investigated, chloride concentration mainly affected the severity of attack rather than morphology. Additionally, the chloride concentration was critical to the nucleation of a pit and slightly affected the propagation through the tested range. The

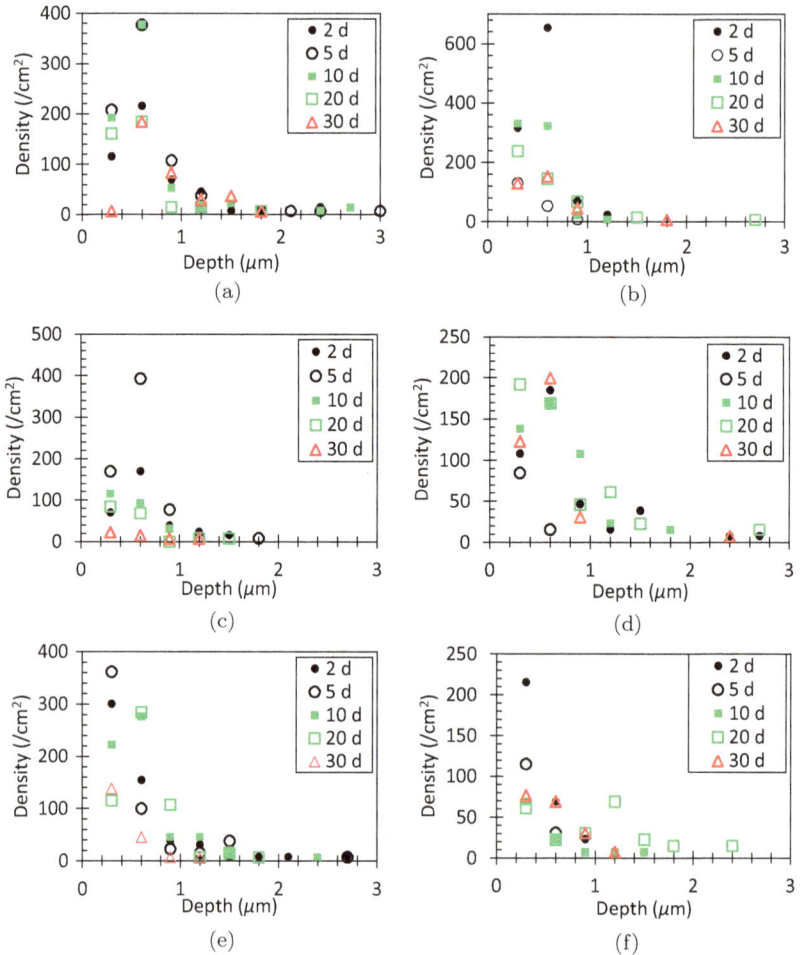

Figure 12. Relationships of pit depth and density: (a) as-received 304L exposed to the solution A, (b) weld 304L exposed to the solution A, (c) as-received 304L exposed to the solution B, (d) weld 304L exposed to the solution B, (e) as-received 304L exposed to the solution C, and (f) weld 304L exposed to the solution C.

chloride concentration was more important in disrupting the film than promoting the propagation of pits. The change of behavior was attributed to the varying response of the matrix and the intermetallic with chloride ions, but these variations are yet to be understood. Further work in this area needs to be performed to elucidate the influence of chloride concentration

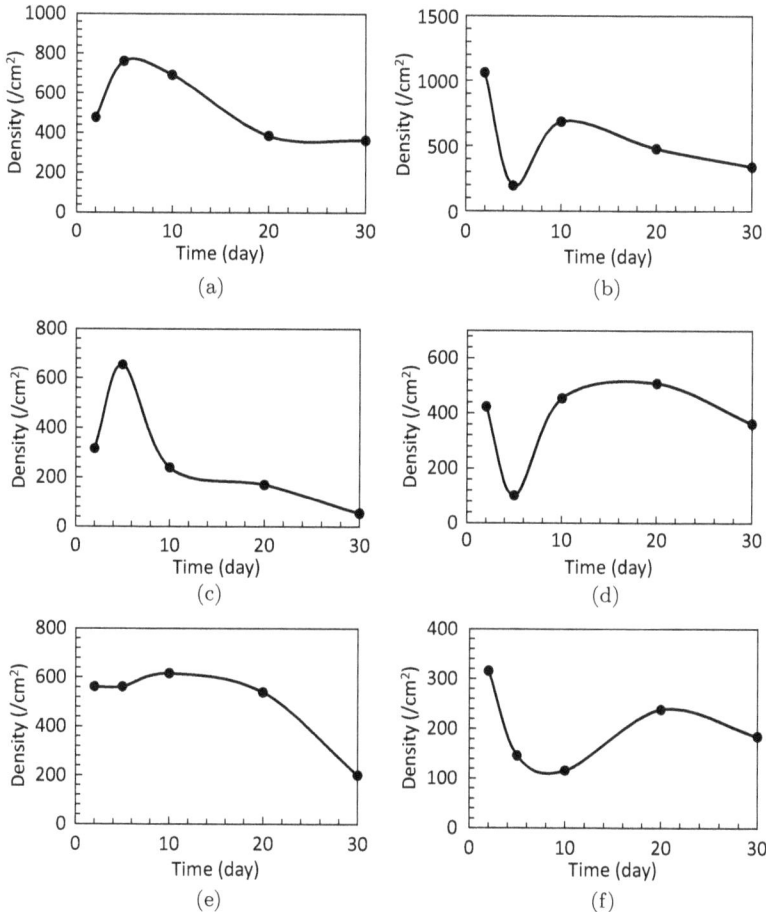

Figure 13. Pit density evolution with time: (a) as-received 304L exposed to the solution A, (b) weld 304L exposed to the solution A, (c) as-received 304L exposed to the solution B, (d) weld 304L exposed to the solution B, (e) as-received 304L exposed to the solution C, and (f) weld 304L exposed to the solution C.

upon corrosion mode determination and also the extent to which that corrosion manifests.

Compared with the effect of chloride concentration, the effect of temperature on the damage accumulation is smaller. A higher temperature solution is expected to produce more damage according to its electrochemical testing (shown in Figures 8 and 9). However, the influence of temperature is difficult to conclude, since in some cases the damage was more severe with the higher temperature solution, coinciding with the accepted understanding, i.e., higher temperature leads to more pitting susceptibility (referring to as-received 304L exposed to solutions B and C, Figures 13(c) and 13(e)). Moreover, the higher pitting density also occurred on the specimen exposed to the lower temperature solution (referring to weld 304L exposed to solutions B and C, Figures 13(d) and 13(f)).

5.3. *Conclusions*

(1) The welding process depresses the corrosion resistance, which is suggested by the OCP measurements, potentiodynamic polarization scans, and EIS measurements. Compared with as-received 304, weld 304L shows the depressed OCP values, accelerated anodic dissolutions, and declined film resistances.

(2) The metastable property of pitting corrosion is presented by the fluctuations (drop and recovery) of the OCP measurements, and the evolution of pit density for the immersion testing up to 720 h.

(3) No dramatic change on pitting corrosion between the as-received and weld 304L can be observed from the statistics of pitting during the test duration up to 720 h.

(4) The pitting corrosion resistance is significantly affected by the chloride concentration and slightly affected by the temperature under the investigated conditions.

(5) The micro-characteristics show that the pit tip has a little metallic element and some salt species. The oxygen diffusion layer around the pit region is on the scale of a micrometer.

6. Fundamentals of CISCC [3]

This section discusses the mechanism of CISCC, including three criteria: sufficiently large tensile stress, susceptible material and aggressive environment, in Section 6.1, and reviews the previous experimental studies relative to the canister corrosion in Section 6.2.

6.1. *SCC mechanism*

An almost uniform conclusion is that no unifying mechanisms of SCC exist. However, many studies have revealed the fundamentals of SCC and suggested the tendency of structure failure for each SS at its application environment. The relation between localized corrosion (e.g., pitting and crevice corrosion) and SCC under many applications has been addressed. CISCC occurs from the combined effect of pitting corrosion under tensile stress [41]. The fundamental steps in the overall process of crack development involve pit initiation, pit growth (dominated by electrochemical mechanism), the transition from a pit to a crack, short crack growth, and long crack growth (dominated by metal separation).

The susceptibility to chloride cracking in SSs is a function of the compositions of crystal phases (e.g., ferrite and austenite phases) and the presence of Cr-depleted zones around Cr–C precipitations [67]. With the materials used on the welded canister, the welding process modifies the microstructure of SS and sensitizes the metal; accordingly, the resistance to pitting corrosion and SCC is changed.

Pitting is recognized as the precursor to SCC as which provides the combination of local aggressive solution chemistry and a stress-concentrating feature [68–74]. In predictive schemes, the pit-to-crack transition is based on the phenomenological requirements that the pit depth must be greater than a threshold depth, corresponding to a threshold stress intensity factor, and that the CGR should exceed the pit growth rate [75]. Accordingly, SCC only occurs over a very narrow potential range close to the repassivation potential of corrosion, as at higher potentials the localized corrosion velocity will always exceed that of crack propagation.

Establishing the necessary conditions for a pit to transfer to a fatigue crack is essential for lifetime prediction. The recognized criteria for this transition consist of two parts, both of which are considered to be necessary for the onset of cracking. First, the pit must grow to a critical size, causing the stress intensity factor to equal the threshold value. This critical size varies depending on the loading, i.e., a smaller pit leads to cracking under higher stress levels. Second, the CGR must exceed the pit growth rate. When the requirements are met, a fracture mechanics approach can be utilized for determining the time to failure. All levels of damage may result in a significant reduction in SCC life: pit depths as shallow as $20\,\mu$m can initiate cracks; however, the deepest pits do not always instigate the cracks. In addition to depth, pit surface area, size, shape, and proximity to other pits are also determined to be critical factors in when and where a crack develops.

The details of how SCC emerges from a pitting corrosion precursor have been clarified through unique 3-D images, showing the early stages of cracks evolving from pits in 3NiCrMoV steam turbine disc steel [76]. The images have demonstrated that SCC invariably originates from corrosion pits [77] and cracks that have nucleated on pit walls can grow around the pit and coalesce to form a complete through-crack [76]. The defect nucleation is a result of high corrosive conditions, and the origin of the crack is located on a stress concentration region [78]. Once the crack is initiated at the corrosion area, it will propagate at a fast rate under the conditions of metal dissolution and WRS. The study of pitting and crack nucleation at the early stages of SCC under ultra-low elastic load suggests that SCC emanates from the defect on the sample surface, and the preferential SCC initiation sites are at the shoulders rather than at the bottoms of the surface defect [79]. The likelihood of SCC propagation from an area of pitting corrosion should increase with time as the depth of metal loss increases [80]. Metal separation is faster than electrochemical dissolution during the process of propagation.

The crack propagation conditions must be fully satisfied otherwise the propagation would not occur. Below the limiting RH that pitting corrosion does not occur has not found any SCC propagated in type 304L SS after 125 days with simulated residual stress [41]. According to the competing theory [80], even if the applied stress intensity factor (K) exceeds the threshold K for SCC, the crack cannot initiate in a pit if the CGR is more rapid than the rate of pit growth. It is reasonable to expect that pitting corrosion will slow down with time as a corroded area deepens and the diffusion distance increases.

6.2. *Experimental studies and past experience*

A study that compares the natural conditions (Miyakojima Island, Japan) with the accelerated conditions (60°C, 95% RH, NaCl mist) supports the importance of the realistic environmental conditions for the canister failure [81]. The SCC propagation rate in SS304/304L specimens under natural conditions was about 1.2×10^{-12} to 1.8×10^{-11} m/s and that under the accelerated test was about 1.0×10^{-10} to 3.5×10^{-9} m/s; besides, the SCC propagation rates under both conditions were independent of the stress intensity factor K in the test range. Accordingly, the propagation rates have two orders difference under two conditions. Under the natural conditions, it will take about 25–375 years to penetrate through a 15-mm canister wall; while under the accelerated conditions, it will take about 50 days to 48 years.

Naturally, the environment changes with time. The canister surface may contact various types of air at different times, which has low temperature with high or low humidity, and relatively high temperature with high or low humility. In other words, there is repetition of wet and dry conditions throughout the time. The repetition seems to be another susceptible factor accelerates CISCC; however, the lab tests with the repetition of wet and dry conditions have demonstrated that it does not necessarily accelerate CISCC even when the surface chloride concentration has reached the concentration level to cause a failure at the constant condition [82].

The tests considering the real canister surface environment (e.g., the actual temperature, humidity, and salt deposit at different exposure times) are critical to predict the failure of an in-service canister, however, such tests are rare. More laboratorial tests have only focused on one factor to study the susceptibility to CISCC.

The investigation of the susceptibility of as-received and sensitized SS304L U-bend specimens along with welded SS304L specimens to SCC in marine atmosphere conditions shows that cracks can occur in sensitized specimens with 1 and $10\,g/m^2$ sea salt deposition, they can occur in as-received specimens with $10\,g/m^2$ salt depositions, and the interdendritic attack can occur on welded specimens all at $45°C$ [83]. If sea-salt concentration is below $1\,g/m^2$, cracks rarely occur on SS304L specimens.

Although previous studies have agreed that salt concentration would affect the susceptibility to CISCC, a multitude of studies have confirmed the corrosion resistance is critically related to the types of salts but not sensitive to the concentration of ions. The corrosivity of salt deposits was controlled by equilibrium chloride concentration in the electrolyte of surface formed by the absorption of water vapor. In a lab test of mixing chlorides, $MgCl_2$ and $CaCl_2$ were fast to cause SS specimen failure; sodium chloride was the slowest. The results are considered as the solubility and chloride activity between the different salts as $MgCl_2$ and $CaCl_2$ are more soluble than NaCl. $MgCl_2$ is the sea-salt constituent responsible for promoting low-temperature CISCC in types 304L and 316L SSs [84]. Although $MgCl_2$ reacted faster than $CaCl_2$, most studies [85, 86] commonly confirmed that $CaCl_2$ is the most likely cause of CISCC of all sea salts at room temperature as which forms more concentrated solutions at a given RH than $MgCl_2$.

At low chloride concentration ($< 0.5\,g/L$), sulfide salts show the synergistic effect on CISCC especially in SS304L and SS316 as the TG cracking was observed [87]. Sulfides are generally detrimental to pitting resistance and crevice corrosion resistance in all grades of SS. The susceptibility of CISCC increases with a high sulfur content, as there is a higher density of sulfide inclusions, which act as initiation sites for pitting and stress corrosion cracks [88]. In the absence of hydrogen sulfide, these relatively

Table 7. RH ranges over which CISCC was observed in the case of sea-salt, $MgCl_2$, and $CaCl_2$ deposits (summarized in [84]).

Wetted deposit type/Cl^- deposition density	Temperature/test duration	RH of maximum CISCC susceptibility	RH range over which CISCC was observed	
			304L	316L
$MgCl_2/2500\,\mu g/cm^2$	25°C/24 months	30%	25–50%	25–50%
	50°C/2 weeks	30%	20–50%	20–50%
	70°C/2 weeks	30%	20–90%	20–80%
Sea-salt/$2500\,\mu g/cm^2$	25°C/24 months	30%	25–50%	25–40%
	50°C/2 weeks	30%	20–50%	25–50%
	70°C/2 weeks	30%	20–80%	20–50%
$MgCl_2/26{,}000\,\mu g/cm^2$	20°C/10 weeks	Unknown	None	None
	30°C/10 weeks	Unknown	27–44%	None
	40°C/10 weeks	Unknown	25–67%	None
$CaCl_2/29{,}000\,\mu g/cm^2$	20°C/10 weeks	Unknown	None	None
	30°C/10 weeks	Unknown	17–49%	17–40%
	40°C/10 weeks	Unknown	12–69%	12–39%

low chloride concentrations would not result in cracking at ambient temperature.

According to the previous studies, the RH range over which CISCC occurs narrows with decreasing temperature [89–93]. For instance, Table 7 shows that for the case of $MgCl_2/2500\,\mu g/cm^2$, when temperature decreases from 70°C to 25°C, the RH range changes from 20–90% to 25–50% accordingly. Table 7 summarizes the RH ranges over which CISCC was observed in the case of sea-salt, $MgCl_2$, and $CaCl_2$ deposits [84].

At the same applied stress, the failure time of SS304L at 50°C is much longer than the temperature at 80°C [69]. Figure 14 shows the comparison of test results of SS304L at 50°C and 80°C. At 50°C, the failure time of applied stress of $\sim 400\,\text{MPa}$ is more than 10 times longer than the results at 80°C. Besides, most of the specimens were not failure below 400 MPa at 50°C, indicating the possibility of failure is extremely low at 50°C. Under the constant humidity (35%), the decrease of temperature tends to decrease the electrochemical and diffusion kinetics, resulting in a lower corrosion rate. The lab tests of U-bend specimens exposed to salt fog conditions at 43°C, 85°C, and 120°C have suggested that only 43°C tests exhibited cracking for SS304L [13]. Cracking severity increased with exposure time as more specimens cracked and the crack lengths increased. Temperatures above 85°C did not initiate SCC because they are beyond the limiting RH temperature, indispensable for atmospheric chloride SCC. Another study suggested that

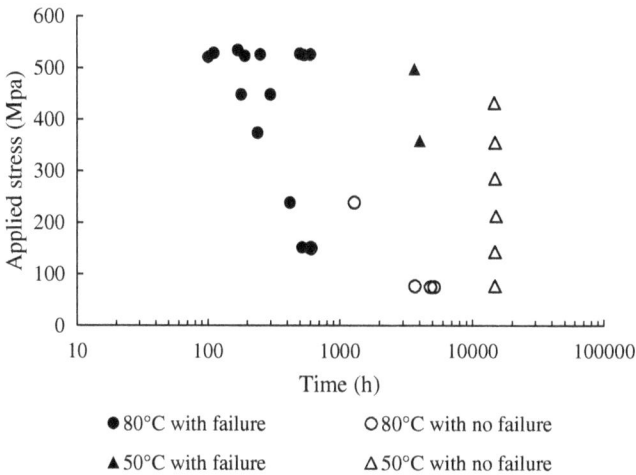

Figure 14. Results of constant-load test of as-received SS304L at 50°C and 80°C with (data from [69]).

the critical RH below which no CISCC to occur at SS304, SS316 is equal to or higher than 15% for sea-salt particles up to 80°C for all test materials [82]. It agrees with the limiting RH principle for the on-site canisters.

7. Experimental Studies of SCC [3]

The objective of this test is to evaluate the CISCC susceptibility of as-received 304L and its weld metal, establish quantitative measurements of CGRs and determine relationships among cracking susceptibility, environmental conditions, and metallurgical characteristics. The specimens investigated in the experiment are described in Section 7.1. The test system and test conditions are introduced in Sections 7.2 and 7.3. SCC CGRs have been identified for the as-received and weld metals in the simulated marine coastal environment (Section 7.4). Extensive characterizations have been performed on material microstructures and stress-corrosion cracks by micro-characteristics imaging and analytical facilities and linked to crack growth test results to help define physical and environmental parameters controlling SCC susceptibility (Section 7.5).

7.1. *Specimen preparation*

The tests used standard 0.5-inch thickness compact tension (CT) specimens with side grooves. The chemical compositions of the CT specimens

Table 8. The chemical compositions of as-received 304L CT specimen used in the investigations (at.%).

Fe	Cr	Ni	C	P	Si	Cu	N	Cb+Ta	Mn
Bal.	18.470	8.190	0.025	0.037	0.260	0.620	0.069	0.013	1.800

S	Mo	Ti	Co	Al	B	V	W	Ta	Sn
0.025	0.400	0.003	0.123	0.006	0.001	0.060	0.029	0.001	0.008

are shown in Table 8, the same as the coupon specimens in Section 5.1.1. The dimensional tolerances and the surface finishes shown in Figure 15 were followed in the specimen preparation. Care was taken in machining to prevent contamination of specimen and notch surfaces that were difficult or impossible to clean. The copper deposit left by electric discharge machining with a copper electrode was cleaned with sandpaper polishing, DI water, and acetone. According to the material certification of the as-received 304L, the yield strength ($\sigma_{\gamma s}$) is 207 MPa, and the ultimate tensile stress (σ_{TS}) is 517 Mpa.

7.2. *SCC test system*

The tests were conducted under the SCC crack growth test system working at high temperature (up to 360°C) and high pressure (up to 2200 psi or 15.2 MPa) with the function of flow chemistry controlling and *in situ* CGR measurement. Figure 16 illustrates the components and structures of the system. The details of the system can be found in the transaction [94]. The system has the features include active constant-stress intensity factor (K) load control, active temperature control, a sensitive crack length measurement apparatus, a recirculating high-temperature water system that controls over all aspects of water chemistry, and continuous monitoring of all pertinent test parameters. The tests were undertaken for well-defined material and at environmental conditions, and have been ensured that the SCC growth rate response was reproducible in the test conditions.

The specimen was tested in the 4-L autoclave which is made of type 316 SS, using the 5000 lbs.-loading frame with 0–5 Hz controllable frequency, which accelerates the cracking propagation. The inner structure of the autoclave is shown in Figure 17. The crack extension can be monitored *in situ* by direct current potential drop (DCPD) with length resolution of about $\pm 1\,\mu$m making it possible to measure extremely low growth rates approaching 10^{-11} m/s (see DCPD setup in Figure 18). The wrapping

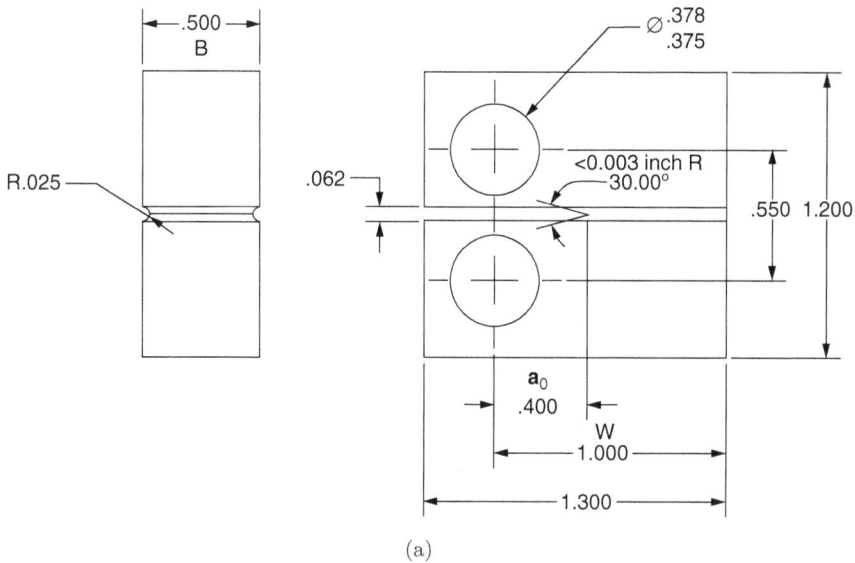

(a)

(b)

Figure 15. Standard 0.5-inch thickness CT specimen: (a) schematic view and (b) solid view.

heating mantle heated the autoclave, and the high-precision proportional-integral-derivative temperature controller controlled the temperature.

7.3. *Test procedure*

The test procedures strictly followed the standard test method suggested in ASTM E1681-03 (2013) [95], which is for determining threshold stress

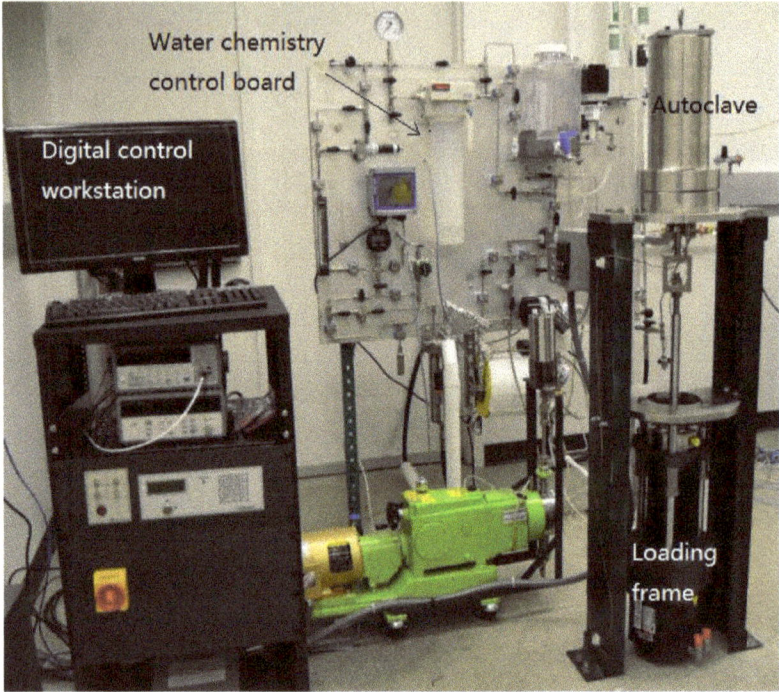

Figure 16. Picture of the SCC test system.

intensity factor for environment-assisted cracking of metallic materials, and the standard test method in ASTM E647-13a (2014) [96], which is for the measurement of CGRs.

Before loading a specimen in the autoclave, the sample thickness B, notch depth a_0, and the distance from the center of holes to the backside of the specimen W (see Figure 15(a)) are all measured and recorded into the data record. Using the sample dimension and the strength of the specimen at the test temperature, in accordance with ASTM E1681-03 (2013) [95], an upper limit on stress intensity factor threshold for environmental assisted cracking (K_{EAC}) value was calculated using the formula:

$$\min\{B, W - a, a\} \geq \frac{4}{\pi} \left(\frac{K_{EAC}}{\sigma_\gamma} \right)^2 \tag{1}$$

where σ_γ is the effective yield strength at the test temperature, and a represents the length of crack (include the notch depth a_0):

$$\sigma_\gamma = \frac{\sigma_{\gamma s} + \sigma_{TS}}{2} \tag{2}$$

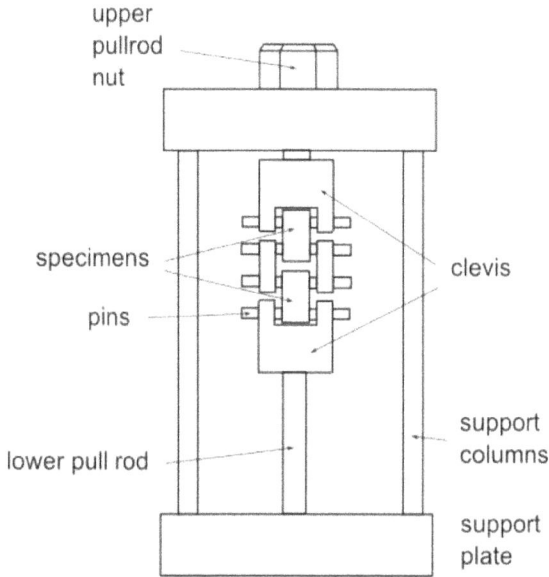

Figure 17. Schematic diagram of autoclave inner structure setup (not scaled).

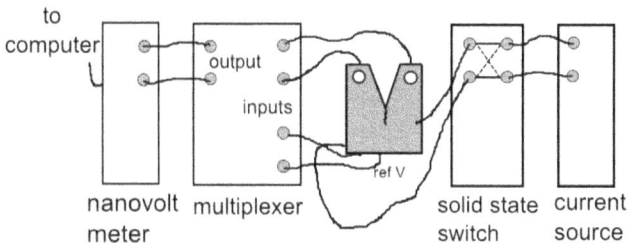

Figure 18. Schematic diagram of DCPD test system setup.

σ_{ys} is the yield strength, σ_{TS} is the ultimate tensile strength. Min$\{B, W - a, a\}$ is the smallest of the specimen thickness, the remaining uncracked specimen width, and the crack length.

The specimen was assembled with the loading frame of the system for fatigue pre-cracking in the room environment before testing in the humid environment. The importance of pre-cracking is to provide a sharpened fatigue crack of adequate size and straightness, which ensures that the effect of the machined starter notch is removed from the specimen K-calibration, and the effects on subsequent CGR data caused by changing crack front shape or pre-crack load history are eliminated. The fatigue pre-crack extended the depth to around 2 mm (see Section 7.4).

The first step in pre-cracking was to cycle the specimen at a relatively high-frequency f (1.5–2 Hz in the test) with a small load ratio ($R = 0.3$ in the test) and maximum stress intensity factor (K_{max}) less than the K level obtained for Eq. (1). As the crack began to grow from the notch, f was reduced while R and K_{max} were increased. Each pre-crack segment can grow beyond the plastic zone created by the previous segment. The increment of the final 1-mm fatigue pre-crack was conducted at a K_{max} value of not more than 60% of the expected K_{EAC}. The direction of cracking was parallel to the groove, within the angle of $\pm10°$.

After pre-cracking, the specimen was disassembled from the loading frame and immersed in the saturated sea-salt solution at room temperature for 10–15 h. Then the specimen was dried in compressed air to make sea-salt deposit on the specimen surface, as well as thoroughly cover the crack-tip region. The specimen was reassembled with the loading frame of the system. A small beaker with DI water of measured weight was placed in the autoclave. The autoclave was sealed by the gasket and heated to the investigated temperature. Note that with some environment–material combinations, preconditioning of the specimen in the investigated environment prior to force or displacement application will greatly influence the resulting K_{EAC} values. Considering the effects of preconditioning, the specimen was immediately exposed to the test environment (i.e., humid environment at the expected temperature) for 8 h before loading. The test has ensured that proper humidity was in the autoclave, and the crack-tip region of the specimen was stayed in the corrosive environment during the test.

Crack transitioning steps were carefully selected to grow the crack in the humid environment by using the following stages: (1) fatigue, (2) corrosion fatigue, and (3) SCC. The pre-cracking was produced by fast cyclic loading till 1 mm before transited to slow cyclic loading to promote SCC. This step was conducted in air. Following, the cyclic loading steps at frequencies of 0.1 Hz down to 0.001 Hz were performed in the humid environment. The final phase involved crack transitioning by very slow cycling. Details about the test steps are shown in the tables of the next section. The post-test specimens were rinsed with DI water, cleaned by ultrasonic with analytical acetone, and then dried by ethanol of analytical grade. The crack surface of each specimen was mechanically abraded and SiC diamond polished to

1 μm. The polished surface was further examined under SEM and transmission electron microscopy (TEM).

7.4. *CGR*

Tables 9 and 10 separately display the test steps for as-received 304L and weld 304L exposed to 15% RH at 70°C. The as-received was applied the K of 9 MPa$\sqrt{}$m for crack propagation, while the weld was applied the lower K of 7 MPa$\sqrt{}$m, due to the yield strength was reduced after welding. The determination of K values was according to Eqs. (8.1) and (8.2). Though the weld was applied at lower K, the CGR was much higher than the as-received at the same test phase, as shown in Figure 19, indicating the welding process largely depressed the resistance to SCC.

7.5. *Microstructure and microchemistry*

The micro-characteristics of the CT specimens were analyzed under SEM/FIB and STEM–energy dispersive spectroscopy (EDS). This section shows the micro-characteristics of the crack tips of weld 304L (Section 7.5.1) and the as-received 304L (Section 7.5.2).

7.5.1. *Weld* 304*L*

Figure 20 displays the main steps of making a TEM foil. Figure 20(a) is an SEM image, showing the overview of the cracking from notch (bottom) to the tip (top). Figure 20(b) is a zoom-in image indicating the tip of the crack. A mark is made by the Pt deposit to illustrate the location of making a TEM foil. Figure 20(c) shows an etched foil ready to be picked up. Figure 20(d) is a complete TEM foil attached to a TEM grid.

Figure 21 shows a STEM image in dark field mode of the cross-section of the crack. A box is marked to present the location of EDS mapping. The crack interface is distinct; however, the crack is contaminated by redeposition. When making a TEM foil in the FIB chamber, while most of the sputtered material is rapidly pumped away into the vacuum system, some sputtered atoms may redeposit onto the freshly milled walls of the specimen. The atoms from the specimen matrix and the Pt deposit, as well as the Ga+ ion beam (Ga+ ion beam is used to hit the sample surface and sputter a small amount of material), may therefore be deposited onto an adjacent region. Although redeposition is impossible to be eliminated, it has been significantly reduced by using low accelerating voltages in our case. Redeposition does not significantly affect the integrity of chemical

Table 9. The pre-cracking and crack-transitioning procedure for SCC crack growth testing of as-received 304L at 9 MPa√m exposed to 15% RH at 70°C. The specimen was immersed in the saturated sea-salt solution at room temperature (RT) and dried in air.

Test phase	Duration (h)	R	Frequency (Hz)	Hold (h)	Humidity	Temperature (°C)	K_{max} (MPa√m)	CGR (m/s)	Crack increment (mm)
1	1	0.3	2	0	air	RT	7	7.06E-08	0.254
2	0.38	0.3	2	0	air	RT	9	7.37E-07	1.016
3	0.18	0.5	2	0	air	RT	9	1.66E-06	1.0922
4	22.5	0.7	0.1	0	15%	70	9	3.45E-09	0.2794
5	81.83	0.7	0.01	0	15%	70	9	9.48E-10	0.2794
6	86.5	0.7	0.001	0	15%	70	9	1.63E-10	0.0508
7	148.28	0.7	0.001	0	15%	70	9	4.76E-11	0.0254
8	170.72	0.7	0.001	0	15%	70	9	—	—
9	24.83	0.7	0.001	0	15%	70	9.5	2.84E-10	0.0254
10	73.5	NA	NA	NA	15%	70	9	9.60E-11	0.0254
11	70.5	NA	NA	NA	15%	70	9	2.00E-10	0.0508

Table 10. The pre-cracking and crack-transitioning procedure for SCC crack growth testing of weld 304L at 7 MPa$\sqrt{m}$ exposed to 15% RH air at 70°C. The specimen was immersed in the saturated sea-salt solution at RT and dried in air.

Test phase	Duration (h)	R	Frequency (Hz)	Hold (h)	Humidity	Temperature (°C)	K_{max} (MPa$\sqrt{m}$)	CGR (m/s)	Crack increment (mm)
1	2.33	0.3	1.5	0	Air	RT	5	2.12E-08	0.1778
2	1.17	0.3	2	0	Air	RT	5	1.21E-07	0.508
3	0.42	0.3	2	0	Air	RT	7	6.77E-07	1.016
4	0.33	0.5	2	0	Air	RT	7	3.60E-07	0.4318
5	1.33	0.7	0.1	0	15%	70	7	5.29E-08	0.254
6	87.5	0.7	0.01	0	15%	70	7	2.42E-10	0.0762
7	107	0.7	0.001	0	15%	70	7	6.59E-11	0.0254
8	62	0.7	0.001	0	15%	70	7	—	—
9	5.17	0.7	0.001	0	15%	70	7.5	1.37E-09	0.0254
10	66.5	0.7	0.001	0	15%	70	7	2.12E-10	0.0508
11	48	NA	NA	NA	15%	70	7	2.94E-10	0.0508

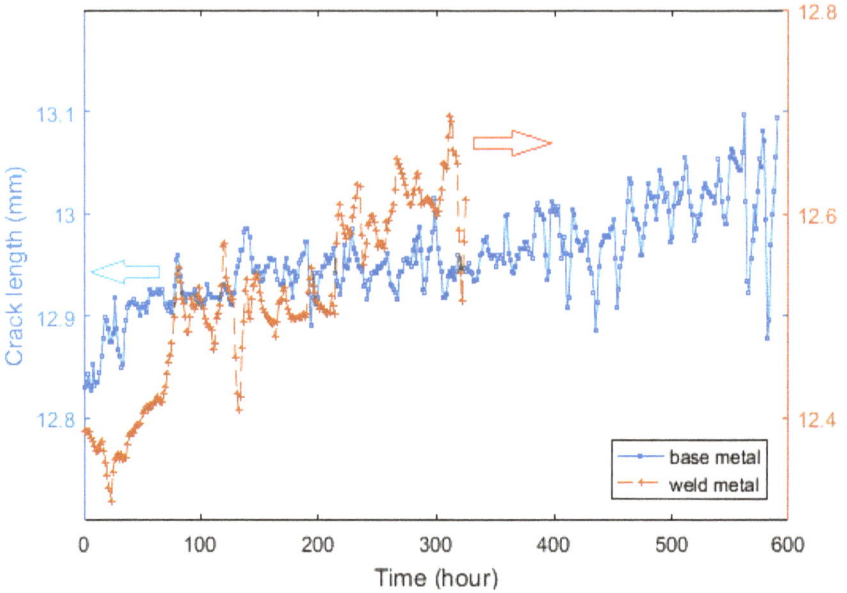

Figure 19. Comparison of CGRs of the as-received 304L and weld.

analysis since the metal elements, such as Fe and Cr, do not redeposit in the crack.

Figure 22 shows a STEM image of the specimen in dark field mode (Figure 22(a)) and the elemental maps (Figures 22(b)–22(g)). The redeposition of Pt layer caused by ion beam trenching is in the crack (Figure 22(b)). Slight of C and O are in the crack. Oxide particles are formed in the branch of the crack, and an oxygen diffusion layer is distinct (Figure 22(g)). The sea-salt elements, Mg and Cl, are detected in the crack (not shown here), being an evidence of the deliquescence of the salt deposit. Only a little metal element and a manifest amount of salt species have been found in the crack.

Figure 23 is the EDS concentration profiles of elements (Pt, C, Si, Cu, Ni, Fe, Cr, and O) measured at the six spots (Spots $1, 2, \ldots, 6$) shown in Figure 22(a). Spots 1 and 2 are in the matrix. Since the content of Cr and C of spot 2 is higher, and the content of Fe and Ni is lower than spot 1, spot 2 is susceptible to be the secondary phase, Cr–C precipitations (note this is a weld specimen). The region of the secondary phase is distinguished in Figures 22(d) and 22(f) by the dashed lines. Spots 3 and 4 are in the crack.

Figure 20. TEM foil preparation of the weld 304L exposed to 15% RH air at 70°C: (a) overview of the cracking from notch (bottom) to the tip (top), (b) the tip of cracking with a Pt layer (marked by Pt deposit), (c) foil has been etched, and (d) the TEM foil is attached to a TEM grid.

There is an extremely low content of metal elements, and an amount of redeposition, including Pt. Spots 5 and 6 are also in the crack. Both of them have an acknowledgeable amount of O and Si. Spot 5 also has significant C. The content of Si and C in the material is trivial. Si and C are susceptible to be the contamination caused by the SiC diamond polishing. The polishing was made before preparing the TEM foil.

Figure 24 is the concentration profiles of C, Fe, Cr, Ni, and O measured by the EDS line scan at the crack interface. The content of O is relatively stable throughout the line, indicating the oxygen diffusion layer at the interface.

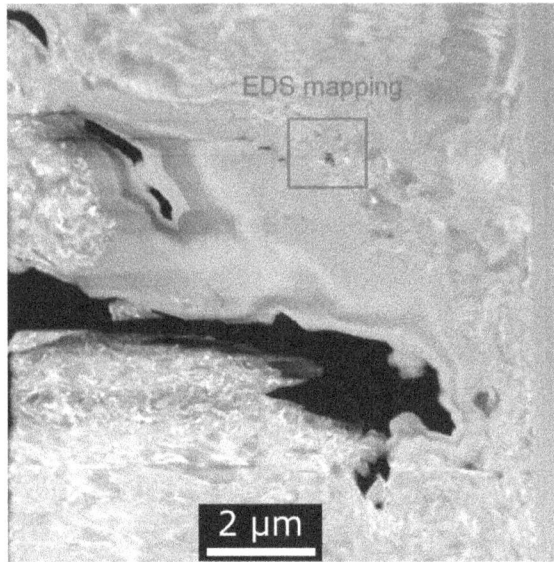

Figure 21. STEM dark field image of the cross-section of the crack with an STEM-EDS mapping area indicated by a box.

7.5.2. *As-received* 304*L*

The steps of making a TEM foil is the same as described in Section 7.5.1. The TEM foil is also made at the tip of the crack. Figure 25 is a STEM image in dark field mode of the crack interface between the matrix and the crack. Note that a 200-nm-thick redeposition layer was made using the ion beam trenching process and accumulated on the edge of the crack. Ga (Figure 25(d)) represents the region of redeposition. Besides, a 40-nm-thick oxygen diffusion layer is measured (Figure 25(b)). To better illustrate the element profiles at the crack interface, the STEM-EDS line scan was conducted on the line in Figure 25(a). Accordingly, Figure 26 shows the EDS concentration profiles of O, Cr, Fe, Ni, and Ga. Between the distance 200 and 240 nm, a peak of O is distinct. The content of Fe, Ni, and Cr drops. Besides, the other location with a little redeposition on the TEM foil was also analyzed (shown in Figure 27). At this location, the redeposition does not affect the integrity of chemical analysis. The EDS mapping was measured at the crack interface. Figure 28 displays the EDS elemental maps of O, Fe, and Ga. There is a little Ga (Figure 28(d)). Similar to Figure 25(b), the oxygen diffusion layer is also found at the interface, and the thickness is about 50 nm.

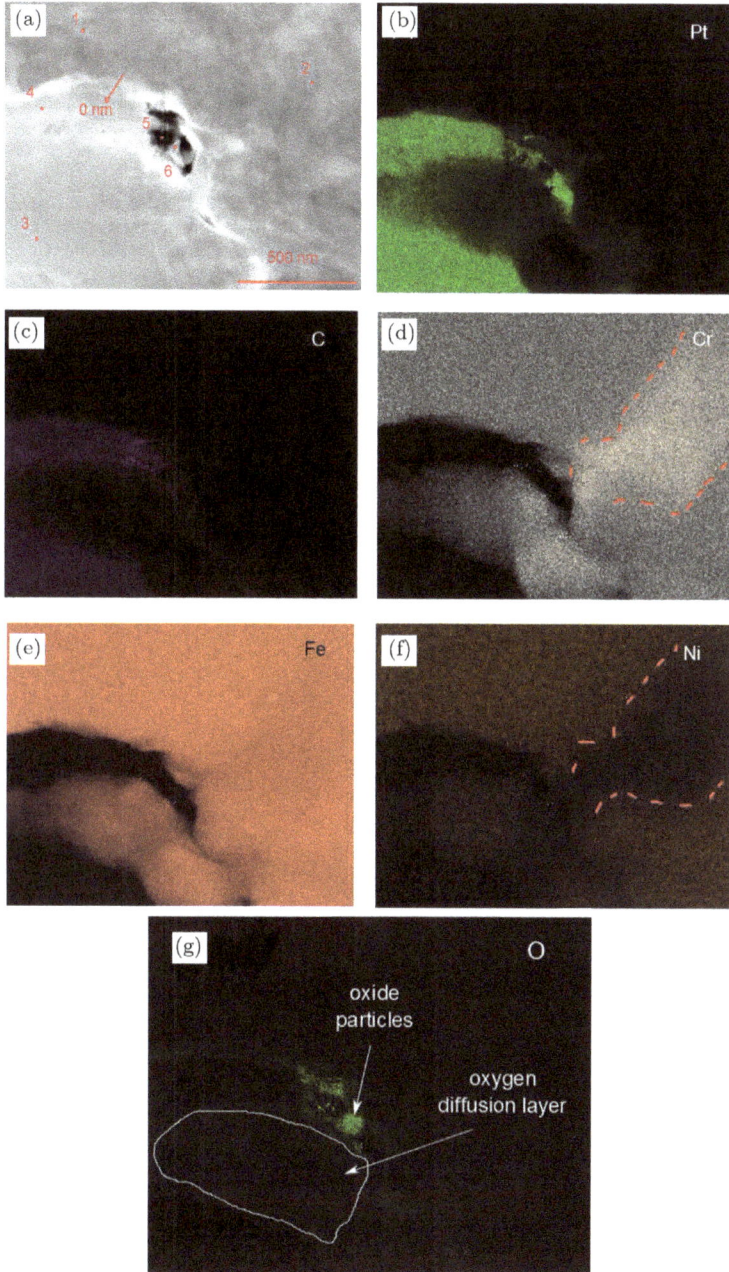

Figure 22. (a) STEM dark field image. Elemental maps of (b) Pt, (c) C, (d) Cr, (e) Fe, (f) Ni, and (g) O.

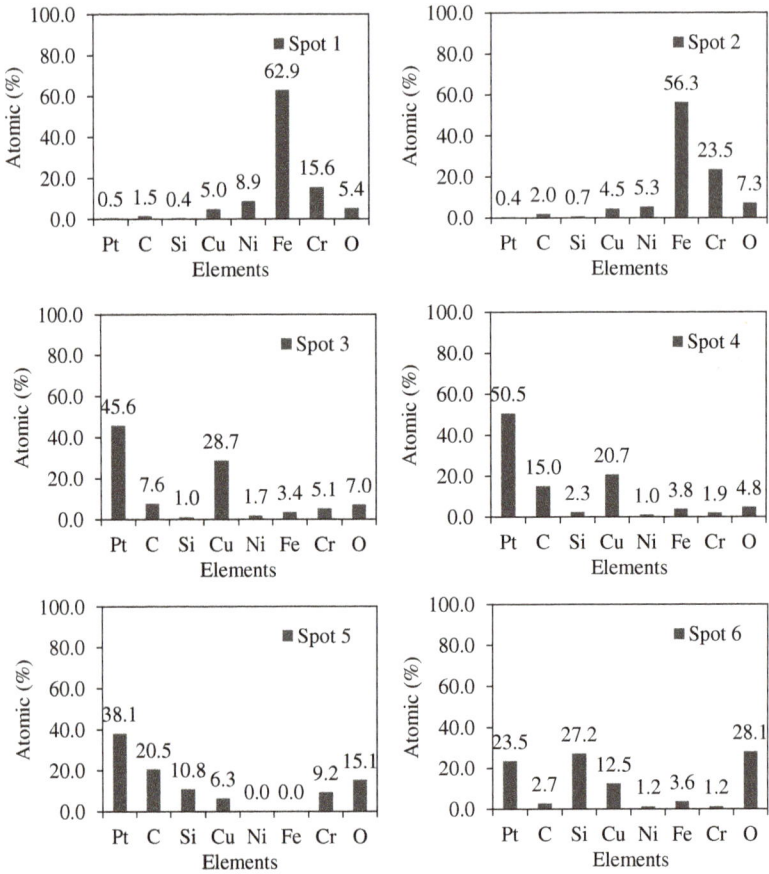

Figure 23. STEM-EDS concentration profiles (atomic normalized) measured at the spots in Figure 22(a).

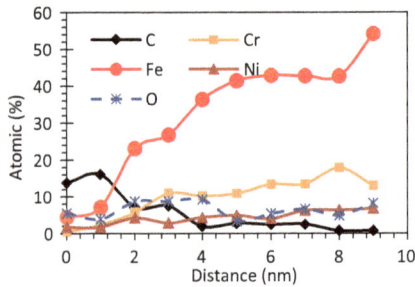

Figure 24. Corresponding STEM-EDS concentration profiles of C, Cr, Fe, Ni, and O measured at the line in Figure 22(a).

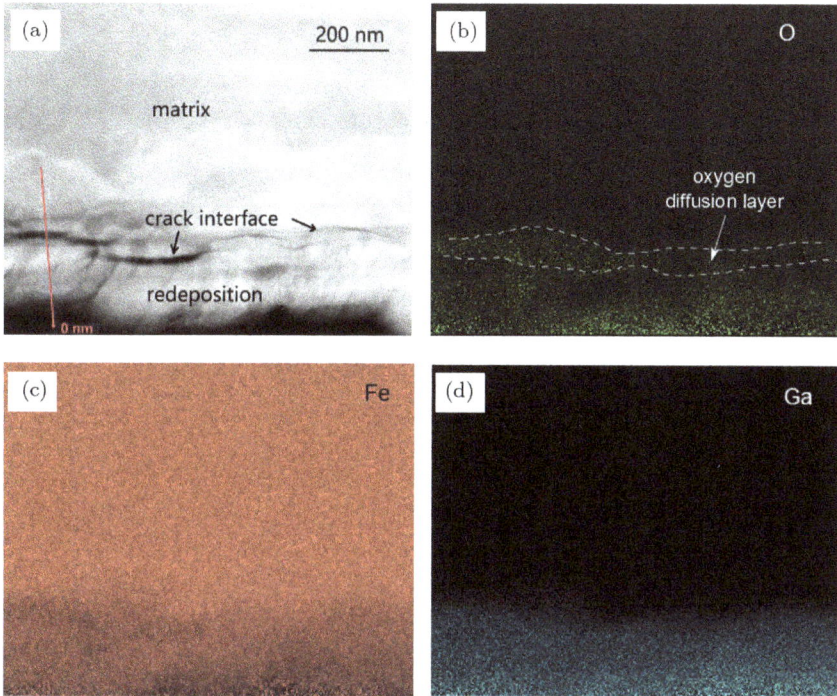

Figure 25. (a) STEM dark field image with an STEM-EDS line scan indicated with a red line. EDS elemental maps of (b) O, (c) Fe, and (d) Ga.

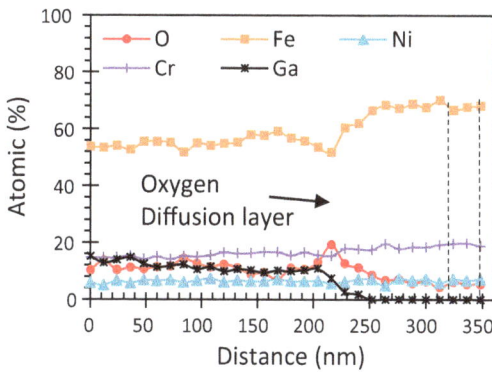

Figure 26. Corresponding EDS concentration profiles of O, Cr, Fe, Ni, and Ga measured at the line of Figure 25(a).

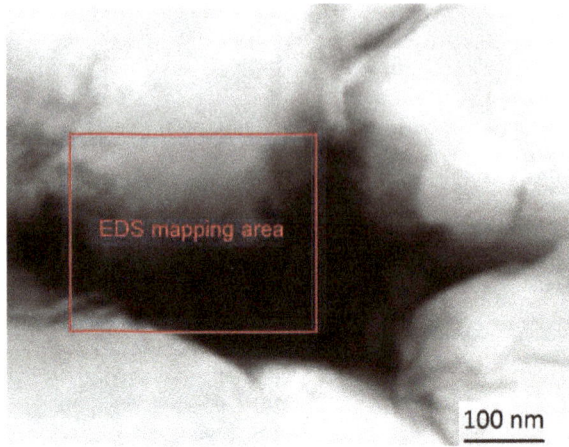

Figure 27. STEM dark field image of the cross-section of the tip of crack with a STEM-EDS mapping area indicated with a red box.

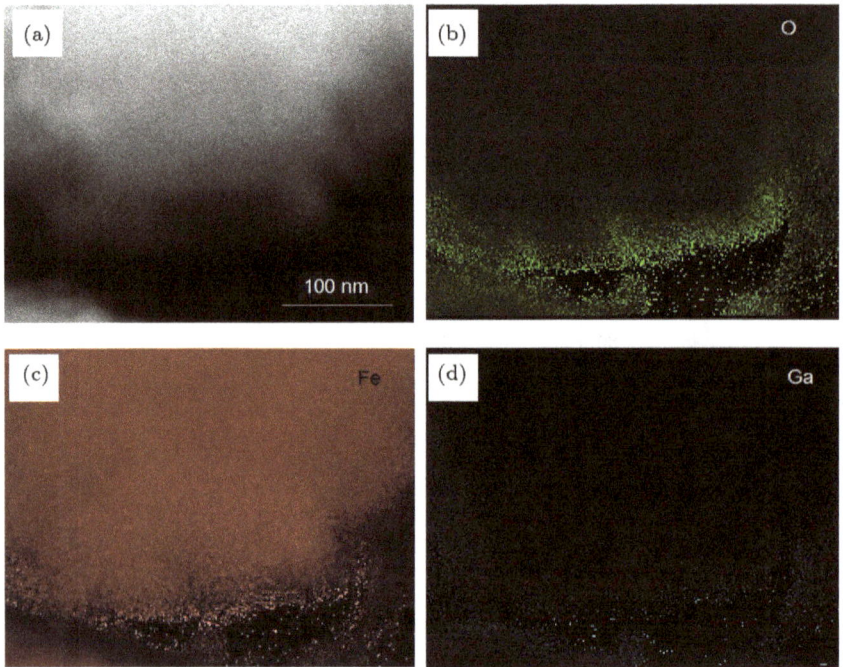

Figure 28. EDS elemental maps of the area in Figure 27. (a) STEM dark field image, (b) O, (c) Fe, and (d) Ga.

References

1. P. C. Durst, Guidance for independent spent fuel dry storage installations (NGSI-SBD-001), United States Department of Energy, Washington, DC (2012).
2. Y. Xie, S. Guo, A. Leong, J. Zhang, and Y. Zhu, Corrosion behaviour of stainless steel exposed to highly concentrated chloride solutions, *Corros. Eng. Sci. Technol.* **52**(4), 283–293 (2017).
3. Y. Xie, Chloride-induced stress corrosion cracking in used nuclear fuel welded stainless steel canisters, Doctoral dissertation, The Ohio State University, Columbus, OH (2016).
4. Dry spent fuel storage designs: NRC approved for general use, U.S.NRC, 3 November (2016). [Online] Available at: https://www.nrc.gov/waste/ spent-fuel-storage/designs.html, Accessed on 3 January 2017.
5. National Research Council, Dry cask storage and comparative risks, Chapter 4, In *Safety and Security of Commercial Spent Nuclear Fuel Storage: Public Report*, Washington, DC, The National Academies Press, pp. 60–74 (2006), DOI:10.17226/11263.
6. Calvert Cliffs stainless steel dry storage canister inspection (1025209), Electric Power Research Institute, Palo Alto, CA (2014).
7. Standard Practice for the Preparation of Substitute Ocean Water, D1141-98, ASTM International, West Conshohocken, PA (2013).
8. National Trends Network, National Atmospheric Deposition Program, (2014), [Online] Available at: http://nadp.sws.uiuc.edu/ntn/, Accessed on 1 May 2015.
9. C. R. Bryan and D. G. Enos, Analysis of dust samples collected from an unused spent nuclear fuel interim storage container at hope creek, Delaware (SAND2015-1746), Sandia National Laboratories (2015).
10. M. Wataru, H. Takeda, K. Shirai, and T. Saegusa, Thermal hydraulic analysis compared with tests of full-scale concrete casks, *Nucl. Eng. Des.* **238**(5), 1213–1219 (2008).
11. S. R. Suffield, J. M. Cuta, J. A. Fort, B. A. Collins, H. E. Adkins, and E. R. Siciliano, Thermal modeling of NUHOMS HSM-15 and HSM-1 storage (PNNL-21788), Pacific Northwest National Laboratories, Richland, WA (2012).
12. J. E. Truman, The influence of chloride content, pH and temperature of test solution on the occurrence of stress corrosion cracking with austenitic stainless steel, *Corros. Sci.* **17**(9), 737–746 (1977).
13. L. Caseres and T. M. Mintz, Atmospheric stress corrosion cracking susceptibility of welded and unwelded 304, 304L and 316L austenitic stainless steels commonly used for dry cask storage containers exposed to marine environments (NUREG/CR-7030), United States Nuclear Regulatory Commission, San Antonio, TX (2010).
14. R. C. Newman and W. R. Whitney, Understanding the corrosion of stainless steel, *Corrosion* **57**(12), 1030–1040 (2001).

15. S. Ramya, T. Anita, H. Shaikh, and R. K. Dayal, Laser Raman microscopic studies of passive films formed on type 316LN stainless steels during pitting in chloride solution, *Corros. Sci.* **52**, 2114–2121 (2010).

16. A. L. Almarshad and D. Jamal, Electrochemical investigations of pitting corrosion behaviour of type UNS S31603 stainless steel in thiosulfate-chloride environment, *J. Appl. Electrochem.* **34**(1), 67–70 (2004).

17. K. T. Kudo, G. Shibata, G. Okamoto, and N. Sato, Ellipsometric and radiotracer measurements of the passive oxide film on Fe in neutral solution, *Corros. Sci.* **8**, 809–814 (1968).

18. G. Okamoto, Passive film of 18-8 stainless steel structure and its function, *Corros. Sci.* **13**, 471–489 (1973).

19. J. R. Galvele, Transport processes in passivity breakdown — II. Full hydrolysis of the metal ions, *Corros. Sci.* **21**, 551–579 (1981).

20. J. Y. Xiong, M. Y. Tan, and M. Forsyth, The corrosion behaviors of stainless steel weldments in sodium chloride solution observed using a novel electrochemical measurement approach, *Desalination,* **327**, 39–45 (2013).

21. M. Ibrahim, S. S. Abd El Rehim, and M. M. Hamza, Corrosion behaviors of some austenitic stainless steels in chloride environments, *Mater. Chem. Phys.* **115**, 80–85 (2000).

22. G. Frankel, Pitting corrosion of metals a review of the critical factors, *J. Electrochem. Soc.* **145**, 2186–2198 (1998).

23. J. Dong, J. Zhou, H. Wang, J. Ling, and L. Shi, An investigation of pitting initiation mechanism of 1Cr12Ni2W1Mo1V steel after induction hardening, *J. Mater. Sci.* **35**, 2653–2657 (2000).

24. H. H. Strehblow, Mechanisms of pitting corrosion, In *Corrosion Mechanisms in Theory and Practicle*, New York, Marcel Dekker, pp. 201–237 (1995).

25. T. P. Hoar, D. C. Mears, and G. P. Rothwell, The relationships between anodic passivity, brightening and pitting, *Corros. Sci.* **5**, 279–289 (1965).

26. A. R. Brooks, C. R. Clayton, K. Doss, and Y. C. Lu, On the role of Cr in the passivity of stainless steel, *J. Electrochem. Soc.* **133**, 2459–2464 (1986).

27. I. Matsushima, Localized corrosion of iron and steel, In *Uhlig's Corrosion Handbook*, John Wiley & Sons, Inc., Hoboken, NJ, pp. 615–619 (2011).

28. Z. Szklarska-Smialowska, *Pitting Corrosion of Metals*, National Association of Corrosion Engineers (1986).

29. Y. S. Sato and H. Kokawa, Preferential precipitation site of sigma phase in duplex stainless steel weld metal, *Scr. Mater.* **40**, 659–663 (1999).

30. M. B. Cortie and E. Jackson, Simulation of the precipitation of sigma phase in duplex stainless steels, *Metall. Mater. Trans. A* **28**, 2477–2484 (1997).

31. A. A. P. Sidharth, Effect of pitting corrosion on ultimate strength and buckling strength of plate — A review, *Digest J. Nanomater. Biostruct.* **4**, 783–788 (2009).

32. L. Li, C. F. Dong, K. Xiao, J. Z. Yao, and X. G. Li, Effect of pH on pitting corrosion of stainless steel welds in alkaline salt water, *Construct. Build. Mater.* **68**, 709–715 (2014).

33. A. Dhanapal, S. Rajendra Boopathy, and V. Balasubramanian, Influence of pH value, chloride ion concentration and immersion time on corrosion

rate of friction stir welded AZ61A magnesium alloy weldments, *J. Alloys Compounds* **523**, 49–60 (2012).

34. K. Laha, K. S. Chandravathi, P. Parameswaran, K. Bhanu Sankara Rao, and S. L. Mannan, Characterization of microstructures across the heat-affected zone of the modified 9Cr-1Mo weld joint to understand its role in promoting type IV cracking, *Metall. Mater. Trans. A* **38A**, 58–68 (2007).

35. H. Park, J. K. Kim, B. H. Lee, S. S. Kim, and K. Y. Kim, Three-dimensional atom probe analysis of intergranular segregation and precipitation behavior in Ti–Nb-stabilized low-Cr ferritic stainless steel, *Scr. Mater.* **68**, 237–240 (2013).

36. Y. Xie, J. Zhang, T. Aldemir, and R. Denning, Multi-state Markov modeling of pitting corrosion in stainless steel exposed to chloride-containing environment, *Reliab. Eng. Syst. Safe.* **172**, 239–248 (2018).

37. M. Hashimoto, S. Miyajima, and T. Murata, A spectrum analysis of potential fluctuation during passive film breakdown and repair on iron, *Corros. Sci.* **33**, 917–925 (1992).

38. K. P. Wong and R. C. Alkire, Local chemistry and growth of single corrosion pits in aluminum, *J. Electrochem. Soc.* **137**, 3010–3015 (1990).

39. H. P. Leckie and H. H. Uhlig, *J. Electrochem. Soc.* **113**, 1262–1267 (1966).

40. D. Asaduzzaman, C. M. Mustafa, and M. Islam, Effects of concentration of sodium chloride solution on the pitting corrosion behavior of AISI-304L austenitic stainless steel, *Chem. Ind. Chem. Eng. Quart.* **17**, 477–483 (2011).

41. R. Melchers and R. Jeffrey, The critical involvement of anaerobic bacterial activity in modelling the corrosion behaviour of mild steel in marine environments, *Electrochim. Acta* **54**, 80–85 (2008).

42. J. Galvele, Tafel's law in pitting corrosion and crevice corrosion susceptibility, *Corros. Sci.,* **47**, 3053–3067 (2005).

43. R. M. Pidaparti, L. Fang, and M. J. Palakal, Computational simulation of multi-pit corrosion process in materials, *Comput. Mater. Sci.* **41**, 255–265 (2008).

44. J. C. Walton, Mathematical modeling of mass transport and chemical reaction in crevice and pitting corrosion, *Corros. Sci.* **30**, 915–928 (1990).

45. J. C. Walton, G. Cragnolino, and S. K. Kalandros, A numerical model of crevice corrosion for passive and active metals, *Corros. Sci.* **38**, 1–18 (1996).

46. P. Marcus and J. Oudar, *Corrosion Mechanisms in Theory and Practice*, Marcel Dekker, Inc., New York (1995).

47. G. A. Henshall, Modeling pitting corrosion damage of high-level radioactive-waste containers using a stochastic approach, *J. Nucl. Mater.* **195**, 109–125 (1992).

48. D. G. Harlow and R. P. Wei, Probabilities of occurrence and detection of damage in airframe materials, *Fatig. Fract. Eng. Mater. Struct.* **22**, 427–436 (1999).

49. D. G. Harlow and R. P. Wei, A probability model for the growth of corrosion pits in aluminum alloys induced by constituent particles, *Eng. Fract. Mech.* **59**, 305–325 (1998).

50. N. Murer and R. G. Buchheit, Stochastic modeling of pitting corrosion in aluminum alloys, *Corros. Sci.* **69**, 139–148 (2013).
51. R. B. Mears and U. R. Evans, The probability of corrosion, *Trans. Farad. Soc.* **31**, 527–542 (1935).
52. G. A. Henshall, W. G. Hasley, W. L. Clarke, and R. D. McCright, Modeling pitting corrosion damage of high level radioactive waste containers, with emphasis on the stochastic approach, Lawrence Livermore National Laboratory (1993).
53. A. Valor, F. Caleyo, L. Alfonso, D. Rivas, and J. M. Hallen, Stochastic modeling of pitting corrosion: A new model for initiation and growth of multiple corrosion pits, *Corros. Sci.* **49**, 559–579 (2007).
54. A. Turnbull, Mathematical modeling of localized corrosion, In *Modelling Aqueous Corrosion from Individual Pits to System Management*, Springer Science+Business Media, Dordrecht, Manadon, Plymouth, UK (1993).
55. T. Shibata and T. Takeyama, Pitting corrosion as a stochastic process, *Nature* **260**, 315–316 (1976).
56. C. Gabrielli, F. Huet, M. Keddam, and R. Oltra, A review of the probabilistic aspects of localized corrosion, *Corrosion* **46**, 266–278 (1990).
57. E. Williams, C. Westcott, and M. Fleischmann, Stochastic models of pitting corrosion of stainless steels I. Modeling of the initiation and growth of pits at constant potential, *J. Electrochem. Soc.* **132**, 1796–1804 (1985).
58. B. Baroux, The kinetics of pit generation on stainless steels, *Corros. Sci.* **28**, 969–986 (1988).
59. A. Valor, F. Caleyo, J. M. Hallen, and J. C. Velázquez, Reliability assessment of buried pipelines based on different corrosion rate models, *Corros. Sci.* **66**, 78–87 (2013).
60. W. J. Quadakkers, J. Żurek, and M. Hänsel, Effect of water vapor on high temperature oxidation of FeCr alloys, *J. Miner. Metals Mater. Soc.* **61**, 44–50 (2009).
61. F. K. Sahrani, M. Aziz, Z. Ibrahim, and A. Yahya, Open circuit potential study of stainless steel in environment containing marine sulphate-reducing bacteria, *Sains Malaysiana* **37**, 359–364 (2008).
62. B. Little, P. Wagner, K. Hart, R. Ray, D. Lavoie, K. Nealson, and C. Agui, The role of metal-reducing bacteria in microbiologically influenced corrosion, In *NACE Corros. 97*, Houston, TX (1997), Paper No. 215.
63. H. M. Ezuber, Influence of temperature on the pitting corrosion behavior of AISI 316L in chloride–CO2 (sat.) solutions, *Mater. Des.* **59**, 339–343 (2014).
64. D. Macdonald, The point defect model for the passive state, *J. Electrochem. Soc.* **139**, 3434–3449 (1992).
65. C. Exartier, S. Maximovitch, and B. Baroux, Streaming potential measurements on stainless steels surfaces: Evidence of a gel-like layer at the steel/electrolyte interface, *Corros. Sci.* **46**, 1777–1800 (2004).
66. C. M. Abreu, M. J. Cristóbal, R. Losada, X. R. Nóvoa, G. Pena, and M. C. Pérez, High frequency impedance spectroscopy study of passive films formed on AISI 316 stainless steel in alkaline medium, *J. Electroanal. Chem.* **572**, 335–345 (2004).

67. M. A. Streicher, Austenitic and ferritic stainless steels, In *Uhlig's Corrosion Handbook*, 3rd edition, John Wiley & Sons, Inc., Hoboken, NJ, pp. 657–693 (2011).
68. A. Turnbull, S. Zhou, L. Orkney, and N. McCormick, Visualization of stress corrosion cracks emerging from pits, *Corrosion* **62**(7), 555–558 (2006).
69. J.-I. Tani, M. Mayuzumi, and N. Hara, Stress corrosion cracking of stainless-steel canister for concrete cask storage of spent fuel, *J. Nucl. Mater.* **379**, 42–47 (2008).
70. A. Sjong and L. Eiselstein, Marine atmospheric SCC of Unsensitized stainless steel rock climbing protection, *J. Fail. Anal. Prev.* **8**, 410–418 (2008).
71. A. Kosaki, Evaluation method of ccorion lifetime of conventional stainless steel canister under oceanic air environment, *Nucl. Eng. Des.* **238**, 1233–1240 (2008).
72. J. Sedriks, *Corrosion of Stainless Steels*, J. Wiley and Sons, New York, NY (1979).
73. S. M. Bruemmer, R. N. Jones, J. R. Divine, and A. B. Johnson, *Evaluating the Intergranular SCC Resistance of Sensitized Type 304 Stainless Steel in Low-Temperature Water Environments*, ASTM Special Technical Testing Publication 821, Philadelphia, PA (1984).
74. R. A. Cottis and R. C. Newman, *Stress Corrosion Cracking Resistance of Duplex Stainless Steels, Volume 94*, H. a. S. O. T. Report, Ed., HSE, London (1995).
75. G. Engelhardt and D. D. Macdonald, Unification of the deterministic and statistical approaches for predicting localized corrosion damage. I. Theoretical foundation, *Corros. Sci.* **46**(11), 2755–2780 (2004).
76. D. A. Horner, B. J. Connolly, S. Zhou, L. Crocker, and A. Turnbull, Novel images of the evolution of stress corrosion cracks from corrosion pits, *Corros. Sci.* **53**, 3466–3485 (2011).
77. A. Turnbull and S. Zhou, Pit to crack transition in stress corrosion cracking of a steam turbine disc steel, *Corros. Sci.* **46**(5), 1239–1264 (2004).
78. J. A. Escobar, A. F. Romero, and J. Lobo-Guerrero, Failure analysis of submersible pump system collapse caused by assembly bolt crack propagation by stress corrosion cracking, *Eng. Fail. Anal.* **60**, 1–8 (2016).
79. L. K. Zhu, Y. Yan, L. J. Qiao, and A. A. Volinsky, Stainless steel pitting and early-stage stress corrosion cracking under ultra-low elastic load, *Corros. Sci.* **77**, 360–368 (2013).
80. C.-S. Chen, T. Shinohara, and S. Tsujikawa, Applicability of the competition concept in determining the stress corrosion cracking behavior of austenitic stainless steels in chloride solutions, *Zairyo-to-Kankyo/Corros. Eng.* **46**, 313–320 (1997).
81. A. Kosaki, An example of corrosion estimation of metal cask, In: The 2002 meeting for Weathering Technology, Japan Weathering Test Center (2002).
82. M. Mayuzumi, J. Tani, and T. Arai, Chloride induced stress corrosion cracking of candidate canister materials for dry storage of spent fuel, *Nucl. Eng. Des.* **238**(5), 1227–1232 (2008).

83. T. Mintz, L. Caseres, X. He, J. Dante, G. Oberson, D. Dunn, and T. Ahn, Atmospheric salt fog testing to evaluate chloride-induced stress corrosion cracking of type 204 stainless steel (ML120720549), In *Corrosion 2012*, Salt Lake City, Utah (2012).

84. B. Cook, S. B. Lyon, N. P. C. Stevens, M. Gunther, G. McFiggans, R. C. Newman, and D. L. Engelberg, Assessing hte risk of under-deposit chloride-induced stress corrosion cracking in austenitic stainless steel nuclear waste containers, *Corros. Eng., Sci. Technol.* **49**, 529–534 (2014).

85. J. W. Oldfield and B. Todd, *Br. Corros. J.* **26**(3), 173–182 (1991).

86. A. Turnbull and S. Zhou, Impact of temperature excursion on stress corrosion cracking of stainless steels in chloride solution, *Corros. Sci.* **50**(4), 913–917 (2008).

87. D. R. McIntyre and C. P. Dillon, *Guidelines for Preventing Stress Corrosion Cracking in the Chemical Process Industries*, MTI Publication (1985).

88. N. R. Smart, Literature review of atmospheric stress corrosion cracking of stainless steels (NR3090/043), Serco Assurance (2007).

89. S. Shoji and N. Ohnaka, Effect of relative humidity and chloride type on stainless-steel room temperature atmospheric corrosion cracking, *Boshoku Gijutsu* **38**, 111–119 (1989).

90. S. Shoji and N. Ohnaka, Effect of relative humidity and chloride type on atmospheric stress corrosion cracking of stainless steels at room temperature, *Boshoku Gijutsu* **38**, 92–97 (1989).

91. S. Shoji, N. Ohnaka, Y. Furutani, and T. Saito, Effects of relative humidity on atmospheric stress corrosion cracking of stainless steels, *Boshoku Gijutsu* **35**, 559–565 (1986).

92. A. Iversen and T. Prosek, Atmospheric stress corrosion cracking of austenitic stainless steels in conditions modelling swimming pool halls (paper No. 1142), In *Eurocorr*, Freiburg im Breisgau, Germany, September (2007).

93. T. Prosek, A. Iversen, C. Taxen, and D. Thierry, Low temperature stress corrosion cracking of stainless steels in the atmosphere in presence of chloride deposits, *Corrosion* **65**, 105–117 (2009).

94. Y. Xie and J. Zhang, High temperature and pressure stress corrosion cracking system, *Transactions of the American Nuclear Society* **109**, 10–14 (2013).

95. E1681-03, *Standard Test Method for Determining Threshold Stress Intensity Factor for Environment-Assisted Cracking of Metallic Materials*, ASTM International, West Conshohocken, PA (2013).

96. E647-13a, *Standard Test Method for Measurement of Fatigue Crack Growth Rates*, ASTM International, West Conshohocken, PA (2014).

Chapter 7

Crystalline Wasteform Phases for ^{137}Cs and ^{90}Sr: Structure and Stability

Hongwu Xu

Earth and Environmental Sciences Division
Los Alamos National Laboratory, Los Alamos
New Mexico 87545, USA
hxu@lanl.gov

1. Introduction

Disposal of the waste resulting from burning nuclear fuel is currently a major problem facing the nuclear industry and is a daunting challenge, both technologically and politically, that must be addressed for the sustainable and expanded use of nuclear energy. Treatment/immobilization of legacy waste from nuclear weapon activities is another related issue for human health and environmental protection. In the past decades, to develop effective strategies and technologies for nuclear waste disposal, extensive studies have been performed [1]. The main immobilization technologies that have been commercially utilized include vitrification, cementation, and bituminization, among which vitrification is the most common and viable. However, each of these technologies has its own advantages and disadvantages for immobilizing different types of wastes. For example, vitrification can produce durable waste glasses incorporating large amounts of various radionuclides with relatively small volumes but is expensive. On the other hand, cementation is inexpensive but is only suitable for disposal of low- and intermediate-level wastes. Moreover, there are some wastes that are difficult to be disposed (e.g., high radionuclide mobility) by either of these technologies, which require development of other wasteforms and immobilization technologies.

Through incorporating radioactive atoms into a crystal structure, crystalline wasteform phases are attractive, particularly for high-level waste disposal [2], due to their high stability and durability at geological repository conditions. Development of such phases for radionuclide immobilization was inspired by nature, where analog natural minerals with high radionuclide abundance can survive throughout geologic times. The basic idea is to identify crystal structures that can accommodate radioactive ions and remain stable for millions of years. Such crystalline wasteform phases usually have the following advantageous attributes: (1) radiation tolerance, (2) thermal or thermodynamic stability, and (3) chemical durability or leach resistance. For example, zircon ($ZrSiO_4$), which contains radioactive uranium (U) and thorium (Th) with concentrations ranging from 10 ppm to 1 wt.%, can be retained after many geologic processes such as erosion, transport, and metamorphism (zircon has thus been used extensively for U–Pb radiometric dating) [3]. Hence, zircon-based wasteforms have been developed to immobilize actinides including plutonium (Pu) (though they may be less durable than originally thought due to the damaging effects of α radiation) [4, 5]. A further approach based on the "nature-inspired" idea was the development of "synroc" (synthetic rock), which can incorporate various radionuclides (actinides and fission products) through different crystalline phases in this ceramic wasteform [6, 7]. Since its invention in 1978, various forms of synroc have been developed to immobilize particular components in the intermediate- and high-level wastes. In addition, more effective ceramic processing technologies, especially hot isostatic pressing (HIP), were used to increase waste loadings and/or cope with difficult wastes such as volatile radionuclides (e.g., ^{137}Cs, ^{99}Tc, and ^{129}I).

Since there are different kinds of radionuclides with dissimilar properties, different crystalline wasteform phases have been or need to be developed. Notably, radionuclides can be categorized into short-lived and long-lived radionuclides, separated by a half-life of ∼30 years [1]. Common short-lived radionuclides in nuclear waste include ^{137}Cs, ^{90}Sr, ^{134}Cs, ^{3}H, etc., most of which are fission products of fissile radionuclides in nuclear fuel. Important long-lived radionuclides are ^{14}C, ^{99}Tc, ^{129}I, ^{237}Np, and Pu isotopes, from fuel fabrication, nuclear reactor operation, and decommissioning. The activity of nuclear waste is initially dictated by that of short-lived, fission products (^{137}Cs, ^{90}Sr), and upon their radionuclide decay, the radiotoxicity of long-lived radionuclides becomes dominant (actinides and their daughter products account for most of the radiotoxicity of nuclear waste after the first 500 years of disposal). A major goal of the synroc development is to immobilize all types of radionuclides within one multi-phase, ceramic wasteform, though they are fixed into the crystal lattices of different phases.

Because of the different sizes, valences, and chemical characters of radioactive atoms/ions, there are several mechanisms by which radionu-clides are incorporated into host crystalline phases. These mainly include direct substitution of isovalent species on a particular structural site (e.g., substitution of Zr^{4+} by Pu^{4+} in zircon, $ZrSiO_4$); and coupled substitutions on multiple sites to satisfy the requirement of charge neutrality (e.g., sub-stitution of Ba^{2+} by $2Cs^+$ in barium hollandite, $Ba_{1.24}Al_{2.48}Ti_{5.52}O_{16}$). One challenge facing the development of crystalline wasteform phases is to increase their waste loadings (a lesser issue for glass wasteform), which is usually dictated by the differences in characteristics between substituting species and the type of host structures of crystalline phases. For exam-ple, host and guest cations with same valences and similar sizes may form complete solid solutions (e.g., $USiO_4$–$ThSiO_4$ solid solution [8]), which result in high waste loadings. Similarly, an open crystal structure tends to accommodate more guest radionuclides (e.g., into its open tunnels) than a dense structure, even for those radionuclides that have a different valence than the substituted host ion. Hence, the structural characteris-tics of radionuclide-substituted crystalline phases determine the amounts of waste loadings. More broadly, they also underlie the thermodynamic stability of these phases and their relations with associated compounds in the systems.

The theme of this chapter is structure, stability, and their relation-ship of crystalline wasteform phases for ^{137}Cs and ^{90}Sr. Because of the diverse structure types of crystalline wasteform phases and different prop-erties of various radionuclides, I will focus on the phases for ^{137}Cs and ^{90}Sr only. ^{137}Cs and ^{90}Sr are fission products of uranium and occur as major short-lived, heat-generating radionuclides in nuclear wastes [9]. Since ^{137}Cs transmutes via beta decay to ^{137}Ba with a half-life of 30.17 years, and since ^{90}Sr beta decays with a half-life of 29.8 years to ^{90}Y, which in turn beta decays to the stable ^{90}Zr with a half-life of 64 h, it is important that their crystalline wasteform phases remain stable over these decay periods and beyond. In other words, the crystal structures need to accommodate both the parent ^{137}Cs and ^{90}Sr and their daughter products ^{137}Ba, ^{90}Y, and ^{90}Zr over the long term. In recent years, the effects of chemical transmutation on the stability of radionuclide-bearing phases have been extensively studied using density functional theory (DFT) calculations; novel daughter phases

were predicted to form upon radionuclide decay in the parent phases —
a phenomenon termed as "radioparagenesis" [10–14], though it still needs
direct experimental verification.

^{137}Cs and ^{90}Sr are present in the radioactive waste effluents resulting
from reprocessing of spent nuclear fuels and from weapon activities. Thus
efficient removal of ^{137}Cs and ^{90}Sr from the effluents is an important step
prior to their safe, long-term storage in solid wasteforms. The ^{137}Cs/^{90}Sr
separation can be achieved using physico-chemical processes such as co-
precipitation, solvent extraction, and ion exchange [15]. Of these methods,
effective separation using an inorganic ion-exchanger (e.g., crystalline silicotitanate, CST, phases) is particularly attractive, as the Cs/Sr-loaded
microporous phase can then be heat-treated *in situ* to form a thermally
stable and chemically durable ceramic [16]. This waste disposal strategy
may reduce cost compared with other methods and minimize the risk of
environmental contamination. Similar approach has been used in thermal
conversion of Cs/Sr-sorbed hexagonal tungsten oxide bronze (HTB) adsor-
bent materials to leach resistant ceramics [17, 18].

Hence, the contents of this chapter include the structures and stability of both microporous ion exchangers and dense wasteform phases for
^{137}Cs/^{90}Sr disposal. It should be pointed out that during the past decades,
there have been considerable efforts that were directed toward the development of improved processes/materials for the removal and disposal of ^{137}Cs
and ^{90}Sr from nuclear waste streams. Thus this chapter is by no means a
complete review of the field. Rather, it mainly summarizes the studies by
this author and his colleagues in recent years. In addition, this chapter
does not deal with radiation effects in nuclear wasteforms, and the reader
may refer to several excellent reviews in this topic [19]. The major experimental techniques used are synchrotron X-ray and neutron diffraction for
structural studies and high-temperature oxide-melt drop-solution calorimetry for stability measurements. Since one of the techniques, i.e., oxide-melt
calorimetry, is rather novel, I'll first give a brief introduction of these techniques before describing the crystalline phases used for ^{137}Cs/^{90}Sr disposal.
Moreover, as aqueous durability is an important criterion for assessing the
performance of any given wasteform, its measurement methods and the
results of the described dense phases will be discussed. Lastly, since crystalline wasteforms are mostly polyphasic (e.g., synroc), the stability relations of Cs/Sr host phases will be considered in relation to other phases in
the ceramic assemblies.

2. Experimental Techniques

2.1. *Powder synchrotron X-ray and neutron diffraction and Rietveld analysis*

Powder diffraction, especially with X-rays, is the most widely used technique for identification and characterization of crystalline solids [20]. With the advent of the Rietveld method and the advances in light sources (synchrotron X-ray and neutron) and detector technology, the Rietveld analysis of powder diffraction data has become a primary method for deriving structural information of crystalline phases, including lattice constants, atomic coordinates, and atomic displacement parameters [21]. In particular, the ease of combination of powder diffraction configurations with sample environmental cells (temperature, pressure, solution, etc.) makes it a powerful tool for *in situ*, time-resolved examination of phase transformations, chemical reactions, and other processes [22]. In addition, many materials (synthetic and natural) exist only as nanocrystalline phases, for which single-crystal X-ray or neutron diffraction is not feasible, leaving powder diffraction the only viable experimental technique to derive atomic parameters. X-ray and neutron diffraction are complementary, as the scattering power of X-ray by an atom scales with its number of electrons (atomic number Z) while that of neutron varies in an irregular fashion with Z. Hence, compared with X-ray diffraction, neutron diffraction is more sensitive to light elements [23] and is advantageous for distinguishing neighboring elements in the periodic table. For structural studies of crystalline wasteform phases for Cs/Sr, both X-ray and neutron diffraction (as well as other techniques such as transmission electron microscopy) have been used. In particular, to examine their structural behavior at elevated temperatures (to simulate the radionuclide decay heat, which is particularly important for ^{137}Cs and ^{90}Sr due to the generated high heat upon their decay), *in situ* heating diffraction experiments were conducted.

2.2. *High-temperature oxide-melt drop-solution calorimetry*

High-temperature oxide-melt calorimetry is a powerful tool for thermodynamic studies and has been used to study a variety of minerals and materials [24–26]. Through measuring the heat flow between a sample and a heat sink maintained at a constant temperature (e.g., 700°C), this method determines the enthalpies of chemical reactions that may be of direct interest (e.g., crystallization) or represent steps in a thermodynamic cycle needed to obtain the enthalpy of interest (e.g., the standard enthalpy of formation of a compound from its constituent elements). Molten lead borate ($2PbO \cdot B_2O_3$) or sodium molybdate ($3Na_2O \cdot 4MoO_3$) has been used as the solvent; the former is a good solvent for silicate samples while the latter is more efficient for transition metal and rare earth compounds. The advantages of calorimetry at high temperature lie in rapid and reproducible reactions, which facilitate high-precision measurements of their released or absorbed heats.

This type of calorimeters was traditionally custom-built but is now commercially available (Setaram AlexSys calorimeter, named after Alexandra Navrotsky who pioneered this technique). The calorimeter consists of twinned sample chambers that are each surrounded by a Pt-PtRh thermopile connecting to a massive metallic block maintained at a constant temperature. The twinned design increases measurement productivity and minimizes the effects of small drifts in temperature. For stability studies of crystalline phases for Cs/Sr disposal, both microporous and dense wasteform phases have been investigated. More specifically, their standard enthalpies of formation from constituent oxides ($\Delta H_{f,ox}{}^0$) and elements ($\Delta H_{f,el}{}^0$) were determined. The obtained stability relations can be related to the structural behavior characterized by X-ray/neutron diffraction.

2.3. *Aqueous durability/leach test*

Aqueous durability of a wasteform phase is typically measured using standard or modified American Society of Testing Materials (ASTM) tests, e.g., the product consistency test (PCT) and Materials Characterization Center (MCC)-1 leach test [27]. Experimental conditions such as solution pH, sample-to-solution ratio, temperature, and time can be varied. The solutions may be buffered or unbuffered. Either powdered or pelletized specimens are used, and their surface areas/porosity can be measured by Brunauer–Emmett–Teller (BET) and geometrical methods. For modified tests, solution samples can be removed at different time intervals. The leached samples are characterized using scanning/transmission electron microscopy (SEM/TEM) and X-ray diffraction (XRD), and the concentrations of the solutions are analyzed by atomic absorption spectrometry (AAS) and inductively coupled plasma (ICP) spectroscopy. The leach rates of the radionuclides can thus be obtained. Detailed analyses of the leach data may shed light on the roles of solution chemistry versus diffusion or passivation in controlling the dissolution behavior of wasteform phases. Combined with analyses of the leachates and the structural and energetic characteristics of the wasteform phase, these provide insights into the leaching mechanisms, which may involve the competition between complexation and hydrolysis as well as the interplay between dissolution kinetics and thermodynamic stability.

3. Microporous Ion-Exchangers for Cs/Sr Separation

3.1. *Microporous titanosilicates for Cs separation*

Microporous titanosilicates or CSTs are a novel class of framework materials that potentially have many industrial applications as catalysts, sorbents, and ion-exchangers (e.g., [28, 29]). Unlike the extensively studied and utilized aluminosilicate zeolites whose framework is composed of $[SiO_4]$ and $[AlO_4]$ tetrahedra, most titanosilicates consist of both tetrahedrally and octahedrally coordinated framework cations ($[SiO_4]$ tetrahedra and $[TiO_6]$ octahedra). Because of the different coordinations of Si^{4+} and Ti^{4+}, microporous titanosilicates offer a variety of unique framework topologies. The novel pore structures in titanosilicates and the new chemistry associated with Ti result in certain superior physical and chemical properties of these materials over zeolites. One example related to nuclear waste disposal is that some titanosilicates have significantly higher selectivities for Cs^+ than for Na^+. Thus they can potentially be used as ion exchangers for removal of Cs^+ from radioactive aqueous wastes with high Na^+ concentrations [30].

Xu *et al.* [31] investigated the structures and energetics of two series of microporous silicotitanate phases in the Na_2O–Cs_2O–SiO_2–TiO_2–H_2O system: (1) the $(Na_{1-x}Cs_x)_3Ti_4Si_3O_{15}(OH)\cdot nH_2O$ ($n = 4 - 5$) phases with a cubic structure (space group $P\bar{4}3m$, an analog of the mineral pharmacosiderite $KFe_4(OH)_4(AsO_4)_3\cdot{\sim}6H_2O$) [32, 33], and (2) the $(Na_{1-x}Cs_x)_3Ti_4Si_2O_{13}(OH)\cdot nH_2O$ ($n = 4 - 5$) phases with a tetragonal structure (space group $P4_2/mcm$) [34, 35]. The framework structures of the two series are similar in the sense that they both are comprised of $[SiO_4]$ tetrahedra and $[Ti_4O_4]$ cubane clusters. However, as the tetragonal series (Ti:Si = 2:4) is relatively silica-deficient compared to the cubic series (Ti:Si = 4:3), the alternating arrangement of $[SiO_4]$ and $[Ti_4O_4]$, which occurs along all the three unit-cell axes in the cubic structure (Figure 1(a)), only occurs along the a and b axis in the tetragonal structure (Figure 1(b)). In the c direction of the tetragonal structure, the $[Ti_4O_4]$ cubane clusters are directly linked to each other (Figure 1(c)). As a result, there is only one type of extra-framework site in the cubic structure, whereas there are two

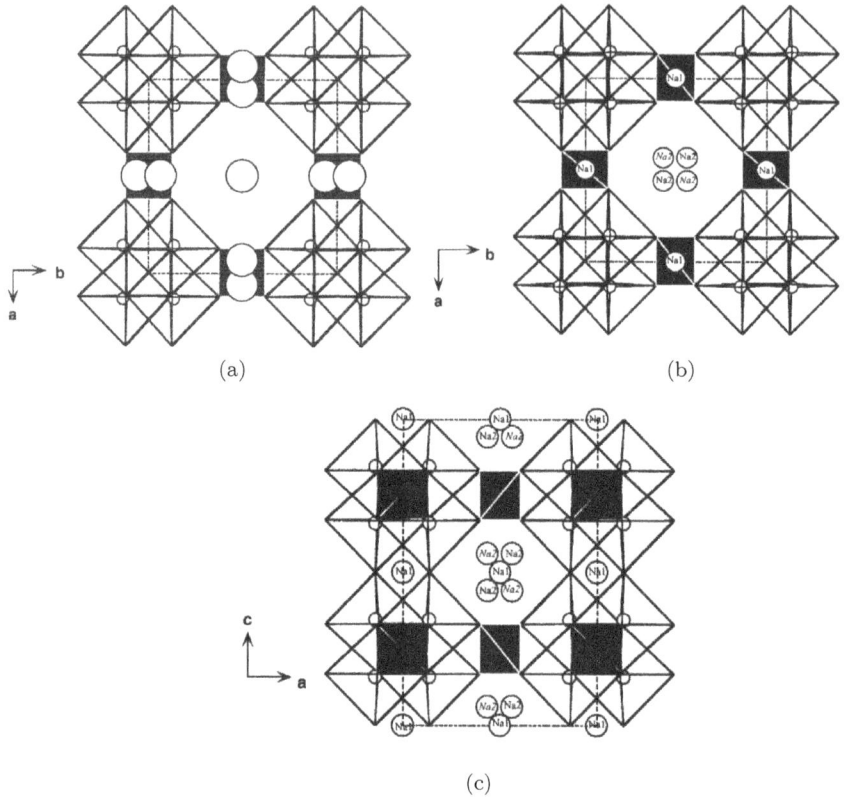

Figure 1. Projections of the crystal structures of (a) $Cs_3Si_3Ti_4O_{15}(OH)\cdot 4H_2O$ down [001] and (b, c) $Na_2Ti_2O_3SiO_4(OH)\cdot 2H_2O$ down (b) [001] and (c) [010] [31]. Spheres represent Cs or Na ions; [TiO_6] octahedra and [SiO_4] tetrahedra are plotted in black and white, respectively. Dash lines outline the unit cell. Structural data of (a) are taken from Harrison *et al.* [33], and those of (b) and (c) from Poojary *et al.* [34].

symmetrically distinct positions in the tetragonal structure, one of which is occupied by Na^+ and cannot be exchanged by Cs^+.

Standard enthalpies of formation for these two series of silicotitanates have been determined by drop-solution calorimetry into molten 2PbO· B_2O_3 at 974 K. The enthalpies of formation from the oxides for the cubic series become more exothermic as Cs/(Na+Cs) increases, whereas those for the tetragonal series become less exothermic (Figure 2). This result indicates that the incorporation of Cs^+ in the cubic phase is somewhat thermodynamically favorable, whereas that in the tetragonal phase

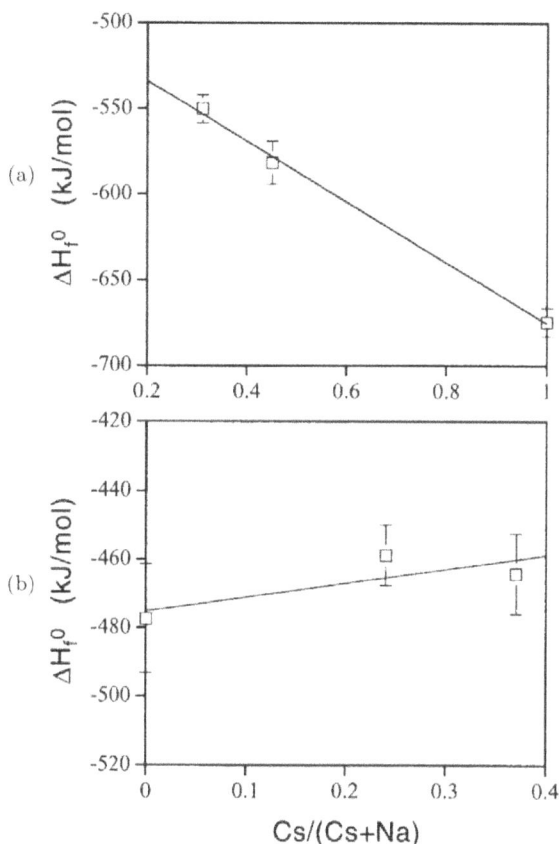

Figure 2. Enthalpies of formation from the oxides for (a) the cubic $(Na_{1-x}Cs_x)_3 Ti_4Si_3O_{15}(OH) \cdot nH_2O$ phases and (b) the tetragonal $(Na_{1-x}Cs_x)_3 Ti_4Si_2O_{13}(OH) \cdot nH_2O$ phases as a function of composition [31].

is thermodynamically unfavorable and kinetically driven. In addition, the cubic phases are more stable than the corresponding tetragonal phases with the same Cs/Na ratios. These disparities in the energetic behavior between the two series are attributed to their differences in both local bonding configuration (i.e., the occurrence of Ti–O–Ti linkages in the tetragonal structure but not in the cubic structure) and degree of hydration.

To determine the effects of K–Cs substitution on the structure and energetics of the cubic pharmacosiderite analog, a complete series of solid solutions with compositions $(K_{1-x}Cs_x)_3Ti_4Si_3O_{15}(OH) \cdot nH_2O$ ($n = 4 - 6$, $0 \leq x \leq 1$) was synthesized using hydrothermal (the K endmember) and

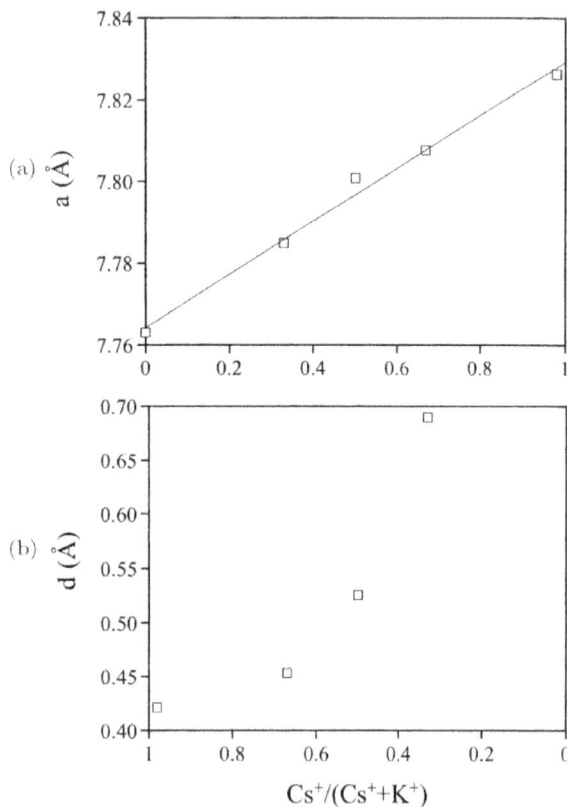

Figure 3. Variation of (a) cell parameter a and (b) the displacement of Cs^+ from the ring center, (0, 1/2, 1/2) of pharmacosiderite analogues $(K_{1-x}Cs_x)_3Ti_4Si_3O_{15}(OH)\cdot nH_2O$ as a function of composition [36]. The lines in (a) are least-square fits to the data. The Cs^+ displacement d is equal to lattice parameter a times the x coordinate of Cs. Errors are smaller than the size of the symbol.

ion-exchange (all other compositions) methods [36]. The Rietveld analysis of synchrotron XRD data shows that the unit-cell parameter a increases linearly with increasing Cs content (Figure 3(a)). In the structure, K^+ is situated in the center of the eight-membered titanosilicate ring, whereas Cs^+ is displaced from the ring center, and the displacement increases with higher K/(Cs+K) ratio (Figure 3(b)). Because unit-cell parameter/volume and the size of titanosilicate ring scale with the Cs content, this behavior implies that the position of Cs^+ is closely related to the ring size. Although the titanosilicate rings for all the compositions appear to be large enough to enclose Cs^+ (as evidenced by the $K^+ \rightarrow Cs^+$ exchange in which Cs^+ penetrates the rings into the cages) [29], the repulsion between Cs^+ and the surrounding Ti^{4+}/Si^{4+} presumably prohibits the occupancy of the ring centers by Cs^+. Since the smaller the ring the larger the repulsion, Cs^+ displaces further with decreasing $Cs/(Cs+K)$.

The enthalpies of formation of $(K_{1-x}Cs_x)_3Ti_4Si_3O_{15}(OH)\cdot nH_2O$ from oxides, determined by drop-solution calorimetry, become more exothermic with increasing $Cs/(Cs+K)$, suggesting a stabilizing effect of $K^+ \rightarrow Cs^+$ on the pharmacosiderite structure (Figure 4). This behavior is similar to that observed in the Na-substituted pharmacosiderite phases, $(Na_{1-x}Cs_x)_3Ti_4Si_3O_{15}(OH)\cdot nH_2O$, whose $\Delta H_{f,ox}{}^0$ decreases with increasing $Cs/(Cs+Na)$ (Figure 2(a)). Calculation of the enthalpy of the $K^+ \rightarrow Cs^+$ exchange reaction based on the measured formation enthalpies indicates that the Cs^+ uptake in $(K_{1-x}Cs_x)_3Ti_4Si_3O_{15}(OH)\cdot nH_2O$ is probably thermodynamically (rather than kinetically) driven, as occurred in $(Na_{1-x}Cs_x)_3Ti_4Si_3O_{15}(OH)\cdot nH_2O$.

3.2. *Microporous niobate phases for Sr separation*

In contrast to the conventional, tetrahedral zeolitic frameworks and the mixed tetrahedral–octahedral titanosilicate frameworks described above, a class of octahedral microporous phases, $Na_2Nb_{2-x}M_xO_{6-x}(OH)_x\cdot H_2O$

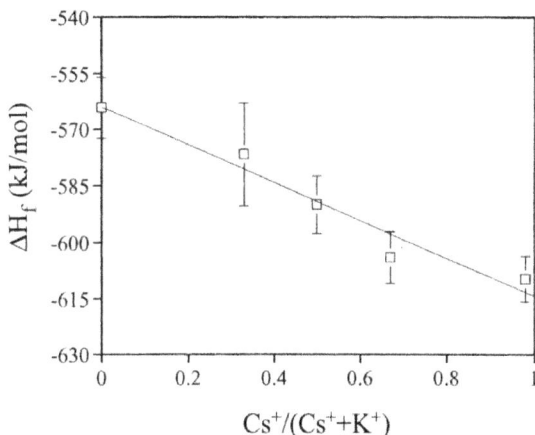

Figure 4. Variation of the standard enthalpies of formation of pharmacosiderite analogues $(K_{1-x}Cs_x)_3Ti_4Si_3O_{15}(OH)\cdot nH_2O$ from the constituent oxides as a function of composition [36]. The line represents a least-square fit to the data.

(M = 2Ti, Zr; $0 \leq x \leq 0.4$), was synthesized using hydrothermal methods [37–39]. These phases, named Sandia octahedral molecular sieves (SOMS), possess a framework structure composed of [NbO$_6$], [MO$_6$], and [NaO$_6$] octahedra linked by corner- or edge-sharing. In the structure, Nb and Ti/Zr occupy the same framework positions, and the remaining Na resides in the channels. This structure is unusual in the sense that Na, which is typically an extraframework cation, also participates in the formation of the framework. Figure 5 shows the structure of the archetype phase, Na$_2$Nb$_2$O$_6 \cdot$H$_2$O. As is shown, [NbO$_6$] octahedra are connected to form double chains that run parallel to [010], whereas [NaO$_6$] into layers parallel to (001), both via edge sharing. The [NaO$_6$] layers alternate with the layers containing [NbO$_6$] double chains along the c axis, forming a three-dimensional (3-D) network. Ion exchange experiments show that the SOMS phases are highly selective for large alkaline earth cations over alkali cations. For example, the distribution coefficient (K_d), which is defined as the ratio of cation adsorbed onto the ion exchanger to the cation remaining in the solution, for the Na$_2$Nb$_{0.6}$Ti$_{0.4}$O$_{5.6}$(OH)$_{0.4} \cdot$H$_2$O phase is about 10^5 for Sr^{2+} but only 95 for K$^+$ [37]. Hence, these materials can potentially be used for separation of radioactive ^{90}Sr from aqueous nuclear wastes (as well as contaminated reactor cooling water or groundwater). Moreover, upon heating, the Sr-exchanged SOMS phases dehydrate and convert to

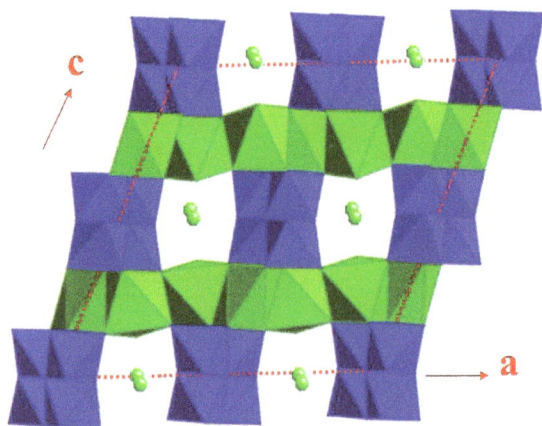

Figure 5. Crystal structure of the archetype phase, Na$_2$Nb$_2$O$_6 \cdot$H$_2$O, projected down [010] [39]. Blue octahedra represent [NbO$_6$] units, green octahedral represent [NaO$_6$] octahedra, and green spheres represent extra framework Na ions. Red dash lines outline the unit cell.

thermally stable and chemically durable perovskites [38]. Thus, these perovskites may serve as permanent ceramic host phases for ^{90}Sr in radioactive
waste management.

The structures and energetics of a series of $Na_2Nb_{2-x}Ti_xO_{6-x}$
$(OH)_x \cdot H_2O$ $(0 \le x \le 0.4)$ phases have been studied using powder XRD and
drop-solution calorimetry [40]. With increasing Ti content, unit-cell parameter b is virtually unchanged, c decreases slightly, while a first decreases up
to $x = \sim 0.3$ and then remains approximately constant thereafter (Figure 6).
As a result, the cell volume V exhibits the same fashion of variation as a.
These general trends may be attributed to the anisotropic nature of the
structure (Figure 5). Replacing portion of Nb^{5+} cations in $Na_2Nb_2O_6 \cdot H_2O$
with the smaller Ti^{4+} decreases the overall volume of $[Nb/TiO_6]$ chains and

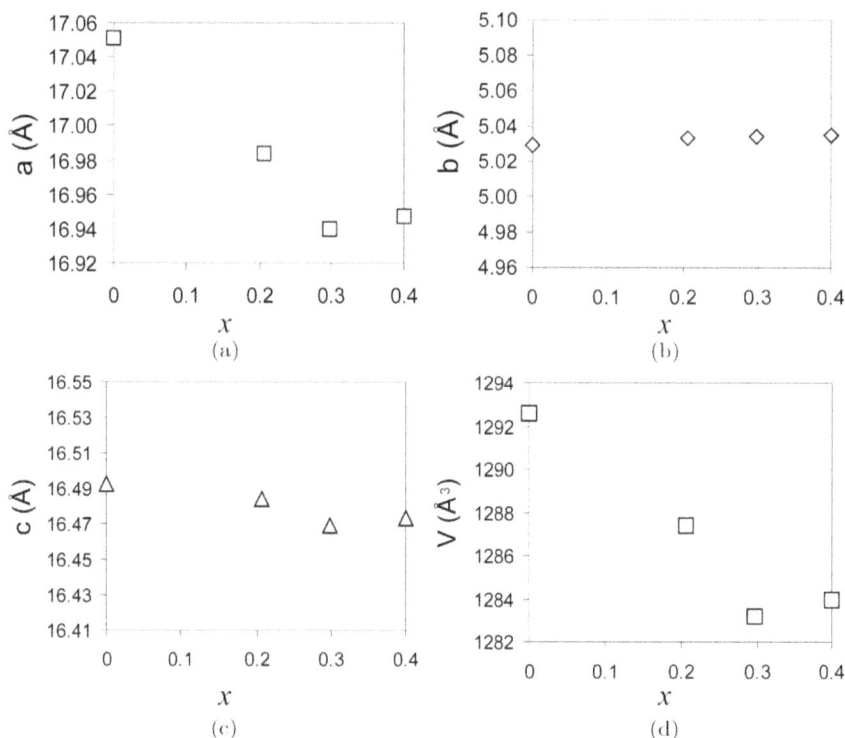

Figure 6. Variation of cell parameters (a) a, (b) b, (c) c, and (d) V of
$Na_2Nb_{2-x}Ti_xO_{6-x}(OH)_x \cdot H_2O$ as a function of composition [40]. Errors are
smaller than the size of symbols. For ease of comparison, a, b, and c are plotted on the same scale.

thus of the structure. However, this volume reduction is not reflected along the b-axis, since a shortening of the edge-sharing chains along b would dramatically increase the repulsion between neighboring Nb^{5+}/Ti^{4+} cations. Instead, the structure shrinks mainly along the a-axis, due to the larger structural flexibility in this direction. When x exceeds ~ 0.3, however, further increasing Ti content does not result in further decrease in a. Rather, a and thus V remain roughly constant. This dimensional conservation is probably due to the geometrical constraints from the $[NaO_6]$ layers, which alternate with the layers containing $[Nb/TiO_6]$ double chains along the c-axis (Figure 5).

Their enthalpies of formation from oxides, determined by drop-solution calorimetry in $3Na_2O \cdot 4MoO_3$ solvent at $974\,K$, display complex variations with composition. As shown in Figure 7, with increasing Ti/Nb, $\Delta H_{f,ox}^0$ values become less exothermic up to $x = \sim 0.2$, but it then becomes more exothermic thereafter. In other words, increasing $Nb^{5+} + O^{2-} \rightarrow Ti^{4+} + OH^-$ substitution first destabilizes the SOMS structure and then stabilizes it with respect to the constituent oxides. This behavior may result from a complex interplay among several competing structural variations with changing Ti/Nb, including the amount of OH^- present in these phases, the strain energy due to the Nb^{5+}/Ti^{4+} repulsion within $[Nb/TiO_6]$ chains and the mismatch between the $[NaO_6]$ and $[Nb/TiO_6]$ layers, the extent of framework Nb^{5+}/Ti^{4+} short-range order and the degree of extraframework Na^+ ordering.

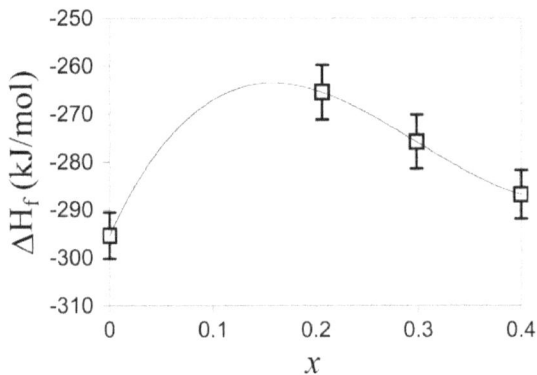

Figure 7. Variation of the enthalpies of formation of $Na_2Nb_{2-x}Ti_xO_{6-x}(OH)_x \cdot H_2O$ from the constituent oxides at $298\,K$ as a function of composition [40]. The curve represents a third-order polynomial fit to the data.

4. Crystalline Wasteform Phases for Cs/Sr Immobilization

4.1. *Titanate hollandites for Cs immobilization*

Titanate hollandites, which have the compositions $A_x(Ti^{4+},B)_8O_{16}$, $(A = Ba^{2+}, Cs^+, Sr^{2+}, Rb^+,$ etc.; $B = Al^{3+}, Fe^{3+}, Mg^{2+}, Ti^{3+}, Ga^{3+}, Cr^{3+}, Sc^{3+},$ etc.) [41–44], comprise a family of potential wasteform phases for immobilization of radioactive ^{137}Cs (and its daughter product ^{137}Ba) due to their high thermal and aqueous stability. In the hollandite structure, $[TiO_6]$ and $[BO_6]$ octahedra are linked via edge- and corner-sharing to form a 3-D framework (Figure 8(a)), in which Ti^{4+} and B^{3+} (such as Al^{3+} and Fe^{3+}) or B^{2+} (such as Mg^{2+}) are disordered over one or two crystallographically distinct sites. A^+ or A^{2+} cations occupy the box-shaped cavity sites, each coordinated by eight oxygen atoms, within tunnels parallel to the *c*-axis (Figure 8(b)). Depending on the mean radius ratio of cations in the A and Ti/B sites, the structure may adopt a tetragonal (space group $I4/m$) or monoclinic (space group $I2/m$) symmetry. The ability of the hollandite structure to accommodate both Cs^+ and Ba^{2+} over its tunnel sites is particularly useful for ^{137}Cs immobilization, as ^{137}Cs transmutes to ^{137}Ba through beta decay with a half-life of about 30 years. It is important that the structure remains stable over this decay period and beyond.

To study their structural behavior at elevated temperatures (as encountered in hot environments due to the $^{137}Cs \rightarrow {}^{137}Ba$ decay), Xu *et al.* [45] conducted *in situ* neutron diffraction experiments of a representative tetragonal phase, $Ba_{1.24}Al_{2.48}Ti_{5.52}O_{16}$, in the temperature range 300–1173 K. The Rietveld analyses of the obtained data show that on heating, unit-cell parameters *a* and *c* increase at similar rates (Figure 9). This isotropic nature of thermal expansion can be explained by the increased volume and regularity of $[(Ti,Al)O_6]$ octahedra and the widening of the $Ti/Al–O_2–Ti/Al$ angle (O_2 is corner-shared by two $[(Ti,Al)O_6]$ octahedra from neighboring $[(Ti,Al)O_6]$ double chains) with increasing temperature. Practically, this property is advantageous, as hollandite-based ceramic or composite products would, upon heating, be less likely to form microcracks. The amplitudes of thermal vibration for Ba, Ti/Al, and O increase with increasing temperature; however, the rate of increase for Ba is much larger. This behavior is due to the occupancy of the box-shaped cavity site by Ba, which has weaker interactions with its neighboring atoms, compared with those for framework Ti/Al and O. On the other hand, the opening of the oxygen-coordinated cavity box is smaller than the size of Ba (Figure 10(a)), even

(a)

(b)

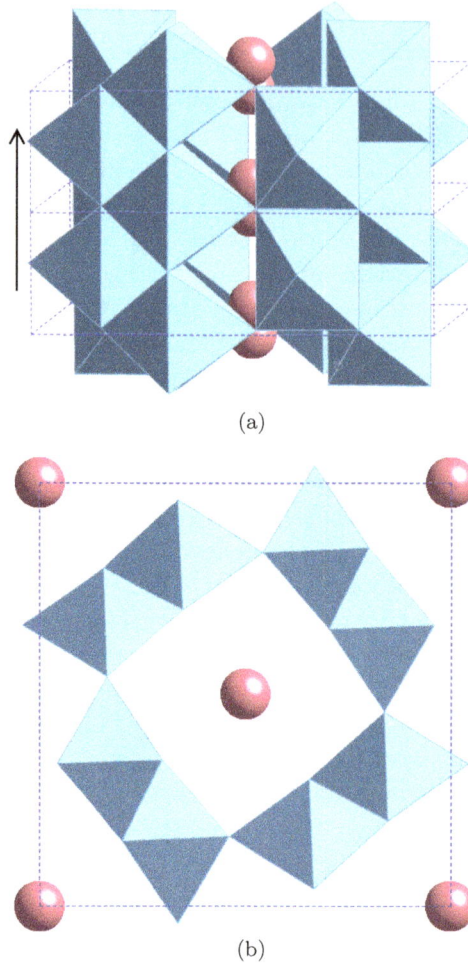

Figure 8. (a) Crystal structure of hollandite, $A_x(Ti^{4+},B)_8O_{16}$ ($A = Ba^{2+}$, Cs^+, etc.; $B = Al^{3+}$, Fe^{3+}, etc.). (b) Projection of the structure along the **c**-axis showing the cross-section of the tunnels [45]. Blue octahedra represent $[(Ti^{4+},B)O_6]$ units, pink balls represent A cations, and dashed lines outline unit cells. Note for each pair of Ba sites shown in (a), only one of them can be occupied by Ba.

at high temperature (Figure 10(b)), preventing evaporation of Ba (or its parent Cs, which is even larger than Ba) from the hollandite structure. These characteristics render titanate hollandites potentially robust wasteforms for Cs/Ba.

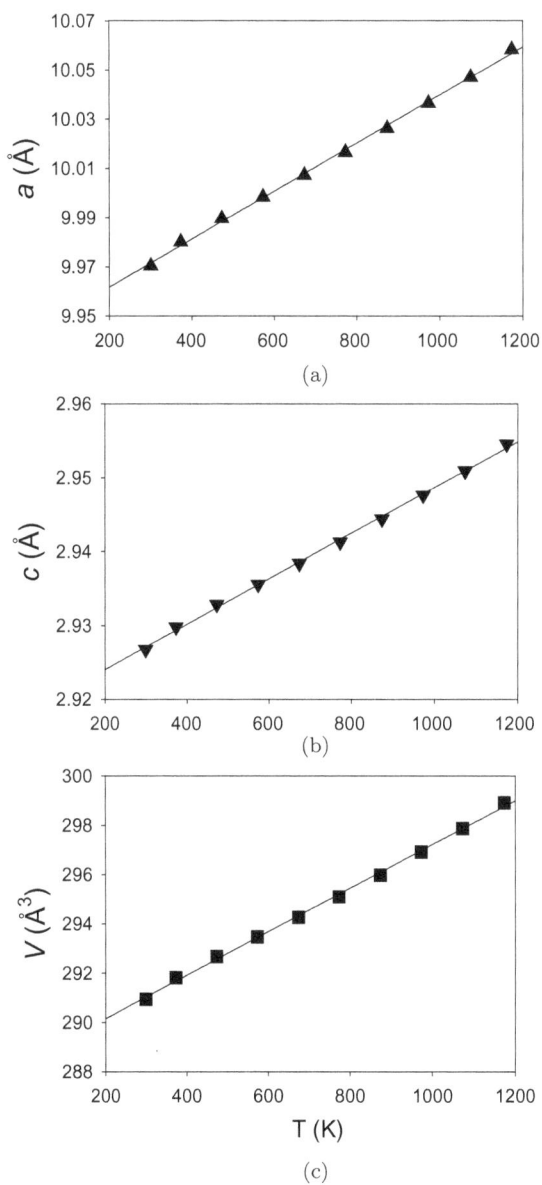

Figure 9. Variation of (a) cell parameters a, (b) c, and (c) cell volume V of $Ba_{1.24}Al_{2.48}Ti_{5.52}O_{16}$ hollandite as a function of temperature [45]. The lines are the best fits to the data.

(a)

(b)

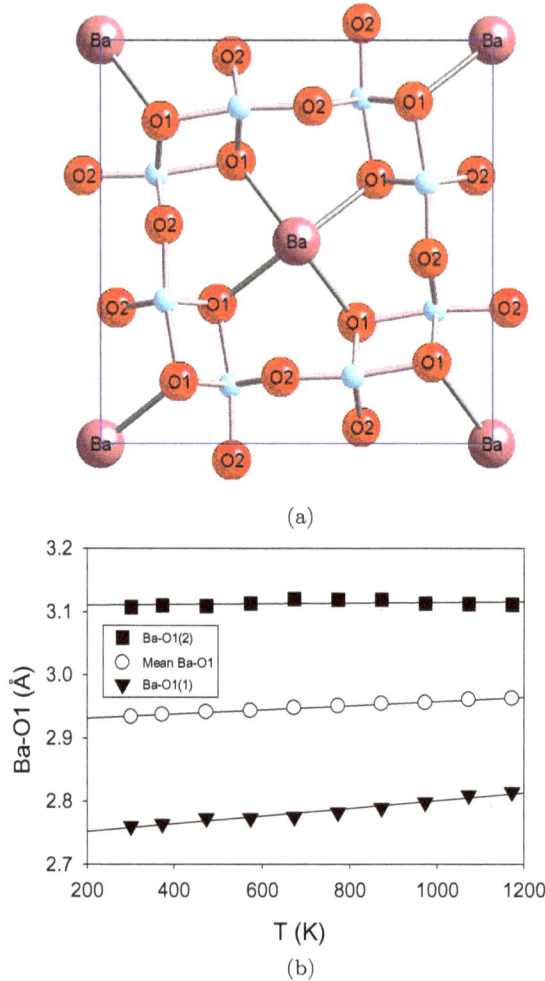

Figure 10. (a) Schematic illustrating the local bonding environments of Ba in the structure of $Ba_{1.24}Al_{2.48}Ti_{5.52}O_{16}$ hollandite. Note that for clarity, the atoms are drawn smaller; (b) Variation of Ba–O1 bond lengths as a function of temperature [45]. The lines are the best fits to the data and serve as the trend guides to eyes.

The high thermal stability of $Ba_{1.24}Al_{2.48}Ti_{5.52}O_{16}$ hollandite is consistent with its exothermic enthalpy of formation from constituent oxides, measured by drop-solution calorimetry into molten $2PbO\cdot B_2O_3$ at 974 K [46]. The determined $\Delta H_{f,ox}^{0}$ is -223.48 ± 9.11 kJ/mol, and taking into account of the entropy of formation (dominated by the configurational component; the vibrational component can be neglected, as all the involved phases are solids), the Gibbs free energy of formation is calculated as $\Delta G_{f,ox}^0 = \Delta H_{f,ox}^0 - T\Delta S_{conf}^0 = -209.25 \pm 5.73\,\text{kJ/mol}$. Thus, this hollandite is thermodynamically more stable than its constituent oxides at standard conditions.

To determine the effects of ionic substitutions on the stability of this phase, Xu *et al.* [47] and Wu *et al.* [48] synthesized three hollandite phases, $Ba_{1.18}Cs_{0.21}Al_{2.44}Ti_{5.53}O_{16}$, $Ba_{1.17}Rb_{0.19}Al_{2.46}Ti_{5.53}O_{16}$, and $Ba_{1.14}Sr_{0.10}Al_{2.38}Ti_{5.59}O_{16}$, which have the same Al/Ti (B cations) stoichiometries but differ in A cation species (Cs^+, Rb^+ and Sr^{2+}). The Rietveld analysis of synchrotron XRD data shows that they adopt the tetragonal structure (space group $I4/m$), and their cell parameters increase with increasing cation size ($Sr^{2+} \rightarrow Rb^+ \rightarrow Cs^+$). Enthalpies of formation of these hollandites from the oxides, measured by drop-solution calorimetry, are similarly exothermic, consistent with the occurrence of extensive cation substitutions in hollandites (Figure 11(a)). Although the large exothermic $\Delta H_{f,ox}^0$ values of the three substituted hollandites are an indication of their higher stability relative to binary oxides, for assessment of their suitability as wasteforms, one must also consider their stability relative to other competing phase assemblages including ternary oxides. A conceivable phase assemblage contains $BaTiO_3$ perovskite, $SrTiO_3$ perovskite (for Sr-substituted hollandite), and other oxides can be evaluated by thermodynamics, which is shown using the following reactions:

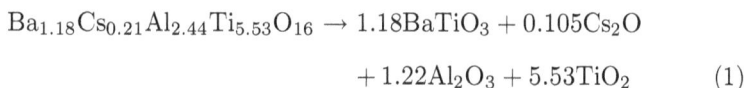

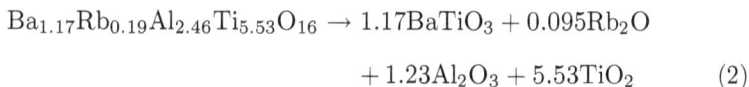

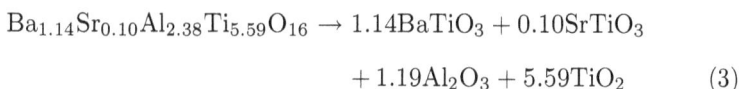

$$Ba_{1.18}Cs_{0.21}Al_{2.44}Ti_{5.53}O_{16} \rightarrow 1.18BaTiO_3 + 0.105Cs_2O$$

$$+ 1.22Al_2O_3 + 5.53TiO_2 \qquad (1)$$

$$Ba_{1.17}Rb_{0.19}Al_{2.46}Ti_{5.53}O_{16} \rightarrow 1.17BaTiO_3 + 0.095Rb_2O$$

$$+ 1.23Al_2O_3 + 5.53TiO_2 \qquad (2)$$

$$Ba_{1.14}Sr_{0.10}Al_{2.38}Ti_{5.59}O_{16} \rightarrow 1.14BaTiO_3 + 0.10SrTiO_3$$

$$+ 1.19Al_2O_3 + 5.59TiO_2 \qquad (3)$$

The enthalpies of the above reactions at standard conditions are derived to be 32.2 ± 13.5, 26.9 ± 13.4, and $14.2 \pm 12.0\,\text{kJ/mol}$ (Figure 11(b)). These results indicate that Cs- and Rb-hollandite are energetically stable at room temperature with respect to $BaTiO_3$, Cs_2O or Rb_2O, Al_2O_3, and TiO_2, since their reaction enthalpies are endothermic. On the other hand,

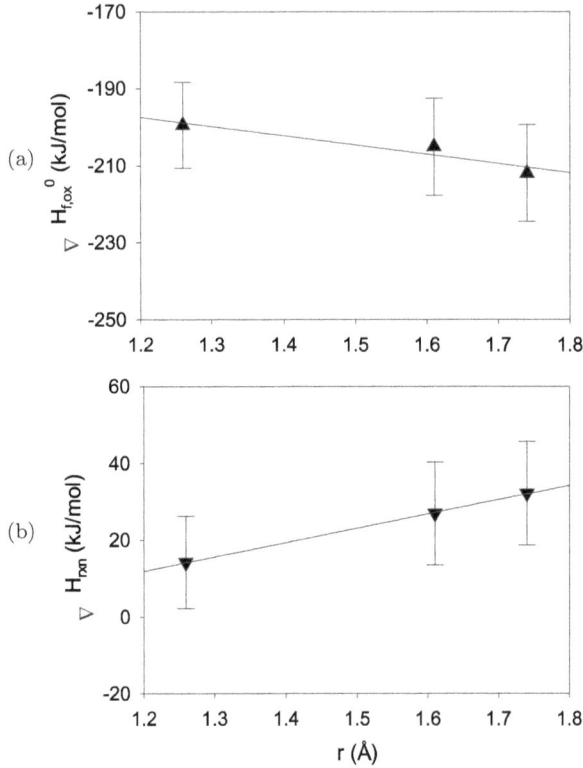

Figure 11. Variations of (a) enthalpies of formation from the constituent oxides at 298 K ($\Delta H^{\circ}_{f,ox}$) and (b) enthalpies of reactions (1), (2) and (3) (ΔH_{rnx}) for the hollandite phases, $Ba_{1.18}Cs_{0.21}Al_{2.44}Ti_{5.53}O_{16}$, $Ba_{1.17}Rb_{0.19}Al_{2.46}Ti_{5.53}O_{16}$, and $Ba_{1.14}Sr_{0.10}Al_{2.38}Ti_{5.59}O_{16}$, as a function of ionic radii of Cs^{+}, Rb^{+}, and Sr^{2+} [48].

Sr-hollandite may be comparable in stability to $BaTiO_3$, $SrTiO_3$, Al_2O_3, and TiO_2, as the reaction enthalpy of its decomposition is closer to zero within experimental uncertainties. These stability relations are in general consistent with phase assemblages in Synroc, where Cs^{+}, Rb^{+}, and Ba^{2+} are incorporated into hollandite phases, whereas Sr^{2+} occurs in titanate perovskite.

In addition to thermodynamic stability, aqueous durability is an important criterion for assessing the performance of a wasteform phase. Previous leach tests of hollandite phases show that Cs release rates can vary significantly, depending on their compositions, structures, and synthesis/processing methods [49–52]. For example, MCC-1 leach tests show that Al-substituted hollandite wasteforms, synthesized by HIP, exhibit much lower Cs leach rates ($< 0.06\,\mathrm{g}\cdot(\mathrm{m}^2\cdot\mathrm{day})^{-1}$) than their Mg-substituted counterparts ($\sim 2.0\,\mathrm{g}\cdot(\mathrm{m}^2\cdot\mathrm{day})^{-1}$) during a 28-day period [51].

4.2. *Pollucite analogs for Cs immobilization*

Pollucite ($CsAlSi_2O_6$) has long been proposed as a solid host for radioactive Cs immobilization because of its relatively high stability under hydrothermal conditions [53–55]. It has a 3-D framework structure composed of corner-sharing $[SiO_4]$ and $[AlO_4]$ tetrahedra with Cs^+ occupying the cavity sites (Figure 12(a)). Substituting Ti^{4+} for Al^{3+} over the tetrahedral sites, accompanied by the incorporation of extra O^{2-} anions into the structure to achieve its charge neutrality [56] (Figure 12(b)), produces a series of substituted pollucites with the general formula $CsTi_xAl_{1-x}Si_2O_{6-0.5x}$, $0 \leq x \leq 1$. These pollucites may be present as a potential host for Cs in the wasteforms from heat treatment of Cs-loaded microporous silicotitanate ion-exchangers. Rietveld analysis of powder synchrotron XRD data indicates that the parent $CsAlSi_2O_6$ pollucite has a tetragonal structure (space group $I4_1/a$), whereas all other compositions are cubic (space group $Ia3d$) [57]. The increased symmetry for the Ti-substituted structures is presumably due to the incorporation of additional O^{2-}, which effectively holds open the expanded cubic framework. As expected, unit-cell volume (V) increases with increasing x. However, the variation of V vs. x is not linear; it shows a negative deviation from the ideal linearity when $x = 0.3 - 1$ (Figure 13(a)). The same trend of nonlinearity is also seen in variation of the enthalpy of formation from oxides, determined by drop-solution calorimetry with molten lead borate as the solvent at $974\,\mathrm{K}$ (Figure 13(b)) [58]. Although the $\Delta H_{f,ox}^{0}$ generally decreases with increasing x, there exists a regime exhibiting exothermic heats of mixing within the composition range from $x = 0.3 - 1$. This non-ideal mixing behavior can be attributed to varying degrees of short-range order associated with the framework cations Al^{3+}, Si^{4+}, and Ti^{4+}, across the series. This interpretation has been confirmed by a later X-ray absorption and Raman spectroscopy study of the same series of samples [59].

Although the thermodynamic stability of $CsTi_xAl_{1-x}Si_2O_{6-0.5x}$ pollucites relative to their constituent oxides decreases with increasing $Ti^{4+} +$ $0.5O^{2-} \rightarrow Al^{3+}$ substitution, the stability of this series with respect to the aqueous species appears not to vary significantly with composition. Specifically, the two endmembers $CsAlSi_2O_6$ and $CsTiSi_2O_{6.5}$ have comparable

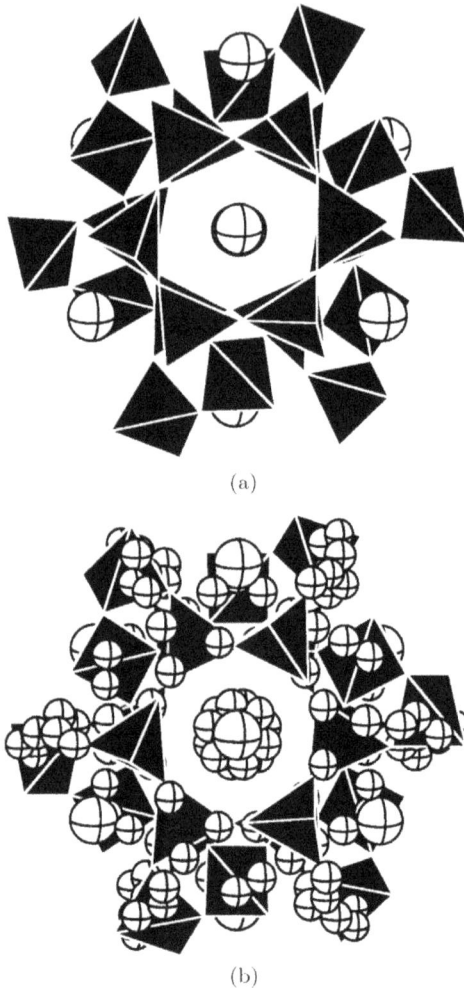

(a)

(b)

Figure 12. Crystal structures of pollucites (a) $CsAlSi_2O_6$ and (b) $CsTiSi_2O_{6.5}$ projected down [1–11, 57]. Large spheres represent Cs^+ ions and small spheres in (b) represent positions of extra oxygens (with an occupancy of 0.060 or 0.023) in excess of 16 oxygens per unit cell; tetrahedra represent $(Si,Al,Ti)O_4$ units.

low rates of cesium leaching [55, 60]. This behavior is probably related to the similar small sizes of their cages in which Cs^+ cations reside, rendering the low mobility of Cs^+ in both structures [60]. Cs release under aqueous conditions also might be controlled by kinetic factors. As demonstrated in the studies on two series of microporous titanosilicate phases [31], kinetic

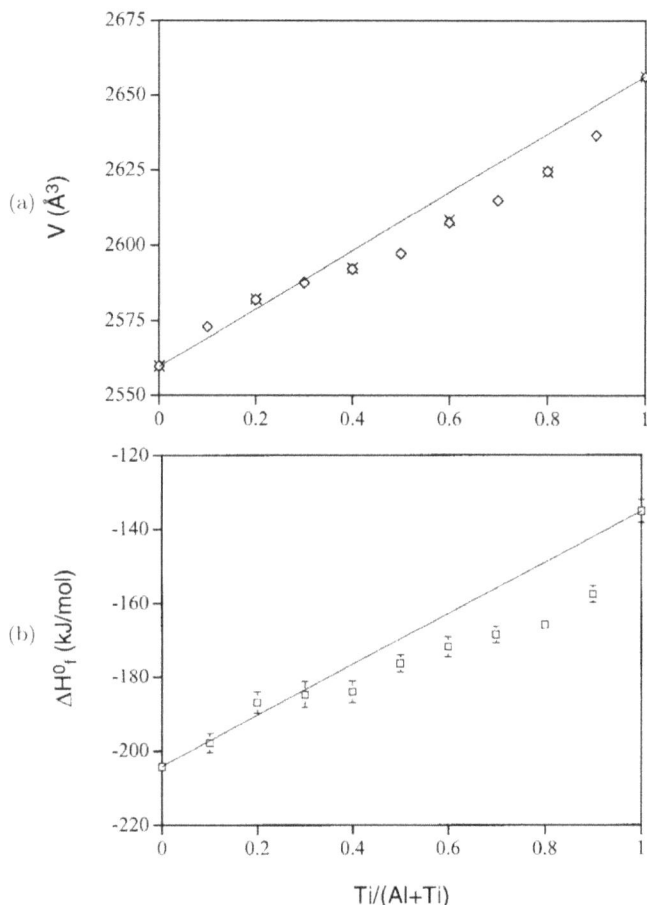

Figure 13. Variation of (a) cell volumes (V) and (b) enthalpies of formation from oxides at 298 K ($\Delta H^{\circ}_{f,ox}$) for $CsTi_xAl_{1-x}Si_2O_{6-0.5x}$ ($0 \leq x \leq 1$) pollucites as a function of composition [58]. In (a), crosses represent synchrotron X-ray data; diamonds represent conventional X-ray data. Errors are smaller than the size of symbols. In (b), line links data points of two endmembers.

factors do seem to play significant roles in the uptake of Cs^+ in titanosilicate structures. In addition, high-temperature neutron diffraction experiments of $CsTiSi_2O_{6.5}$ shows that this phase is thermally stable up to at least 1073 K (the highest temperature tested) (Figure 14). The cubic symmetry of the structure means that its thermal expansion is isotropic, thereby minimizing the chance of microcracking in pollucite-based ceramic or composite products at elevated temperatures.

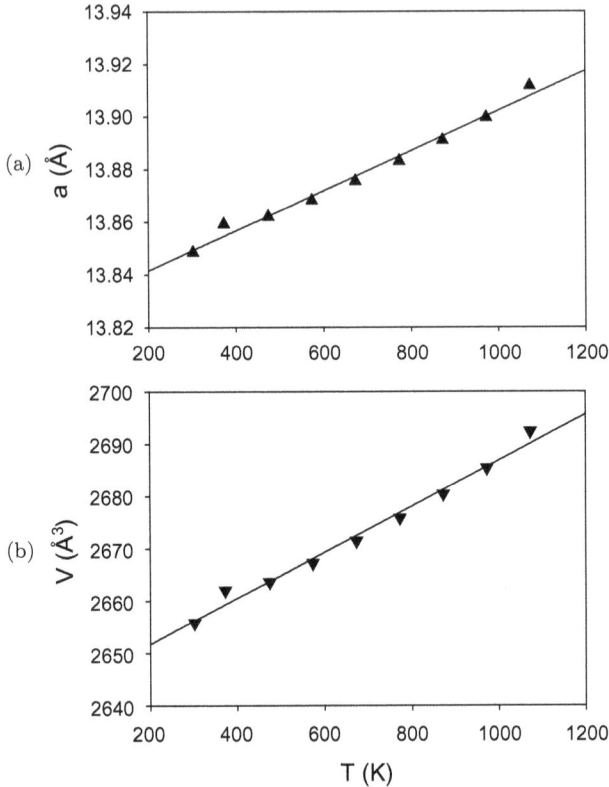

Figure 14. Variation of (a) cell parameter a and (b) cell volume V of $CsTiSi_2O_{6.5}$ pollucite as a function of temperature. The lines are the best fits to the data.

As stated earlier, because ^{137}Cs transmutes to [137]Ba through the beta decay, it is important to study the structure and stability of Ti-substituted pollucites, especially $CsTiSi_2O_{6.5}$, upon Cs → Ba substitution. Two series of Ba/Ti-substituted samples, $Cs_xBa_{(1-x)/2}TiSi_2O_{6.5}$ and $Cs_xBa_{1-x}TiSi_2O_{7-0.5x}$, ($x = 0.9$ and 0.7), were synthesized by high-temperature crystallization from their respective precursors [61]. The Rietveld analysis of synchrotron XRD data reveals that while $Cs_xBa_{(1-x)/2}TiSi_2O_{6.5}$ samples are phase-pure, $Cs_xBa_{1-x}TiSi_2O_{7-0.5x}$ samples contain $Cs_{3x/(2+x)}Ba_{(1-x)/(2+x)}TiSi_2O_{6.5}$ pollucites (i.e., also two-Cs-to-one-Ba substitution) and a secondary phase, fresnoite ($Ba_2TiSi_2O_8$). Thus the $Cs_xBa_{1-x}TiSi_2O_{7-0.5x}$ series is energetically less favorable than $Cs_xBa_{(1-x)/2}TiSi_2O_{6.5}$. To study the stability systematics of

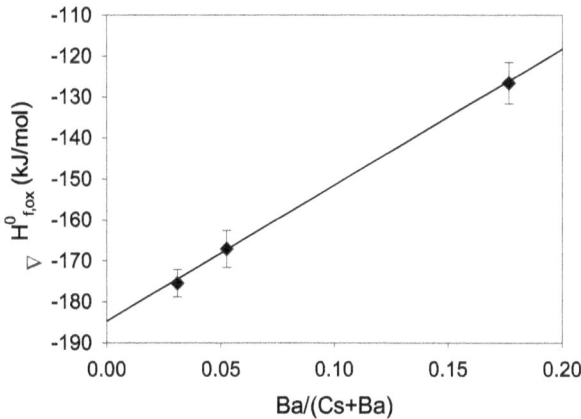

Figure 15. Enthalpies of formation from the constituent oxides at 298 K ($\Delta H^{o}_{f,ox}$) of the pollucite phases $Cs_{0.9}Ba_{0.05}TiSi_2O_{6.5}$, $Cs_{0.7}Ba_{0.15}TiSi_2O_{6.5}$, and $Cs_{0.94}Ba_{0.03}TiSi_2O_{6.49}$ as a function of Ba/(Cs+Ba) [61].

$Cs_xBa_{(1-x)/2}TiSi_2O_{6.5}$ pollucites, their enthalpies of formation from oxides were derived from drop-solution calorimetric measurements with the lead borate solvent at 973 K. The results show that with increasing Ba/(Cs+Ba) ratio, the thermodynamic stability of these phases decreases with respect to their component oxides (Figure 15). Hence, from the energetic viewpoint, continued Cs $\rightarrow$ Ba transmutation tends to destabilize the parent silicotitanate pollucite structure. Nevertheless, the Ba-substituted pollucite co-forms with fresnoite ($Ba_2TiSi_2O_8$, which incorporates the excess Ba and has high leach resistance) [62], thereby still providing viable ceramic wasteforms for all the Ba decay products.

The Ba-to-Cs substitution has also been attempted on the parent pollucite $CsAlSi_2O_6$ and its Fe analog $CsFeSi_2O_6$, though inconsistent results were obtained. Rodriguez *et al.* [63] reported synthesis of $Cs_xBa_{(1-x)/2}AlSi_2O_6$, $Cs_xBa_{(1-x)/2}Al_xFe_{1-x}Si_2O_6$ and $Cs_xBa_{1-x}Al_x$ $Fe_{1-x}Si_2O_6$ with $1 \geq x \geq 0.7$ by using a hydrothermal method. Their Rietveld analysis of XRD data indicated different effects of the three types of substitutions on the pollucite structure: (1) Ba^{2+} for Cs^+ cation has little effect on cell dimensions, (2) intermediate concentrations of Ba^{2+} and Fe^{3+} substitution result in net minor expansion due to Fe^{3+} addition, and (3) large Ba and Fe substitutions lead to overall framework contraction. On the other hand, Vance *et al.* [64] found only small levels of Ba-to-Cs substitution of up to ~0.07 formula units in $Cs_{(1-2x)}Ba_xAlSi_2O_6$ pollucites prepared by sol–gel methods and sintering at 1400°C; extra Ba tended to form $BaAl_2Si_2O_8$. Moreover, they found no evidence of Ba substitution in $CsFeSi_2O_6$ via cation vacancies or Fe^{2+} formation, where Ba appeared to enter a Fe-silicate glass phase.

4.3. Wadeite-type phases for Cs immobilization

Although classified as a cyclosilicate, wadeite, $K_2ZrSi_3O_9$, possesses a framework structure composed of three-membered $[Si_3O_9]$ rings of $[SiO_4]$ tetrahedra and isolated $[ZrO_6]$ octahedra, via corner-sharing, with K^+ occupying the cavities [65] (Figure 16). The wadeite structure is very tolerant of ionic substitutions, not only in its framework sites but also over the interstitial positions. Balmer et al. [27] synthesized a Cs-substituted wadeite, $Cs_2ZrSi_3O_9$, using a sol–gel method. Rietveld analysis of powder XRD data confirmed that this compound is isostructural with the prototype wadeite $K_2ZrSi_3O_9$ (space group $P6_3/m$). The aqueous durability of $Cs_2ZrSi_3O_9$ was measured using the PCT and MCC-1 methods. Figure 17 shows normalized Cs mass losses of $Cs_2ZrSi_3O_9$ as a function of time. The results from both measurements indicate that the Cs loss was high in the first day, decreased dramatically over the next day, and then remained low thereafter. However, the Cs leach rates of $Cs_2ZrSi_3O_9$ depend on the solution conditions. In modified leach tests with buffered (pH = 7) and unbuffered solutions, the Cs release rates are smaller than $1.2 \times 10^{-4} g \cdot (m^2 \cdot day)^{-1}$ over a 7-day period. On the other hand, in unsaturated, unbuffered solutions with a pH of 9–10, the durability was much lower, with 7-day Cs release rates of $\sim 2.2 \times 10^{-3} g \cdot (m^2 \cdot day)^{-1}$. The ability of this phase to retain Cs in aqueous environments is attributed to its condensed ring structure, in which the size of the channel openings is smaller than that of Cs^+. However, dissolution of the framework silicate can occur at higher pHs, resulting in the release of Cs. Thus this wadeite-type phase may be utilized for storage of radioactive ^{90}Cs at certain aqueous conditions.

Poojary et al. [66] synthesized a microporous compound, $K_2(ZrSi_3O_9) \cdot H_2O$, and its ion-exchanged phases with Na and Cs. They found that heating these phases releases the lattice water and transforms them into wadeite-type phases. High-temperature synchrotron XRD of $Cs_2ZrSi_3O_9$ indicates that this phase remains stable up to 1273 K — the highest temperature tested. Rietveld analyses of the obtained data show that on heating, unit-cell parameter a increases much more rapidly than c; the axial coefficients of thermal expansion are $\alpha_a = 16.481 \times 10^{-6}$; $\alpha_c = 2.996 \times 10^{-6}$ (Figure 18). This anisotropy is probably due to the more significant expansion of $[Si_3O_9]$

(a)

(b)

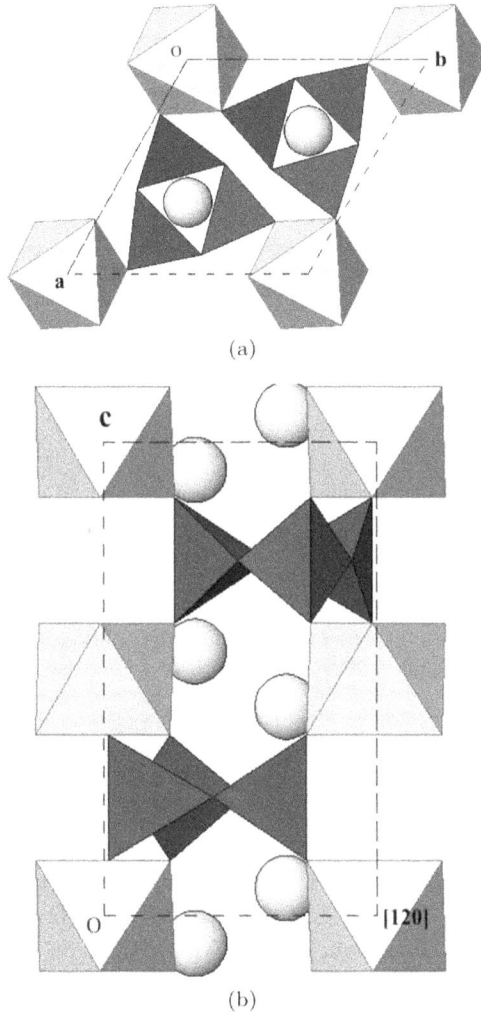

Figure 16. Structure of wadeite-type phases, $A_2BSi_3O_9$, projected down (a) c-axis and (b) a-axis [67]. Drawn here is wadeite $K_2ZrSi_3O_9$. Black tetrahedra represent $[SiO_4]$ units, gray octahedra represent $[ZrO_6]$, and spheres represent K^+ cations. Dashed lines outline the unit cell.

rings along (001) via Si-O-Si widening and Si–O lengthening. Hence, extraction of Cs^+ from aqueous wastes by this ion-exchanger and conversion of the Cs-loaded ion-exchanger to the wadeite analog via *in situ* heating are a potential method for radioactive Cs remediation.

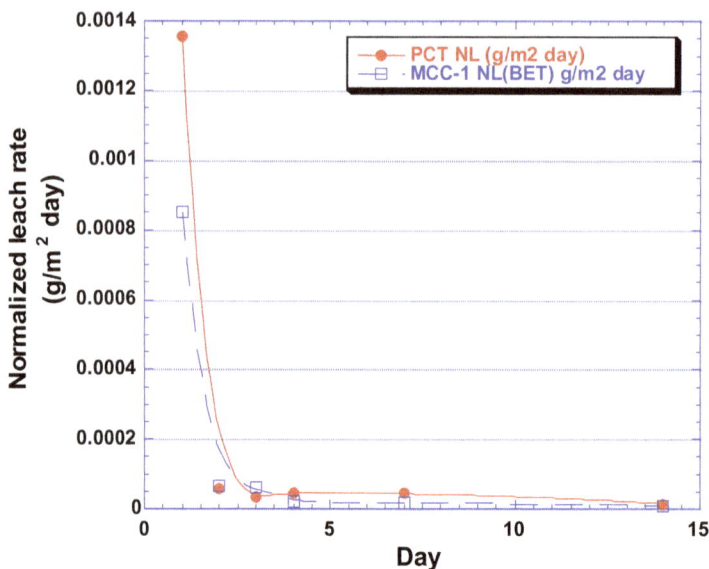

Figure 17. Normalized Cs mass losses of $Cs_2ZrSi_3O_9$ as a function of time measured using the PCT method (powder sample; solution:sample = 185:1) and MCC-1 test (pellet sample) [27].

To determine the energetics of $Cs_2ZrSi_3O_9$ and its stability relations to other wadeite-type phases, Xu *et al.* [61] measured standard enthalpies of formation of several compounds including $K_2ZrSi_3O_9$ and its analogs $K_2TiSi_3O_9$ and $Cs_2ZrSi_3O_9$, using drop-solution calorimetry into $2PbO \cdot B_2O_3$ solvent at 975 K. The obtained values (in kJ/mol) are as follows: $\Delta H_{f,ox}^0(K_2TiSi_3O_9) = -355.8 \pm 3.0$, $\Delta H_{f,el}^0(K_2TiSi_3O_9) = -4395.1 \pm 4.8$, $\Delta H_{f,ox}^0(K_2ZrSi_3O_9) = -374.3 \pm 3.3$, $\Delta H_{f,el}^0(K_2ZrSi_3O_9) = -4569.9 \pm 5.0$, $\Delta H_{f,ox}^0(Cs_2ZrSi_3O_9) = -396.6 \pm 4.4$, and $\Delta H_{f,el}^0(Cs_2ZrSi_3O_9) = -4575.0 \pm 5.5$. Therefore, with increasing the size of the octahedral framework cation $[Ti^{4+}$ (0.605 Å) $\rightarrow Zr^{4+}$ (0.72 Å)] or of the interstitial alkali cation $[K^+$ (1.55 Å) $\rightarrow Cs^+$ (1.78 Å)] [68], the formation enthalpies become more exothermic. This trend is consistent with the general behavior of increasing energetic stability with decreasing ionic potential (z/r) seen in many oxide and silicate systems [69].

4.4. *Perovskite series for Sr immobilization*

Perovskites are arguably the most important family of materials due to their vast diversity in composition, structure, and physical properties. Titanate

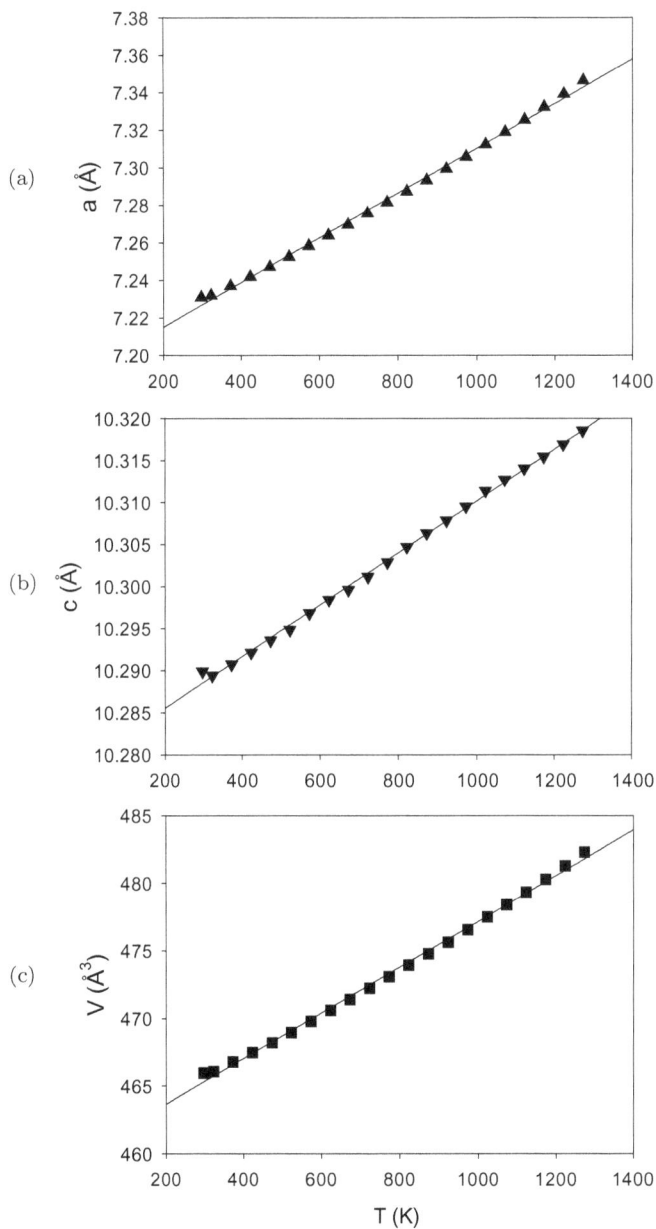

Figure 18. Variation of (a) cell parameters a, (b) c, and (c) cell volume V of wadeite-type phase $Cs_2ZrSi_3O_9$ as a function of temperature. The lines are the best fits to the data.

perovskites occur in Synroc as host phases for Sr. As described earlier, the microporous phases $Na_2Nb_{2-x}Ti_xO_{6-x}(OH)_x \cdot H_2O$ ($0 \leq x \leq 0.4$), which have high selectivities for Sr, transform to $Na_2Nb_{2-x}Ti_xO_{6-0.5x}$ perovskites upon heating. Combined Rietveld analysis of powder neutron and X-ray diffraction data reveals that these perovskite phases adopt the structure of the endmember $NaNbO_3$ with the space group *Pbma* (Figure 19(a)) [70]. However, the substituted structure is deficient in oxygen over its O1 and O3 sites, which compensates the charge imbalance between the substituting Ti^{4+} and Nb^{5+} (Figures 19(b) and 19(c)). Raman spectroscopy indicates that the O^{2-} vacancies are locally associated with Ti^{4+}, resulting in $[TiO_5]$ coordination. There is no evidence for long-range order or superstructure. With increasing Ti content, the orthorhombic structure becomes more like the cubic, as reflected by the smaller differences among its cell parameters (a_p, b_p, and c_p) and the cell angle β_p approaching $90°$ (in terms of the pseudocubic subcell) (Figure 20). Standard enthalpies of formation were determined by drop-solution calorimetry into molten $3Na_2O \cdot 4MoO_3$ solvent at 974 K [40]. As Ti content increases, $\Delta H_{f,ox}^0$ becomes less exothermic (Figure 21), which suggests a destabilizing effect of the $Nb^{5+} \rightarrow Ti^{4+} + 0.5V_O^{\cdot\cdot}$ substitution on the perovskite structure with respect to the constituent oxides. This behavior can be attributed to a number of structural factors, including the occurrence of O^{2-} vacancies and of Ti–O–Nb linkages as well as the size mismatch between Ti^{4+} and Nb^{5+}.

As potential host phases for Sr in the wasteforms from heat treatment of Sr-loaded microporous phases $Na_2Nb_{2-x}Ti_xO_{6-x}(OH)_x \cdot H_2O$, perovskite solid solutions along the $NaNbO_3$–$SrTiO_3$ join were synthesized using the sol–gel and solid-state sintering methods [71]. XRD analysis indicates that as Sr + Ti content increases, the perovskite structure changes from the orthorhombic to tetragonal and to cubic. The enthalpies of formation from the constituent oxides, determined by drop-solution calorimetry into molten $3Na_2O \cdot 4MoO_3$ at 974 K, become less exothermic with increasing Sr+Ti content (Figure 22(a)), suggesting a destabilization effect of the substitution, $Na^+ + Nb^{5+} \rightarrow Sr^{2+} + Ti^{4+}$ on the perovskite structure with respect to the constituent oxides. The trend of decreasing thermodynamic stability with decreasing structural distortion (relative to the ideal cubic structure; assessed by the tolerance factor t) is opposite to that seen in most ABO_3 perovskites (Figure 22(b)). This behavior can be interpreted in terms of the dominance of acid–base chemistry, expressed by the ionic potential ratio of B to A cation $(z/r)_B/(z/r)_A$, (where z is the formal charge and r the ionic

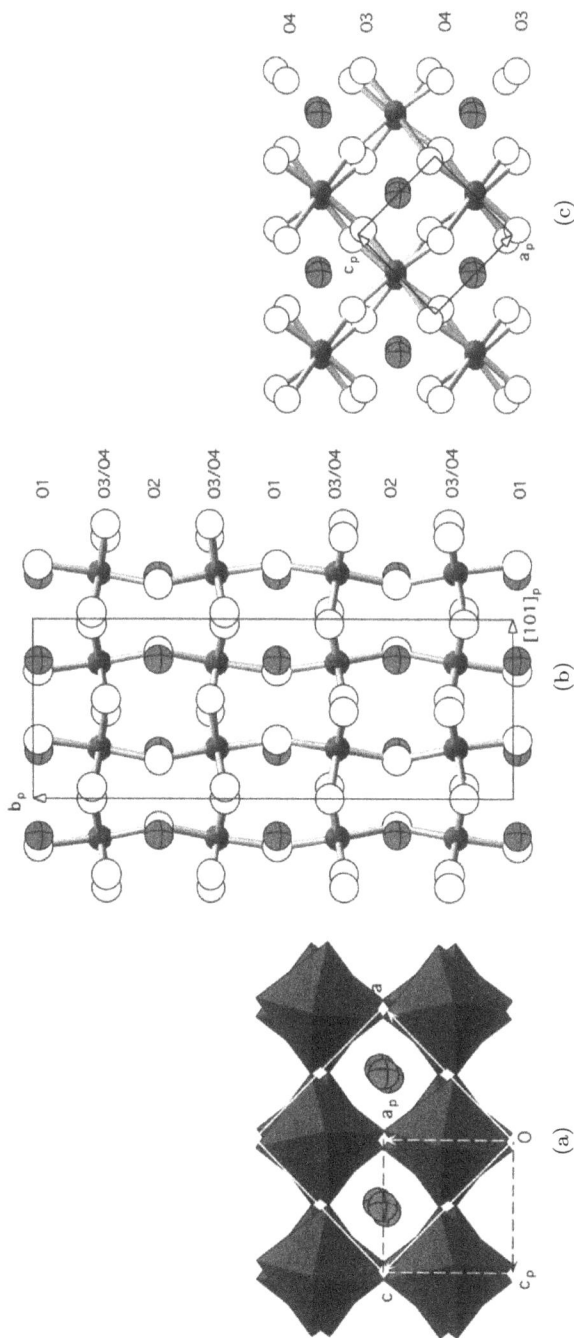

Figure 19. (a) Polyhedral representation of the NaNbO₃ perovskite structure projected down [010]. Octahedra represent [NbO₆] units, and spheres represent Na ions. Solid lines outline the unit cell and dashed lines the pseudocubic subcell. (b, c) Bonding representation of the NaTi$_x$Nb$_{1-x}$O$_{3-0.5x}$ perovskite structure projected along (b) [100] (or [10-1]$_p$) and (c) [010] (or [010]$_p$) [71]. Solid spheres represent B cations (B = Nb or Ti), filled equatorials represent Na ions, and circles represent O atoms. Thin lines outline the pseudocubic subcell. For clarity, atoms in a partial subcell along the projection axis are shown.

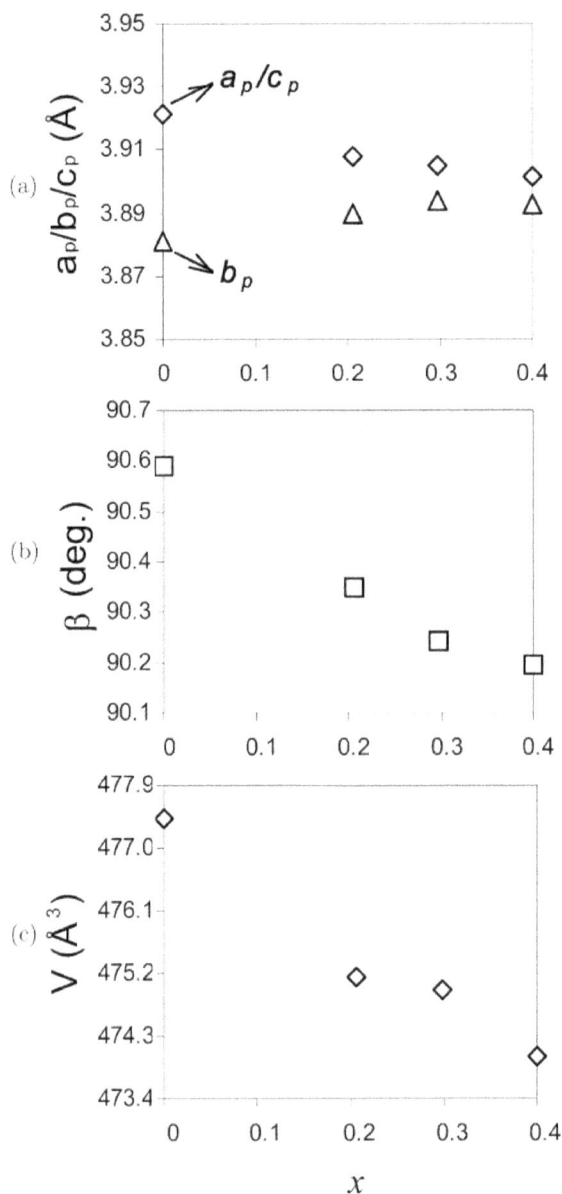

Figure 20. Variation of cell parameters (a) a_p, b_p, c_p, (b) β_p, and (c) cell volume V in terms of the pseudocubic subcell for $Na_2Nb_{2-x}Ti_xO_{6-0.5x}$ perovskites as a function of composition [40]. Errors are smaller than the size of symbols.

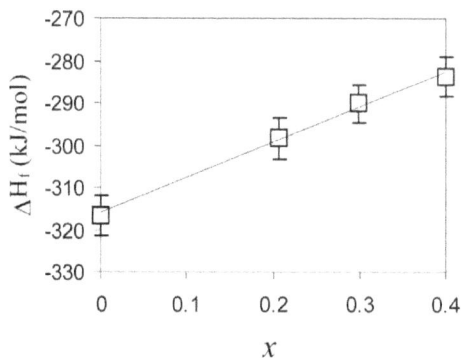

Figure 21. Variation of the enthalpies of formation of $Na_2Nb_{2-x}Ti_xO_{6-0.5x}$ perovskites from the constituent oxides at 298 K ($\Delta H^\circ_{f,ox}$) as a function of composition [40]. The line represents a least-square fit to the data.

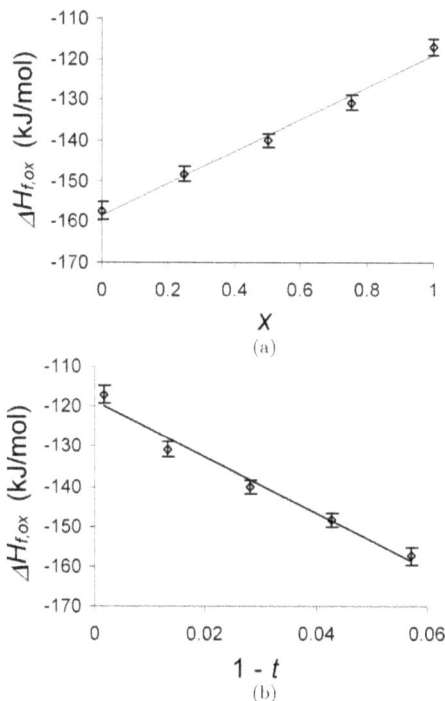

Figure 22. Variation of the enthalpies of formation of $Na_{1-x}Sr_xNb_{1-x}Ti_xO_3$ perovskites from the oxides at 298 K ($\Delta H^\circ_{f,ox}$) as a function of (a) x and (b) $(1-t)$ [71]. t is the tolerance factor for ABO_3 perovskites: $t = (r_A + r_O)/\sqrt{2}(r_B + r_O)$, where r_A, r_B, and r_O refer to the ionic radius of A, B, and O^{2-}, respectively. The lines are the best fits to the data.

radius; A = Na, Sr; B = Nb, Ti) in determining phase stability. Moreover, the enthalpic variation with Sr+Ti content is nearly linear, and thus the enthalpies of the morphotropic transitions across the series are rather small.

Although these substituted perovskites exhibit high thermodynamic stability relative to their constituent oxides, no systematic leach tests have been conducted, except for the $SrTiO_3$ endmember, which does show good leach resistance [72]. Thus their aqueous durability needs to be further investigated. Moreover, in contrast to the $^{137}Cs \rightarrow ^{137}Ba ^{137}Cs$ transmutation studies on the Cs wasteform phases (hollandite and pollucite), there has been essentially no work to determine the effects of $^{90}Sr \rightarrow ^{90}Y \rightarrow ^{90}Zr$ transmutations on the structures and stability of Sr-bearing perovskites and other crystalline phases. Hence, this is also an area requiring future studies.

5. Concluding Remarks

Incorporation of ^{137}Cs and ^{90}Sr into crystalline wasteform phases is potentially a powerful strategy for radioactive Cs/Sr immobilization. Through fixing Cs or Sr at a crystal lattice site, the host phases generally exhibit higher thermodynamic stability and aqueous durability (as well as radiation tolerance) than other types of nuclear wastes (e.g., glass and cement). These characteristics are particularly important for ^{137}Cs and ^{90}Sr, as they generate tremendous amounts of heat via the beta decay and are highly soluble in aqueous environments. Though diverse in chemical composition and crystal structure, the Cs/Sr wasteform phases (the above-described hollandite, pollucite, wadeite, and perovskite as well as other phases such as apatite and sodium zirconium phosphate — NZP) possesses a 3-D framework structure with Cs^+/Sr^{2+} occupying the cavity or tunnel sites. Since 3-D frameworks typically have high structural flexibility via polyhedral tilting/rotation on changing temperature, it is important to study their thermal stability and whether the enclosed Cs^+/Sr^{2+} remain in the frameworks at elevated temperatures. For instance, the opening of the cavity box in aluminotitanate hollandites is smaller than the size of Cs (and its daughter product Ba), even at high temperature, preventing evaporation of Cs/Ba from the hollandite structure. In aqueous environments, whether the 3-D frameworks retain their integrity depends on solution conditions such as pH. An example is the leachability of the wadeite-type phase $Cs_2ZrSi_3O_9$, which shows high durability in solutions of pH = 7 and low resistance at higher pHs due to dissolution of the silicate framework.

^{137}Cs and ^{90}Sr exist in radioactive aqueous wastes from the reprocessing of spent nuclear fuels and weapon activities (such as those at Hanford, WA). To reduce the waste volumes, ^{137}Cs and ^{90}Sr first need to be extracted/concentrated from the wastes prior to their immobilization in solid wasteforms. Microporous titanosilicate and niobate phases can be used to separate Cs/Sr from the waste effluents that are frequently rich in other alkali or alkaline earth cations; titanosilicates are particularly effective for the extraction of Cs while niobates are efficient ion-exchangers for Sr separation. Then the Cs/Sr-loaded microporous phases may be heated and dehydrated *in situ*, resulting in dense ceramic wasteform phases that are stable both thermally and at aqueous conditions. Hollandites, pollucites, and wadeites are potential wasteform phases from heat treatment of Cs-loaded microporous titanates, and perovskites result from thermal conversion of Sr-loaded niobate ion-exchangers. This approach is attractive, as it combines ^{137}Cs/^{90}Sr separation and immobilization into one integrated process and may ultimately reduce the cost for aqueous Cs/Sr waste disposal.

A recent yet still underexplored area in crystalline wasteform materials research is the determination of the effects of chemical transmutation of the radionuclides on the structures and stability of wasteform phases. These effects are particularly important for ^{137}Cs and ^{90}Sr, as they are short-lived; ^{137}Cs and ^{90}Sr transmute via beta decay to ^{137}Ba and ^{90}Y/^{90}Zr with halflives of ~30 years, respectively. A major advantage of the crystalline wasteform phases for ^{137}Cs and ^{90}Sr is the ability of their framework structures to potentially accommodate both the parent ^{137}Cs and ^{90}Sr and their decayed products of ^{137}Ba and ^{90}Y/^{90}Zr over their cavity/tunnel sites. Titanate hollandites (which occur in Synroc) have been shown to host both Cs$^+$ and Ba^{2+} extensively, though pollucites can do so only to limited extents. Synthesis of Ba-substituted $Cs_2ZrSi_3O_9$ wadeites has not been attempted, and neither does that for Y/Zr-substituted titanate perovskites. Although it would be ideal that the crystal structure of a wasteform phase maintains its stability over the entire decay period (i.e., forming a complete substituted solid–solution series), this might be crystal-chemically too demanding, given that the substitution involves cations of different sizes, charges, and even chemical characters. Nevertheless, to fully assess the effects of radionuclide transmutation on the wasteform performance, further studies are needed.

Acknowledgments

I thank many colleagues especially A. Navrotsky, T. M. Nenoff, M. Nyman, M. L. Balmer, Y. Su, C. R. Stanek and B. P. Uberuaga for fruitful collaborations in nuclear wasteform materials research over the years. Some of the presented works and the preparation of this chapter were supported by the laboratory-directed research and development program of Los Alamos National Laboratory (LANL) and the University of California Lab Fees Research Program. The newly reported high-temperature neutron diffraction data of $CsTiSi_2O_{6.5}$ pollucite and synchrotron XRD data of $Cs_2ZrSi_3O_9$ wadeite were collected at the Lujan Neutron Scattering Center of LANL and the National Synchrotron Light Source (NSLS) at Brookhaven National Laboratory, respectively. Use of the Lujan Center and NSLS was supported by the U.S. Department of Energy, Office of Science, and Office of Basic Energy Sciences. LANL is operated by Los Alamos National Security LLC, under DOE Contract DE-AC52-06NA25396.

References

1. M. I. Ojovan and W. E. Lee, *An Introduction to Nuclear Waste Immobilisation*, Elsevier, 315 pp. (2005).
2. W. J. Weber, A. Navrotsky, S. Stefanovsky, E. R. Vance, and E. Vernaz, Materials science of high-level nuclear waste immobilization, *MRS Bull.* **34**, 46–53 (2009).
3. J. M. Hanchar and P. W. O. Hoskin (eds.), *Zircon. Reviews in Mineralogy*, Vol. 53, Mineralogical Society of America, 500 pp. (2003).
4. R. C. Ewing, Nuclear waste forms for actinides, *Proc. Natl. Acad. Sci. USA* **96**, 3432–3439 (1999).
5. I. Farnan, H. Cho, and W. J. Weber, Quantification of actinide α-radiation damage in minerals and ceramics, *Nature* **445**, 190–193 (2007).
6. A. E. Ringwood, S. E. Kesson, N. G. Ware, W. O. Hibberson, and A. Major, Immobilization of high level nuclear reactor wastes in SYNROC, *Nature* **278**, 219 (1979).
7. A. E. Ringwood, S. E. Kesson, N. G. Ware, W. O. Hibberson, and A. Major, The SYNROC process: A geochemical approach to nuclear waste immobilization, *Geochem. J.* **13**, 141 (1979).
8. X. Guo, S. Szenknect, A. Mesbah, N. Clavier, C. Poinssot, D. Wu, H. Xu, N. Dacheux, R. Ewing, and A. Navrotsky, Energetics of uranothorite ($Th_{1-x}U_xSiO_4$) solid solution, *Chem. Mater.* **28**, 7117–7124 (2016).
9. R. Wigeland, T. Bauer, T. Fanning, and E. R. Morris, Separations and transmutation criteria to improve utilization of a geological repository, *Nucl. Technol.* **154**, 95–106 (2006).
10. C. Jiang, C. R. Stanek, N. A. Marks, K. E. Sickafus, and B. P. Uberuaga, Predicting from first principles the chemical evolution of crystalline compounds

due to radioactive decay: The case of the transformation of CsCl to BaCl, *Phys. Rev. B* **79**, 132110 (2009).

11. C. Jiang, C. R. Stanek, N. A. Marks, K. E. Sickafus, and B. P. Uberuaga, Radioparagenesis: The formation of novel compounds and crystalline structures via radioactive decay, *Philos. Mag. Lett.* **90**, 435–446 (2010).

12. B. P. Uberuaga, C. Jiang, C. R. Stanek, K. E. Sickafus, N. A. Marks, D. J. Carter, and A. L. Rohl, Implications of transmutation on the defect chemistry in crystalline waste forms, *Nucl. Instrum. Meth. Phys. Res. B: Beam Interact. Mater. Atoms* **268**, 3261–3264 (2010).

13. M. Sassi, B. P. Uberuaga, C. R. Stanek, and N. A. Marks, Transmutation in $^{90}SrF_2$: A density-functional-theory study of phase stability in ZrF_2, *Phys. Rev. B* **85**, 094104 (2012).

14. C. R. Stanek, B. P. Uberuaga, B. L. Scott, R. K. Feller, and N. A. Marks, Accelerated chemical aging of crystalline nuclear waste forms, *Curr. Opin. Solid State Mater. Sci.* **16**, 126–133 (2012).

15. B. Aguila, D. Banerjee, Z. Nie, Y. Shin, S. Ma, and P. K. Thallapally, Selective removal of cesium and strontium using porous frameworks from high level nuclear waste, *Chem. Commun.*, **52**, 5940–5942 (2016).

16. M. L. Balmer and B. C. Bunker, Inorganic ion exchange evaluation and design — silicotitanate ion exchange waste conversion, Pacific Northwest Laboratory Report No. PNL-10460 (1995).

17. V. Luca, C. S. Griffith, E. Drabarek, and H. Chronis, Tungsten bronze-based nuclear waste form ceramics. Part 1. Conversion of microporous tungstates to leach resistant ceramics, *J. Nucl. Mater.* **358**, 139–150 (2006).

18. C. S. Griffith, F. Sebesta, J. V. Hanna, P. Yee, E. Drabarek, M. E. Smith, and V. Luca, Tungsten bronze-based nuclear waste form ceramics. Part 2: Conversion of granular microporous tungstate–polyacrylonitrile (PAN) composite adsorbents to leach resistant ceramics, *J. Nucl. Mater.* **358**, 151–163 (2006).

19. R. C. Ewing, W. J. Weber, and F. W. Clinard Jr, Radiation effects in nuclear waste forms for high-level radioactive waste, *Prog. Nucl. Energy* **29**, 63–127 (1995).

20. D. L. Bish and J. E. Post (eds.), Modern Powder Diffraction. Reviews in Mineralogy, Vol. 20, Mineralogical Society of America, 369 pp. (1989).

21. H. Xu, P. J. Heaney, and G. H. Beall, Phase transitions induced by solid solution in stuffed derivatives of quartz: A powder synchrotron XRD study of the LiAlSiO4-SiO2 join, *Am. Mineral.* **85**, 971–979 (2000).

22. J. Zhang, A. Celestian, J. B. Parise, H. Xu, and P. J. Heaney, A new polymorph of eucryptite (LiAlSiO4), ε-eucryptite, and thermal expansion of α- and ε-eucryptite at high pressure, *Am. Mineral.* **87**, 566–571 (2002).

23. H. Xu, Y. Zhao, S. C. Vogel, L. L. Daemen, and D. D. Hickmott, Anisotropic thermal expansion and hydrogen bonding behavior of portlandite: A high-temperature neutron diffraction study, *J. Solid State Chem.* **180**, 1519–1525 (2007).

24. A. Navrotsky, Progress and new directions in calorimetry: A 2014 perspective, *J. Am. Ceram. Soc.* **97**, 3349–3359 (2014).

25. H. Xu, Y. Zhang, and A. Navrotsky, Enthalpies of formation of microporous titanosilicates ETS-4 and ETS-10, *Microporous Mesoporous Mater.* **47**, 285–291 (2001).

26. T. Varga, C. Lind, A. P. Wilkinson, H. Xu, C. E. Lesher, and A. Navrotsky, Heats of formation for several crystalline polymorphs and pressure-induced amorphous forms of AMo2O8 (A= Zr, Hf) and ZrW2O8, *Chem. Mater.* **19**, 468–476 (2007).

27. M. L. Balmer, Y. Su, H. Xu, E. R. Bitten, D. E. McCready, and A. Navrotsky, Synthesis, structure determination and aqueous durability of $Cs_2ZrSi_3O_9$, *J. Am. Ceram. Soc.* **84**, 153–160 (2001).

28. M. W. Anderson, O. Terasaki, T. Ohsuna, P. J. O. Malley, A. Phillipou, S. P. MacKay, A. Ferreira, J. Rocha, and S. Lidin, Microporous titanosilicate ETS-10: A structural survey, *Philos. Mag. B* **71**, 813–841 (1995).

29. A. Clearfield, Structure and ion exchange properties of tunnel type titanium silicates, *Solid State Sci.* **3**, 103–112 (2001).

30. R. G. Anthony, C. V. Philip, and R. G. Dosch, Selective adsorption and ion exchange of metal cations and anions with silico-titanates and layered titanates, *Waste Manage.* **13**, 503–512 (1993).

31. H. Xu, A. Navrotsky, M. D. Nyman, and T. M. Nenoff, Thermochemistry of microporous silicotitanate phases in the $Na_2O\text{-}Cs_2O\text{-}SiO_2\text{-}TiO_2\text{-}H_2O$ system, *J. Mater. Res.* **15**, 815–823 (2000).

32. E. A. Behrens, D. M. Poojary, and A. Clearfield, Syntheses, crystal structures, and ion-exchange properties of porous titanosilicates, $HM_3Ti_4O_4(SiO_4)_3 \cdot 4H_2O$ (M = H^+, K^+, Cs^+), structural analogues of the mineral pharmacosiderite, *Chem. Mater.* **8**, 1236–1244 (1996).

33. W. T. A. Harrison, T. E. Gier, and G. D. Stucky, Single-crystal structure of $Cs_3HTi_4O_4(SiO_4)_3 \cdot 4H_2O$, a titanosilicate pharmacosiderite analog, *Zeolites* **15**, 408–412 (1995).

34. D. M. Poojary, R. A. Cahill, and A. Clearfield, Synthesis, crystal structures, and ion-exchange properties of a novel porous titanosilicate, *Chem. Mater.* **6**, 2364–2368 (1994).

35. D. M. Poojary, A. I. Bortun, L. N. Bortun, and A. Clearfield, Structural studies on the ion-exchanged phases of a porous titanosilicate, $Na_2Ti_2O_3SiO_4 \cdot 2H_2O$, *Inorg. Chem.* **35**, 6131–6139 (1996).

36. H. Xu, A. Navrotsky, M. Nyman, and T. M. Nenoff, Crystal chemistry and energetics of pharmacosiderite-related microporous phases in the $K_2O\text{-}Cs_2O\text{-}SiO_2\text{-}TiO_2\text{-}H_2O$ system, *Micropor. Mesopor. Mater.* **72**, 209–218 (2004).

37. M. Nyman, A. Tripathi, J. B. Parise, R. S. Maxwell, W. T. A. Harrison, and T. M. Nenoff, A new family of octahedral molecular sieves: Sodium Ti/Zr^{IV} niobates, *J. Am. Chem. Soc.* **123**, 1529–1530 (2001).

38. M. Nyman, A. Tripathi, J. B. Parise, R. S. Maxwell, and T. M. Nenoff, Sandia octahedral molecular sieves (SOMS): Structural and property effects of charge-balancing the M^{IV}-substituted (M = Ti, Zr) niobate framework, *J. Am. Chem. Soc.* **124**, 1704–1713 (2002).

39. H. Xu, M. Nyman, T. M. Nenoff, and A. Navrotsky, The prototype Sandia Octahedral Molecular Sieve (SOMS) $Na_2Nb_2O_6 \cdot H_2O$: Synthesis, structure and thermodynamic stability, *Chem. Mater.* **16**, 2034–2040 (2004).

40. H. Xu, A. Navrotsky, M. Nyman, and T. M. Nenoff, Octahedral microporous phases $Na_2Nb_{2-x}Ti_xO_{6-x}OH_x \cdot H_2O$ and their related perovskites: Crystal chemistry, energetics and stability relations, *J. Mater. Res.* **20**, 618–627 (2005).

41. V. Aubin-Chevaldonnet, D. Caurant, A. Dannoux, D. Gourier, T. Charpentier, L. Mazerolles, and T. Advocat, Preparation and characterization of $(Ba, Cs)(M, Ti)_8O_{16}$ ($M = Al^{3+}$, Fe^{3+}, Ga^{3+}, Cr^{3+}, Sc^{3+}, Mg^{2+}) hollandite ceramics developed for radioactive cesium immobilization, *J. Nucl. Mater.* **366**, 137–160 (2007).

42. K. R. Whittle, S. E. Ashbrook, G. R. Lumpkin, I. Farnan, R. I. Smith, and S. A. T. Redfern, The effect of caesium on barium hollandites studied by neutron diffraction and magic-angle spinning (MAS) nuclear magnetic resonance, *J. Mater. Sci.* **42**, 9379–9391 (2007).

43. S. E. Kesson and T. J. White, $[Ba_xCs_y][(Ti,Al)_{2x+y}{}^{3+}Ti_{8-2x-y}{}^{4+}]O_{16}$ Synroc-type hollandites. II. Structural chemistry, *Proc. R. Soc. Lond. A Math. Phys. Sci.* **408**, 295–319 (1986).

44. R. W. Cheary, Caesium substitution in the titanate hollandites Ba_xCs_y $(Ti_{y+2x}{}^{3+}Ti_{8-2x-y}{}^{4+})O_{16}$ from 5 to 400 K, *Acta Crystallogr.* **B47**, 325–333 (1991).

45. H. Xu, G. C. C. Costa, C. R. Stanek, and A. Navrotsky, Structural behavior of $Ba_{1.24}Al_{2.48}Ti_{5.52}O_{16}$ hollandite at high temperature: An in-situ neutron diffraction study, *J. Am. Ceram. Soc.* **98**, 255–262 (2014).

46. G. C. C. Costa, H. Xu, and A. Navrotsky, Thermochemistry of barium hollandites, *J. Am. Ceram. Soc.* **96**, 1554–1561 (2013).

47. H. Xu, L. Wu, J. Zhu, and A. Navrotsky, Synthesis, characterization and thermochemistry of Cs-, Rb- and Sr-substituted barium aluminium titanate hollandites, *J. Nucl. Mater.* **459**, 70–76 (2015).

48. L. Wu, J. Schliesser, B. F. Woodfield, H. Xu, and A. Navrotsky, Heat capacities, standard entropies and Gibbs energies of Sr-, Rb- and Cs-substituted barium aluminotitanate hollandites, *J. Chem. Thermodyn.* **93**, 1–7 (2016).

49. M. L. Carter, E. R. Vance, D. R. G. Mitchell, J. V. Hanna, Z. Zhang, and E. Loi, Fabrication, characterization and leach testing of hollandite, $(Ba,Cs)(Al,Ti)_2Ti_6O_{16}$, *J. Mater. Res.* **17**, 2578–2589 (2002).

50. F. Angeli, P. McGlinn, and P. Frugier, Chemical durability of hollandite ceramic for conditioning cesium, *J. Nucl. Mater.* **380**, 59–69 (2008).

51. M. L. Carter, A. L. Gillen, K. Olufson, and E. R. Vance, HIPed tailored hollandite waste forms for the immobilization of radioactive Cs and Sr, *J. Am. Ceram. Soc.* **92**, 1112–1117 (2009).

52. T. Suzuki-Muresan, J. Vandenborre, A. Abdelouas, B. Grambow, and S. Utsunomiya, Studies of (Cs,Ba)-hollandite dissolution under gamma irradiation at 95°C and at pH 2.5, 4.4 and 8.6, *J. Nucl. Mater.* **419**, 281–290 (2011).

53. S. A. Gallagher and G. J. McCarthy, Preparation and X-ray characterization of pollucite ($CsAlSi_2O_6$), *Inorg. Nucl. Chem.* **43**, 1773–1777 (1981).

54. S. Komameni and W. B. White, Stability of pollucite in hydrothermal fluids, *Sci. Basis Nucl. Waste Manage.* **3**, 387–396 (1981).

55. K. Yanagisawa, M. Nishioka, and N. Yamasaki, Immobilization of cesium into pollucite structure by hydrothermal hot-pressing, *J. Nucl. Sci. Technol.* **24**, 51–60 (1987).

56. M. L. Balmer, Q. Huang, W. Wong-Ng, R. S. Roth, and A. Santoro, Neutron and X-ray diffraction study of the crystal structure of $CsTiSi_2O_{6.5}$, *J. Solid State Chem.* **130**, 97–102 (1997).

57. H. Xu, A. Navrotsky, M. L. Balmer, and Y. Su, Crystal chemistry and phase transitions in substituted pollucites along the $CsAlSi_2O_6$-$CsTiSi_2O_{6.5}$ join: A powder synchrotron XRD study, *J. Am. Ceram. Soc.* **85**, 1235–1242 (2002).

58. H. Xu, A. Navrotsky, M. L. Balmer, Y. Su, and E. R. Bitten, Energetics of substituted pollucites along the $CsAlSi_2O_6$-$CsTiSi_2O_{6.5}$ join: A high-temperature calorimetric study, *J. Am. Ceram. Soc.* **84**, 555–560 (2001).

59. N. J. Hess, Y. Su, and M. L. Balmer, Evidence of edge-sharing TiO_5 polyhedra in Ti-substituted pollucite, $CsTi_xAl_{1-x}Si_2O_{6+x/2}$, *J. Phys. Chem. B* **105**, 6805–6811 (2001).

60. M. L. Balmer, Y. Su, I. F. Grey, A. Santoro, R. S. Roth, Q. Huang, N. Hess, and B. C. Bunker, The Structure and properties of two new silicotitanate zeolites, *Mater. Res. Soc. Symp. Proc.* **465**, 449–455 (1997).

61. H. Xu, M. E. Chavez, J. N. Mitchell, T. J. Garino, H. L. Schwarz, M. A. Rodriguez, D. X. Rademacher, and T. M. Nenoff, Crystal structure and thermodynamic stability of Ba/Ti-substituted pollucites for radioactive Cs/Ba immobilization, *J. Am. Ceram. Soc.* **98**, 2634–2640 (2015).

62. T. J. Park, M. J. Davis, P. Vullo, T. M. Nenoff, J. L. Krumhansl, and A. Navrotsky, Thermochemistry and aqueous durability of ternary glass forming Ba-titanosilicates: Fresnoite ($Ba_2TiSi_2O_8$) and Ba-titanite ($BaTiSiO_5$), *J. Am. Ceram. Soc.* **92**, 2053–2058 (2009).

63. M. Rodriguez, T. J. Garino, D. X. Rademacher, X. Zhang, and T. M. Nenoff, The synthesis of Ba- and Fe-substituted $CsAlSi_2O_6$ pollucites, *J. Am. Ceram. Soc.* **96**, 2966–2972 (2013).

64. E. R. Vance, D. J. Gregg, G. J. Griffiths, P. R. Gaugliardo, and C. Grant, Incorporation of Ba in Al and Fe pollucite, *J. Nucl. Mater.* **478**, 256–260 (2016).

65. D. E. Henshaw, The structure of wadeite, *Min. Mag.* **30**, 585–595 (1955).

66. D. M. Poojary, A. I. Bortun, L. N. Bortun, and A. Clearfield, Syntheses and X-ray powder structures of $K_2(ZrSi_3O_9) \cdot H_2O$ and its ion-exchanged phases with Na and Cs, *Inorg. Chem.* **36**, 3072–3079 (1997).

67. H. Xu, A. Navrotsky, M. L. Balmer, and Y. Su, Crystal-chemical and energetic systematics of wadeite-type phases $K_2TiSi_3O_9$, $K_2ZrSi_3O_9$ and $Cs_2ZrSi_3O_9$, *Phys. Chem. Miner.* **32**, 426–435 (2005).

68. R. D. Shannon and C. T. Prewitt, Effective ionic radii in oxides and fluorides, *Acta Cryst. B* **25**, 925–946 (1969).

69. A. Navrotsky and Z. Tian, Systematics in the enthalpies of formation of anhydrous aluminosilicate zeolites, glasses and dense phases, *Chem. Eur. J.* **7**, 769–774 (2001).
70. H. Xu, Y. Su, M. L. Balmer, and A. Navrotsky, A new series of oxygen-deficient perovskites in the $NaTi_xNb_{1-x}O_{3-0.5x}$ system: Synthesis, crystal chemistry and energetics, *Chem. Mater.* **15**, 1872–1878 (2003).
71. H. Xu, A. Navrotsky, M. L. Balmer, and Y. Su, Perovskite solid solutions along the $NaNbO_3$-$SrTiO_3$ join: Phase transitions, formation enthalpies, and implications for general perovskite energetics, *Chem. Mater.* **17**, 1880–1886 (2005).
72. T. Kastrissios, M. Stephenson, P. S. Turner, and T. White, Hydrothermal dissolution of perovskite: Implications for Synroc formulation, *J. Am. Ceram. Soc.* **70**, C-144–C-146 (1987).

Index